PUBLIC SERVICE LIBERALISM

PUBLIC SERVICE LIBERALISM

TELECOMMUNICATIONS AND TRANSITIONS IN PUBLIC POLICY

Alan Stone

PRINCETON UNIVERSITY PRESS PRINCETON, NEW JERSEY

Published by Princeton University Press, 41 William Street,
Princeton, New Jersey 08540
In the United Kingdom:
Princeton University Press, Oxford

Library of Congress Cataloging-in-Publication Data

Stone, Alan, 1931–
Public service liberalism : telecommunications and transitions
in public policy / Alan Stone.
p. cm.
Includes index.
ISBN 0-691-07877-7
1. Telephone—Government policy—United States—History.
2. Telephone—United States—History. I. Title.
HE8819.S76 1991
384.6′0973—dc20 90-22302

This book has been composed in Linotron Caledonia

Princeton University Press books are printed
on acid-free paper and meet the guidelines
for permanence and durability of the Committee
on Production Guidelines for Book Longevity
of the Council on Library Resources

Printed in the United States of America by
Princeton University Press, Princeton, New Jersey

10 9 8 7 6 5 4 3 2 1

To the Memory of Jerome Schnur

There are matters in which the interference of law is required, not to overrule the judgment of individuals respecting their own interest, but to give effect to that judgment.

The same principle which points out . . . cases to which the principal objection to government interference does not apply, extends also to a variety of cases, in which important public services are to be performed.

—John Stuart Mill

Contents

List of Tables xi

Preface xiii

1. Liberalism Revised 3
2. The Telephone and the Public Service Idea 23
3. Protection of the Newborn 51
4. Structural Liberalism: The Issues of Economic Structure 84
5. The Progressive Impulse and the Telephone 122
6. Public Service Liberalism and the New Political Economy 165
7. The Administrative State and Public Service Liberalism 205
8. The Contraction of the World 238
9. The End of the Old Deal 271

Index 287

List of Tables

2.1 Telephones in the United States 39

5.1 Bell and independent telephones in selected years 131

5.2 Effect of competition on minimum rates 134

5.3 New independent telephone companies established per year 1894–1902 136

6.1 Bell and independent companies and stations on January 1, 1909 176

Preface

Writing at the end of the 1950s, Daniel Bell proclaimed that ideology in the sense of sweeping reductionist ideas applicable to every public problem was dead. "Ideology makes it unnecessary for people to confront individual issues on their individual merits. One simply turns to the ideological vending machine, and out comes the prepared formulae."[1] Bell had in mind fascism and communism, the dynamic ideologies of the 1930s. But while fascism never made a major revival in the postwar West and communism's intellectual collapse was even more complete than Bell could have contemplated, ideology was not dead. The New Left and the Green Movement moved to occupy the intellectual space vacated on the left. The persistent economic backwardness of the socialist bloc coupled with the economic problems that began to reach the agenda in the West during the late 1960s and early 1970s, in turn, triggered the revival of a conservatism that challenged the welfare state consensus. Deregulation, privatization, competition, and laissez-faire were, in this view, the keys to both freedom and economic progress.

The new conservatism scored important political successes in the United Kingdom and the United States and clearly influenced policymakers not only in the West, but even in the economically troubled socialist bloc and the Third World. Intellectually the new conservatism stood heads and shoulders above its competitors in the early 1990s, drawing from the thought of such luminaries as Nobel laureates Milton Friedman, F. A. Hayek, and George Stigler.

It is the contention of this book that the new conservatism is wrong. Rather, there is an ideology termed public service liberalism, based on the work of John Stuart Mill, that better addresses public problems than the laissez-faire ideology and the sanctimonious muddle that modern liberalism has become. It is premised on a marked preference for private property and liberty of contract but also provides an important role for state action. It recognizes that there are important social values that laissez-faire, with its paramount focus on economic efficiency, cannot adequately meet. And it satisfies Daniel Bell's injunction that individual issues must be confronted on their individual merits and not in a reductionist manner. Moreover, as I attempt to show, it has been tried and not found wanting.

This book examines the American telephone industry from its incep-

[1] Daniel Bell, *The End of Ideology*, rev. ed. (New York: Free Press, 1962), p. 405.

tion until 1934 to illustrate public service liberalism and how it served a complex structure of values, not just economic efficiency. The government-business interface in the telephone industry was not an unusual exception to a general theory of laissez-faire in the period covered. Rather, during much of that period, government-business relations in the telephone industry was an application of a general conception of public policy—public service liberalism.

Virtually every theory of public policy (with the notable exception of the Green Movement's) looks upon technological progressiveness as one of the keys—perhaps the most important one—to economic progress and a good material life for the masses. This book began as part of a series of studies on how public policy regimes affect the dynamics of technological progressiveness in the telecommunications industry. The research was generously funded by the John and Mary R. Markle Foundation, one of the missions of which is to analyze public policy issues in communications. As the research progressed, I felt that to understand what the future of telecommunications policy should be, the first step should be a close examination of its extraordinary past record, especially in the realm of technological progressiveness. As I delved into that record, I began to see that telecommunications policy was an application of a large conception of public policy—public service liberalism. In my view that conception, and not competing ones, can be the key to the future of American progressiveness, not only in telecommunications but in many other areas as well.

In addition to the generous funding and release time from teaching provided by the John and Mary R. Markle Foundation, I am enormously indebted to other people who read the entire manuscript or portions of it, offering valuable criticism. I gratefully thank James Anderson (Texas A & M University), Benjamin Ginsberg (Cornell University), Edward J. Harpham (University of Texas at Dallas), Hudson Janisch (University of Toronto Law School), and Almarin Phillips (Wharton School, University of Pennsylvania) for their advice and encouragement. Archival research was critical to the writing of this book. Mildred Ettlinger and Robert G. Lewis at the AT&T Archives and Peggy Chronister at the Museum of Independent Telephony were extraordinarily helpful. I learned much from discussions about technology and public policy with Gary D. Bray and E. Subissati of Bell Canada and C. C. Gotlieb of the Department of Computer Science, University of Toronto. As always, my largest debt is to Celeste P. Stone for her efforts in innumerable ways. Without her collaboration it would have not been possible. Gail Ullman and Cindy Hirschfeld of Princeton University Press continuously provided excellent advice and judgment. I would also like to thank, Lois Krieger, copy editor, and, Elizabeth Cunningham, indexer, for their exemplary efforts.

PUBLIC SERVICE LIBERALISM

1

Liberalism Revised

The Counterrevolution in Public Policy

SINCE the mid-1970s a new trend has come to dominate thinking about the political economy in much of the Western world. The Keynesian and mixed economy notions that dominated most post–World War II political-economic thinking have been supplanted by fervent advocacy of deregulation of state-controlled sectors, privatization of state-owned enterprises, and most important, a belief in competition as the only reasonable economic arrangement. In some countries, such as the United States, where state-owned enterprises are few compared to Western Europe, only a part of this framework is emphasized. But throughout the capitalist world and even to some extent in the less developed and socialist nations, the triad of deregulation, privatization, and competition has been in the ascendancy with varying emphases from country to country.

The dominance of these theoretical views in economic and governmental circles ensued for a variety of reasons. First, such major intellectual leaders as Nobel laureates Milton Friedman, Friedrich A. Hayek, and George Stigler and their numerous followers have forcefully advocated these policy prescriptions. But intellectual advocacy will usually not persuade public policymakers or become a dominant ideological trend unless events in the real world seem to call out for new solutions. In the West, oil shocks, rapid inflation, high levels of unemployment, and other troublesome trends shook the consensus in the efficacy of postwar policies. At the same time the stagnation—indeed, backwardness—of the socialist economies led most observers in the West to reaffirm their belief in the dire consequences of state intervention, either in the extreme form of ownership or in the lesser form of excessive regulation. In both circumstances, it was held, the vital ingredient of entrepreneurship was thwarted by the heavy hand of government. The argument continued that a revival of entrepreneurship and progressiveness could come about only through competition. Finally, in the Third World it became clear that those nations adopting capitalism were growing, while those following the socialist path were stagnating or, worse yet, crumbling.

The new ideologists of deregulation, privatization, and competition claimed they were not embarked on a new experiment but rather the rediscovery of an older public policy tradition. Although aware of the period's harshness, they looked wistfully backward to the great economic takeoffs in England and the United States in which laissez-faire was the alleged engine of growth, inventiveness and entrepreneurship. Then during the twentieth century the state undertook more and more control and restraint. Gradually, according to this view, state control took its toll, eroding entrepreneurship, innovation, productivity, and finally, economic growth.[1]

It is the contention of this book that the view of a laissez-faire past with minimal government intervention is profoundly wrong. The enormous energy of the English and American economies in the nineteenth and early twentieth centuries and the extraordinary technological progress and inventiveness of that period were associated with considerable government intervention. The underlying ideological perspective behind the remarkable achievements of the period was not laissez-faire but public service liberalism, a conception we will more fully examine later. Public service liberalism posits a presumption in favor of free competition that can be rebutted by showing that government intervention is necessary to attain social values that the unfettered market may not achieve. This conception, elaborated by John Stuart Mill, was gradually abandoned. But the change in conception was not simply from free competition to state control; it was far more complicated than that. Even today, the ideologues of free competition ignore the large state role in such dynamic economies as Japan, South Korea, and the Federal Republic of Germany.

Public service liberalism, like every other conception of public policymaking, always had its opponents. Socialists, populists, and libertarians for differing reasons opposed an ideology whose centerpiece is a rebuttable presumption in favor of free competition, but a presumption that can readily be invoked when important values in addition to economic efficiency are at stake. But as the story of American telecommunications policies illustrates, the decline of public service liberalism and its replacement by modern liberalism cannot be attributed to these opponents. Rather, it is attributable to policymakers and elites who did not overtly challenge the old order. Instead, during the Progressive Era they used new approaches that undermined public service liberalism to solve policy problems as they arose. The most important set of such problems were those that arose in connection with the need for a rapid, effective mobilization in World War I. Modern liberalism, which then began to supplant public service liberalism, entailed a more positive state role in which the issues were no longer posed in the forms required under pub-

[1] See, for example, Paul W. MacAvoy, *The Regulated Industries and the Economy* (New York: Norton, 1979), chap. 1.

lic service liberalism. Rather, under the exigencies of rapid wartime economic mobilization and drastic changes in the allocation of resources, it was assumed that state intervention and leadership, often of an elaborate kind, were required when there were problems to be solved. No longer was there a presumption in favor of market and voluntary solutions.

The changes in public philosophy were, thus, not dramatic but subtle. The participants did not debate these changes as Abraham Lincoln and Stephen Douglas debated the great issues of slavery and states' rights, with each side knowing precisely where it stood. Rather, like some fundamental changes, the participants during the period from World War I until the Great Depression focused more on specifics than on major confrontations in philosophy. As we shall see in chapter 8, the coming of radio provides an excellent case of how the transition in public philosophy occurred. Then the collapse of the American economy in 1929 shattered the older system of public service liberalism in virtually every sector except the set of industries conventionally grouped together as public utilities. In turn, the burgeoning state role had another important outcome—what political scientist Theodore J. Lowi describes as interest group liberalism.[2] Under it various forces sought to use state intervention in both the legislative and administrative arenas to achieve their ends, often by portraying their private ends as public ones.

Although the New Deal is often considered the beginning of a trend, it is, according to the view advanced here, the culmination. Like measles, modern liberalism began earlier than its most transparent symptoms. The forces that gradually led to the erosion of public service liberalism and the ascendancy of modern liberalism are numerous, but several are critical. Among the most important of these factors was the development of professionalism. Historian Robert Wiebe locates the burgeoning of professionalism at the dawn of the twentieth century. Older professions, such as medicine, raised formal entry barriers in order to bar the unfit, while at the same time establishing professional societies (such as the American Medical Association) to disseminate technical information within the profession and take stands on public policy issues, such as public health, on which government could act. Expanding older professions were joined by newer ones such as economics, teaching, social work, and administration.[3]

The formal entry and/or educational requirements of the new professionals both enhanced their prestige and provided them with confidence in their problem-solving abilities. Instead of relying *primarily* on the

[2] See Theodore J. Lowi, *The End of Liberalism*, 2d ed. (New York: Norton, 1977).

[3] Robert H. Wiebe, *The Search for Order, 1877–1922* (New York: Hill and Wang, 1967), chap. 5.

forces of markets and voluntary agreements to solve problems with government participating when these were insufficient, the new professionals placed primary reliance on specialists engaging in scientific planning to achieve results. Frequently the principal instrument that was expected to carry out the planning was the state, both because of its resources and its ability to impose sanctions on the disobedient. Of course, the new professionals also appreciated the long-standing fear of government excess. Therefore citizen protection was to be balanced with activist government. This would be the function of the new field of administrative law.

The scale of business enterprise also triggered specialized managerial expertise. Thus, professionalism arose in a number of facets of business enterprises, such as accounting, production techniques, and distribution. At the same time a general philosophy of scientific management under the leadership of Frederick W. Taylor and Robert G. Valentine took root.[4]

Many of the managerial elite, including some of the established figures who preceded the rise of professionalism, also embraced planning and cooperation with government. Judge Elbert Gary of the powerful United States Steel Corporation went so far as to call for the abandonment of competition and the allocation of market shares in proportion to capacity.[5] But more important than Judge Gary's pronouncements was the dramatic formation of trade associations in virtually every leading industry after the turn of the century, but especially between 1915 and 1920.[6] These institutions were based on cooperation and the sharing of market information among erstwhile rivals instead of on old-fashioned rivalry. From the government's perspective it was easier to deal with one trade association than with a host of industrialists.[7]

During the Progressive Era important political leaders close to Theodore Roosevelt and Woodrow Wilson embraced the new ideas. Herbert Croly, who deeply influenced Roosevelt, held that corporate planners closely regulated by professional governmental administrators can better attain the national economic interest without sacrificing efficiency than any other system. Although differing with Croly in some ways, Louis Brandeis, one of the strongest influences on Wilson, virtually triggered a scientific management craze through his skillful opposition to proposed railroad rate increases, claiming that scientific management would obvi-

[4] Ibid., pp. 151, 152; see also Samuel Haber, *Efficiency and Uplift: Scientific Management in the Progressive Era, 1890–1920* (Chicago: University of Chicago Press, 1964).

[5] See Ida M. Tarbell, *The Life of Elbert H. Gary* (New York: Appleton, 1925), p. 212.

[6] Robert F. Himmelberg, "The War Industries Board and the Antitrust Question in November 1918," *Journal of American History* 53 (June 1965): 59–74.

[7] See Robert F. Himmelberg, *The Origins of the National Recovery Administration* (New York: Fordham University Press, 1976), pp. 5–8.

ate the need for the rate increases. Whether in industry or public administration, planning and scientific management would supplant both the irrationality of partisan politics and the destructive impulses of class struggle. Bossism and socialism would recede before scientific management and planning.[8]

The event that caused the convergence of many of these trends, as well as modern liberalism's supplanting of public service liberalism, was America's preparedness campaign and entry into World War I. Older mechanisms could not rapidly redirect resources to wartime purposes. As historian Ellis W. Hawley observed: "The mere placing of orders through competitive bidding, so most leaders had concluded, could not bring this about, at least not without excessive costs and numerous delays."[9] By war's end government operated the railroad, telephone, and telegraph industries and exerted firm control over a wide range of other industries, including food and steel. Important representatives of the new professionals and the ideology of modern liberalism, such as Bernard Baruch and Herbert Hoover, assumed important roles in wartime planning, whereas agencies with broad powers, such as the War Industries Board, were staffed with lesser-known professionals confident in their abilities to plan the effective deployment of resources.[10]

Most of the governmental apparatus constructed during the war was dismantled when it ended. And many public leaders sought to return to the old order. Nevertheless, the precedent had been set, and the new professionals and their ideology of modern liberalism were in the ascendancy. After all, the wartime experience confirmed to them the virtues of a theory of public policy premised on state intervention as a first step, not merely as a possibility, after the presumption against it was rebutted. During the period before his ascent to the presidency, Hoover sought to expand the Commerce Department into a super agency that would direct cooperative economic action. Again, the Transportation Act of 1920, although rife with compromise, sought to rationally reshape the railroad

[8] See Oscar Kraines, "Brandeis Philosophy of Scientific Management," *Western Political Quarterly* 13 (March 1960): 191–201; William J. Cunningham, "Scientific Management in the Operation of the Railroads," *Quarterly Journal of Economics* 25 (May 1911): 539–62; Stephen Skowronek, *Building a New American State* (Cambridge: Cambridge University Press, 1982), pp. 268–69, 281–86; Herbert Croly, *The Promise of American Life* (1909; rpt., Cambridge, Mass.: Harvard University Press, 1965), pp. 115, 359, 362; David W. Noble, *The Progressive Mind, 1890–1917* (Chicago: Rand McNally, 1970), pp. 38, 39; and Haber, *Efficiency and Uplift*, pp. 51–62, 103, 113.

[9] Ellis W. Hawley, *The Great War and the Search for a Modern Order* (New York: St. Martin's, 1979), p. 22.

[10] See, generally, Robert D. Cuff, *The War Industries Board* (Baltimore: Johns Hopkins University Press, 1973); Wiebe, *The Search*, p. 170; and Hawley, *The Great War*, pp. 22–26.

industry under the aegis of the Interstate Commerce Commission (ICC); the act marked a significant enlargement of the old style of railroad regulation that lightly intervened in industry affairs in order to assure the attainment of values other than economic efficiency.[11]

Although Hoover's version of modern liberalism—the associative state—emphasized cooperation and was premised on joint government-business agreements, the virtual collapse of the business system in 1929, coupled with numerous scandals in such critical industries as finance and electric utilities, put business on the defensive even before the onset of the New Deal in 1933. Could business managements, with their presumed failures, be accorded the respectability of expertise? During the New Deal the state became the senior partner. Notwithstanding this important change, the New Deal represented, as Robert F. Himmelberg shows, a continuation of the idea of planning through government intervention that sprouted during the war and was nurtured by Hoover and others during the 1920s.[12] The older ideas of public service liberalism were in full-scale retreat. The contrast between early public intervention in telecommunications and policies toward radio in the 1920s and 1930s strikingly illustrate the wide gulf that had been crossed and the submerging of an older public policy tradition in favor of modern liberalism. The realm of public service liberalism would thereafter be limited to a few sectors, such as the traditional public utilities. During the Gerald Ford presidency the older philosophy would begin its retreat in these sectors as well.

Communications and Public Policy

The American communications industry from its origins to the New Deal is a primary example of public service liberalism. Telecommunications was not an exception to a general rule in which the state did not intervene in social and economic affairs. The communications and transportation sectors—the Post Office, telegraph, telephones, railroads, and general roads—constituted a high percentage of the gross national product (GNP). Further, many other industries and activities were subject to substantial government intervention under public service liberalism. The reasons that the state intervened in the private sector differed from industry to industry, situation to situation. Nevertheless, a general theo-

[11] Hawley, *The Great War*, pp. 101, 102; and K. Austin Kerr, *American Railroad Politics 1914–1920* (Pittsburgh: University of Pittsburgh Press, 1968), pp. 143–46.

[12] Himmelberg, *The Origins*, pp. 181–83.

retical construct—public service liberalism—determined when state intervention was appropriate and when it was not. One important class of industries in which the state routinely intervened was public services. Among important public services was communication, including the telegraph and the telephone.

Seeing why public service liberalism required close regulation of these instrumentalities and how they were regulated illuminates public service liberalism generally. In the course of this examination we will see how public service liberalism spurred the development and deployment of new technology and advanced the economy generally; competition is only one route to technological progress. Government intervention *of the proper type* can also spur it.

We begin with the myth of laissez-faire and move from that to a general overview of government intervention in the Anglo-American world from the nineteenth century to the 1930s. Most of the rest of this book is devoted to exploring the development of the government–private sector relationship in the telephone industry within that context, as well as its results. Past styles of public policy, unlike past events, can be recaptured.

The Myth of Laissez-faire

One of the most persistent myths about nineteenth-century political economy in Britain and the United States is that the state virtually did not interfere in economic and social affairs. The persistence of this view is at least partially attributable to the fact that it has received the approval of several eminent figures in the social sciences. According to their views, economic liberalism can be traced to the latter part of the seventeenth century in England and obtained doctrinal justification from Adam Smith in *The Wealth of Nations* (1776). Under its influence, England attained commercial and industrial supremacy in the ninenteenth century. The application of economic liberalism in the United States contributed importantly to pre–Civil War growth and the extraordinary transformation of the economy in the period following the war.

According to this view, the major attributes of economic liberalism were: (1) severe limitations on government coercive activities, (2) a rule of law based on private contractual and property right arrangements, and (3) reliance on market arrangements to enhance welfare. Government's role was limited to assuring the "provision of services that the spontaneous forces of the market may not produce or may not produce adequately."[13] According to F. A. Hayek, one of the most forceful advocates of this conception of traditional English liberalism, the system simultaneously maximized individual liberty and led to unprecedented material prosperity.

Consider also the views of Joseph Schumpeter, one of the giants of twentieth-century economics. He conceived traditional economic liberalism as "the theory that the best way of promoting economic development and general welfare is to remove fetters from the private enterprise economy and to leave it alone."[14] Schumpeter starkly contrasted this conception to the one the word *liberalism* has taken on since the 1930s in which the state has assumed a highly interventionist role. In his view, "The enemies of the system of private enterprise have thought it wise to appropriate its label."[15] Thus, the "enemies" of private enterprise unwittingly paid tribute to its success. The views of these scholars and others have become widely accepted by both their ideological allies and others, helping perpetuate the myth that laissez-faire was the mighty engine that advanced the most dynamic economies in the world through the nineteenth century into the first third of the twentieth.

Yet these views are supported by surprisingly little historical evidence. Nearly every scholar who has carefully studied the years in question in the United States or Britain has observed a remarkable degree of government intervention in economic and social affairs. As economic historian George Rogers Taylor remarked: "Present day writers, and especially those who fear the growing power of the government . . . make nostalgic reference to an earlier period in American history when a policy of laissez faire had general acceptance. . . . The exact time when this idyllic condition is supposed to have prevailed is seldom made more definite than are most references to the 'good old days.' "[16] Taylor and others have found numerous instances of such intervention, but very often at subnational levels. Intervention at local levels rather than at the national one is to be expected, however, when most activities were local or regional, at best, and communication was rudimentary compared to post-1930s standards.

Perhaps the best contemporary evidence of the extent of government intervention during the supposed age of laissez-faire in England comes from Herbert Spencer, the most rigorous advocate of laissez-faire as an

[13] F. A. Hayek, *Studies in Philosophy, Politics and Economics* (Chicago: University of Chicago Press, 1967), p. 165.

[14] Joseph Schumpeter, *History of Economic Analysis* (New York: Oxford University Press, 1954), p. 394.

[15] Ibid.

[16] George Rogers Taylor, *The Transportation Revolution: 1815–1860* (New York: Holt, Rinehart and Winston, 1951), p. 354.

ideology. In 1884 Spencer, looking back at nineteenth-century English public policy, complained bitterly of state intervention during both liberal and Tory regimes. Spencer scrupulously listed the innumerable statutes that enlarged governmental restraint during that period. For example, in 1861 alone Parliament enacted laws pertaining to working, fixed rates of hire for horses, taxes for rural drainage, and many other public works programs.[17] Spencer compiled his lists for year after year, noting the wide variety of subjects and goals to which legislation pertained. He directed his enmity most sharply at the Liberal party: "Such then are the doings of the party which claims the name of Liberal; and which calls itself Liberal as being the advocate of extended freedom."[18]

Spencer was correct from his perspective in complaining about the direction public policy had taken, but he was incorrect in his understanding of the theoretical underpinnings of state intervention during the long period in which laissez-faire was presumed to have reigned. Yet as historian William D. Grampp meticulously showed, the state intervened for many purposes during the so-called heyday of laissez-faire, including the regulation of rates, the redistribution of income through taxation, the settlement of labor disputes, support of employment, the control of food quality, and the regulation of working conditions.[19] Historian Arthur J. Taylor shows that the understanding of English leaders during the nineteenth-century was that laissez-faire achieved economic goals. Government was responsible for the attainment of many other goals, including health and safety, to which laissez-faire was not the appropriate route.[20]

And that is precisely the point, as relevant today as in the nineteenth-century. Laissez-faire is the best path (except under the conditions of market failure) to the achievement of certain economic values, the most important of which is efficiency.[21] But then, as now, economic values constitute only one set of values. Others may be less scientific and more difficult to define, but that does not make them any less important. Reflecting this understanding, Lord T. B. Macaulay commented during the Ten Hours Bill debate in 1846 on laissez-faire: "But you would fall into error if you apply it to the transactions which are not purely commercial. . . . The principle of non-interference is one that cannot be applied without great restriction where the public health or the public morality is

[17] Herbert Spencer, *The Man versus the State* (Indianapolis: Liberty Press, 1981), p. 17.

[18] Ibid., p. 24.

[19] William D. Grampp, *Economic Liberalism* (New York: Random House, 1965), 2:87–91.

[20] Arthur J. Taylor, *Laissez-Faire and State Intervention in Nineteenth Century Britain* (London: Macmillan, 1972), pp. 43–48.

[21] Market failure is "the inability of a system of private markets to provide certain goods, either at all or at the most desirable or 'optimal' levels." David W. Pearce, ed., *The MIT Dictionary of Modern Economics*, 3d ed. (Cambridge, Mass.: MIT Press, 1986), p. 264.

concerned."[22] And as we will see, much else was excepted from laissez-faire, because as constitutional historian G.S.R. Kitson Clark succinctly observed, "If you have any social values, freedom is never a sufficient answer to all problems."[23]

One of the functions of interest group politics was to help define what these "problems" were. Thus, Parliament enacted smoke control legislation in 1819, 1847, and 1863 not because smoke emissions constituted a market failure in the form of externalities, but because factory smoke defaced the countryside.[24] The actions and debates of smoke control may be comprehended by the terminology and conceptual apparatus of contemporary microeconomics, but that was not the way these problems were understood by contemporary observers. When we look at events in the eyes of the participants, a more sophisticated conception of the government–private sector interface emerges, the conception of public service liberalism. The state's role in a capitalist society is not simply intervention versus nonintervention, regulation versus laissez-faire. The choices are wider.

Public Service Liberalism

Two principal underpinnings of the complex approach to public policy that will be called public service liberalism are: (1) an intellectual tradition that can be traced to Adam Smith and other figures and (2) the common law. The first of these is found in the writing of the English political economists whose influence extended far beyond the academy. John Maynard Keynes aptly summarized the influence of such figures as Smith, David Ricardo and John Stuart Mill: "Practical men who believe themselves to be quite exempt from any intellectual influences are usually the slaves of some defunct economist. Madmen in authority, who hear voices in the air, are usually distilling their frenzy from some academic scribbler of a few years back."[25] Their views, in short, resonated in the corridors of government.

The critical importance of the common law in justifying government intervention (as opposed to providing rules for private dispute settlement) is sometimes overlooked. Long before current concerns over judicial activism, common law courts in England, the United States, and

[22] Quoted in Taylor, *Laissez-Faire*, p. 44.

[23] G.S.R. Kitson-Clark, *An Expanding Society: Britain 1830–1900* (Cambridge: Cambridge University Press, 1967), p. 13.

[24] Ibid., p. 158.

[25] John Maynard Keynes, *The General Theory of Employment, Interest and Money* (London: Macmillan, 1936), p. 383.

other common law countries were shaping public policy in the economic and social realms. They did so, however, in the mainstream of a strong common law tradition and not, as in much current activist judicial decision making, as radical departures from common law principles. The distinction, in short, between the nineteenth-century judge and the current activist judge is not between passive judicial behavior and a current predilection for public policymaking. Judges in the Anglo-American legal system have always made public policy. Rather, the distinction is between making public policy within the logic and framework of such centuries-old common law conceptions as a "public service" and the radical departure from such conceptions on the basis of a judge's whims, values, or strained views of "rights."

Let us begin with the political economist strain. The starting point is, of course, Adam Smith, to whom Schumpeter and others point as the progenitor of free market economics. Smith clearly *favored* a free market, but on closer scrutiny his views are far more complicated. Smith proclaimed, in a famous passage, that the pursuit of individual self-interest, as if "by an invisible hand . . . frequently promotes that of the society more effectually than when he really intends to promote it."[26] While most commentators draw an inference favoring the free market from this statement, they tend to ignore the critical presence of the word *frequently*. Smith customarily carefully chose his words and did not use *always*, *often* or *most of the time*. He chose a word connoting many exceptions to the general rule. And, indeed, when we examine Smith's works we find many exceptions to the rule. As economist Jacob Viner observed, Smith's primary interest was, as a practical reformer interested in policies, in forcefully advocating the main point—the adoption of laissez-faire in most circumstances.[27] Accordingly, he nowhere gathered together the exceptions in an integrated form. Nevertheless they constituted part of his overall outlook.

Smith, for example, saw a clear need for government-operated courts to mediate disputes and provide a framework of law for doing so. Government has an obligation to maintain law and order as well as national defense. But Smith went far beyond these elementary government obligations. He advocated the regulation of certain infrastructural services, such as roads, canals, and other public works, or government operation of them, whichever better served the public interest. He favored government protected patent and copyright monopolies for limited periods, tariffs to protect infant industries until they became viable, and temporary

[26] Adam Smith, *An Inquiry into the Nature and Causes of the Wealth of Nations* (1776; rpt., Oxford: Oxford University Press, 1976), 1:456.

[27] Jacob Viner, *The Long View and the Short* (Glencoe, Ill.: Free Press, 1958), p. 232.

monopolies for high-risk enterprises such as trading companies. Smith advocated public regulation in many instances. Thus, banking should be strictly regulated in order to promote a stable monetary framework. Recalling the Great Fire of London, he favored building restrictions. He favored paternalistic legislation that protected children and the poor. And this by no means exhausts Smith's list of governmental obligations; it simply corrects the common mistaken view of Smith's conception of the state's role.[28]

Lurking behind these exceptions were two fundamental points. The more commonplace one is that free markets cannot always operate according to the rules of a perfectly harmonious order. Monopolies, for example, may be inevitable in some circumstances and competition may, accordingly, not be able to function. But more important is the recognition that there are many other values in addition to those that the free market provides. Suppose, for example, that certain persons are too poor to purchase drinking water at prevailing market prices. Should they be permitted to die of thirst? Smith, as we have seen, was unwilling to sacrifice other values on the altar of economic efficiency. *The Wealth of Nations* is replete with such moralizing terms as *abuse* as well as value judgments on a host of subjects. These values are not as easy to spell out as the one generally favoring economic efficiency over inefficiency, but they constituted an important part of Smith's legacy as it was understood in the nineteenth century.

In the face of considerable state intervention during the nineteenth century, it was left to John Stuart Mill to spell out in more detail the system of public service liberalism that had evolved in England. We will look at his synthesis and then consider it in the context of English public policy. In Book V of his *Principles of Political Economy*, Mill delineated a theory of government intervention—the theory of public service liberalism. It begins with a rebuttable presumption in favor of laissez-faire. One should favor laissez-faire for several reasons. First, "There is, or ought to be some space in human existence thus entrenched around and sacred from authoritative intrusion no one who professes the smallest regard to human freedom or dignity will call in question."[29] Second, to be prevented from doing what one is inclined to do tends to erode mental and physical abilities. Thus, costs are imposed not only on the individual, but on society in the form of reduced productivity and innovation. Third, history is replete with examples of governmental abuse of the individual.

[28] See the summaries in ibid., pp. 228–44; and R. L. Couch, "Laissez-Faire in Nineteenth Century Britain: Myth or Reality," *Manchester School of Economics and Social Studies* 35, no.3 (September 1967): 203–10.

[29] John Stuart Mill, *Principles of Political Economy* (London: Longmans, Green, 1929), p. 943.

This tendency can be mitigated by diminishing the powers of government. Fourth, when a large number of duties are imposed on government they are either not done or done very poorly. Government, in short, tends to do things worse than individuals and cooperating groups directly interested in the matters.[30]

Mill, in keeping with the standard beliefs of his time, concluded that "in every instance, the burthen of making out a strong case [should be] not on those who resist, but on those who recommend, government interference. *Laissez-faire*, in short, should be the general practice: every departure from it, unless required by some great good, is a certain evil."[31] The elaboration of the exceptions to the injunction is, as befits Mill, complex and cannot be reduced to a few simple rules. These included what we now term consumer protection and labor legislation. They also included: government operation of "public services" when no private firm is interested in operating them; the protection of patent monopolies for limited periods in order to compensate and reward inventors for the public benefits they bestowed; and the regulation of privately owned public services, including, if necessary, restrictions on entry.[32]

Mill's views were not a statement of some hoped for utopia, but an articulation of the evolution of British public policy. They lauded the British system of political economy as it had evolved. It is within this context that we can understand Herbert Spencer's consternation at the amount of public intervention undertaken by the English government. Spencer was libertarian; Mill was much more in the mainstream of his time. He was a public service liberal. We can thus understand the extraordinary amount of actual intervention that occurred during the so-called heyday of laissez-faire. England, for example, as early as 1832, regulated transportation, and this regulation continued through the nineteenth century. When private ownership in England failed to reasonably meet public service requirements, it was often superseded by local government ownership.[33] The pattern was hardly laissez-faire, but both British Conservatives and Liberals supported it.

One survey of economic intervention in England lists in the area of promotion and regulation of business enterprises

> the railway companies acts from 1823 on . . . the reformed post office (1840) and nationalization of telegraphs, telephones, and broadcasting (1856–1869,

[30] Ibid., pp. 942–48.

[31] Ibid., p. 950. See also Samuel C. Hollander, *The Economics of John Stuart Mill* (Toronto: University of Toronto Press, 1985), 2:684–92.

[32] Mill, *Principles*, pp. 932, 975, 978; and Hollander, *Economics of Mill*, 2:758–63.

[33] See John P. McKay, *Tramways and Trolleys* (Princeton: Princeton University Press, 1976), pp. 19, 20, 91, 184.

> 1878–1911, 1922); inspection and enforcement of standards of amenity, safety, legality and so forth, in steam power, railways, mercantile marine, gas and water supply, weights and measures, food adulteration, patents, bankruptcy, and so on . . . and the facilitation and regulation of limited-liability joint-stock companies (1856–1862). A great variety of land acts notably restricted freedom of contract in a number of ways.[34]

Paralleling the development of national legislation was the growth of an administrative apparatus, often with considerable discretionary power to make policy based on broad grants of authority, and a large body of regulatory laws enacted at the level of local government.[35] And as we will see in our later discussion of the public service principle, the courts, too, were imposing restrictions on enterprise under the common law. Indeed, much of the discussion surrounding the growth of government was conducted in the language and context of the common law.[36]

American State Intervention

American government intervention was at least as extensive as Britain's and had much the same intellectual backdrop. The American Constitution, however, constructed a far more delicate balance of responsibility between the national government and the local ones than existed in England. Consequently, during the nineteenth and the first third of the twentieth century much public policy that was undertaken at the national level in England was undertaken at the state level in the United States. Further, even though judicial policymaking was important in England, the development of judicial review under the Constitution focused a considerable amount of centralized power in the federal courts, especially the Supreme Court, which was lacking in England. That is, the Supreme Court rejected a variety of state policies on the ground that such policies transgressed the American Constitution. While English courts wielded considerable power and shaped public policy under the common law in many ways, they lacked the power to declare policies unconstitutional because they violated a written fundamental law.

[34] J. B. Brebner, "Laissez-Faire and State Intervention in Nineteenth Century Britain," in E. M. Carus-Wilson, ed., *Essays in Economic History* (New York: St. Martin's, 1962), 3:261.

[35] See, for example, Kitson-Clark, *Expanding Society*, pp. 39, 137, 138; William C. Lubenow, *The Politics of Governmental Growth* (Hamden, Conn.: Archon, 1971), p. 82; and Phyllis Deane, *The First Industrial Revolution* (Cambridge: Cambridge University Press, 1965), pp. 218, 219.

[36] Lubenow, *Governmental Growth*, p. 26.

Notwithstanding these important differences, however, each American state (except Louisiana) has operated under a common law system that relied heavily on the elaborated structure of English common law. The predominant influence on the intellectual and public policy leaders of the new nation was English. In historian John R. Nelson's words: "America was heir to a liberal intellectual tradition already a century and a half old at her founding. This tradition and the ideas it contained constituted the liberal world view that was the ground, the substance, and the very language of the early American leaders."[37] England's intellectual and material predominance and progressiveness during the nineteenth century assured that it would be the principal model followed, although not blindly. Indeed, the United States would make important contributions to the development of public service liberalism.

Consider government promotion of new enterprises. Alexander Hamilton, although one of the Constitution's principal architects, was a highly controversial figure. Nevertheless, his path of encouraging manufacturing in the United States through active governmental policy was the one taken. His 1791 *Report on Manufactures* charted the national government's active role in this endeavor. Among the policies he advocated in the capital-short new nation filled with infant industries were duties on foreign manufactures, patent protection, careful inspection of imported products to protect consumers, incentives to lure foreign capital, and the development of a sophisticated banking structure.[38] In the same period at the more important state levels, governments were employing the police power—laws designed to protect public health, safety, morals, and welfare—to intervene in economic and social affairs. Thus, Oscar and Mary Handlin's important study of public policy in Massachusetts describes a variety of measures undertaken to encourage industry and economy, including inspection laws, pilotage regulations, subsidies, educational promotion to develop economic skills, occupational licensing, and the liberalized chartering of new businesses. All of these diverse policies were aimed at advancing the economy, a goal conceived as the common good. Thus, developing a modern transport system within the state served the advancement of all industry and, therefore, the common good.[39]

[37] John R. Nelson, Jr., *Liberty and Property* (Baltimore: Johns Hopkins University Press, 1987), p. 1.

[38] Alexander Hamilton, *Report on Manufactures*, in Jacob E. Cooke, ed., *The Reports of Alexander Hamilton* (New York: Harper & Row, 1964), pp. 115, 148, 168, 175–77.

[39] Oscar Handlin and Mary Flug Handlin, *Commonwealth* (Cambridge, Mass.: Harvard University Press, 1947), pp. 71–75, 171–80, 232; and Bernard Schwartz, *The Rights of Property* (New York: Macmillan, 1965), p. 39.

All of the other states that historians have studied reveal a similar breadth of public policymaking. State policies were designed to promote the development and expansion of industry. Yet at the same time states undertook to assure that business activities operated to promote the common good, rather than to act in conflict with it. On the one hand, states would promote enterprises through regulatory, licensing, subsidy, or other policy instruments. On the other, the activities of these enterprises could be curbed or compelled to operate for the public good through the police power, including, as Chief Justice John Marshall observed, "inspection laws, quarantine laws, health laws of every description."[40] American public service liberalism conceived a kind of stewardship under which unregulated operation would yield many important public values, such as efficient operation, low prices, variety, and product quality. But the state would also regulate and assist enterprises to assure that these goals would be met and to attain other goals as well. When the required regulation was very extensive, the industry or activity was called a public service or public utility. In turn, as we will see in the case of the telephone, this had important policy consequences. But at this juncture, the critical point is that the policy distinction was not between unregulated and regulated activities and enterprises. Rather, the emphasis was on the extent of regulation required.

Let us consider state policy in Georgia, often considered less innovative and active than the principal eastern states. Milton S. Heath's comprehensive study provides details. As early as 1755 Georgia enacted a public roads act to promote the common good. It instituted inspection laws for lumber, meat products, tar, turpentine, leather, and tobacco before the turn of the nineteenth century.[41] Such policies as bank-chartering legislation, promotion of infrastructure, and mixed state-private enterprises followed. When the enormous economic importance of the railroad was recognized in the 1830s, the state undertook a variety of policies to promote and regulate it. At the same time, as the banking industry illustrated, certain industries were subject to strict prescription and penalties for failing to fulfill their particular obligations to serve the public good. Not only were banks and services customarily treated as public utilities subject to a variety of regulations, but such other businesses as cotton warehouses and flour mills and such professionals as physicians, surgeons, and midwives were similarly treated. Finally, extensive intervention was undertaken not only at the state level, but at the

[40] *Gibbons* v. *Ogden*, 9 Wheat. 1, 203 (1824).

[41] Milton Sydney Heath, *Constructive Liberalism* (Cambridge, Mass.: Harvard University Press, 1954), pp. 52–57.

local one as well. Courts, as well as the legislature and executive branch, were involved in the policy process.[42]

Other state studies have reached parallel conclusions to those of the Massachusetts and Georgia cases. James N. Primm's study of Missouri public policy from 1820 to 1860 and Louis Hartz's celebrated examination of Pennsylvania focus on chartering and its implications. Primm, for example, emphasizes the close regulation of banking, public utilities, and industries that were central to Missouri's economic development, such as tobacco and warehouses.[43] Hartz, too, focuses on regulation and the licensing of occupations. Like other states, he emphasizes that the state grant of corporate powers was not freely given. In order to be chartered, a corporation had to show that it would serve a public purpose.[44] All of the states examined support the view described here as public service liberalism, that the state may intervene to support industrial development, often through legislation favoring specific commercial interests such as railroads, canals, and water-powered gristmills.[45] Depending on the specific values sought in each case, government intervention can be short or long-run.

Many of the industries in which such intervention was justified have been variously described as "public services," "public utilities," "public employments," and so forth. In the next chapter we define the concept of a public service as it was understood in the nineteenth century and before. But for now, we need only point out that the conception of public service liberalism here set forth can explain the strong endorsement of public roads and canals that Albert Gallatin, Jefferson's financial expert and a self-proclaimed disciple of Adam Smith, provided in 1808. Without such public effort the "beneficial works for American commerce will not be prosecuted."[46]

For much the same reason New York State commissioners favored in 1811 the construction of the Erie Canal, and all states during this period fixed rates and fees for transportation, including ferries, canals, bridges,

[42] Ibid., pp. 188, 231, 239, 281, 312, 314, 361–67.

[43] James Neal Primm, *Economic Policy in the Development of a Western State: Missouri* (Cambridge, Mass.: Harvard University Press, 1954), pp. 27–34, 93–95, 116–19.

[44] Louis Hartz, *Economic Policy and Democratic Thought* (1948; rpt., Chicago: Quadrangle, 1968), pp. 63–86, 96.

[45] See, for example, Leonard W. Levy, *The Law of the Commonwealth and Chief Justice Shaw* (Cambridge, Mass.: Harvard University Press, 1957), pp. 303–21; and Harry H. Pierce, *The Railroads of New York* (Cambridge, Mass.: Harvard University Press, 1953), pp. 15–17.

[46] Quoted in Joseph Dorfman, *The Economic Mind in American Civilization* (New York: Viking, 1946), p. 328; and Nelson, *Liberty and Property*, pp. 101–26. On Gallatin's reverence for Smith, see Nelson, *Liberty and Property*, p. 206; and Grampp, *Economic Liberalism*, 1:165.

and roads.[47] When the American canal era was in full bloom during the 1820s, rates and tolls were strictly controlled, and public regulatory officials were granted wide discretion in decision making.[48] And because values in addition to those that the free market could provide were needed in the expanding urban settings of the 1850s, cities undertook the operation of water distribution, sewage, and fire control and strictly regulated private firms in the provision of these and other public services.[49]

The industry that was the quintessential public service in the nineteenth century was, of course, the railroad. As early as 1815 New York granted its first railroad charter in which the railroad was compelled to operate for the benefit of the public. Early charters limited rates and net earnings on capital. These restrictions and the imposition of public service obligations were in keeping with James Madison's 1816 admonition that government should undertake measures that would bind the country together by promoting intercourse and improvements.[50] The public service conception of railroads as stewards whose privileges extended only so far as they promoted the public welfare—and which could be withdrawn if they failed to do so—continued through the nineteenth century.[51] As railroads assumed the position of the nation's most dynamic industry, independent commissions assumed much of the responsibility for their continuous regulation, and in 1887 the ICC was formed to undertake national regulation. Although political battles unquestionably shaped the particular powers that legislatures granted to the railroad commissions, it is important not to forget their underlying public service character. As Henry S. Haines, one of the nation's leading railroad analysts, wrote in 1907: "It only performs its accepted service at all by permission of the same sovereign power to which it owes its very existence. . . . The claim upon its service which is of a public character grows out of the service that it may be required to render to the members of the community."[52]

[47] Dorfman, *Economic Mind*, pp. 329, 396.

[48] Harry N. Scheiber, *Ohio Canal Era* (Athens: Ohio University Press, 1969), pp. 247, 248.

[49] Charles W. Cheape, *Moving the Masses* (Cambridge, Mass.: Harvard University Press, 1980), pp. 1–15.

[50] Lewis H. Haney, *A Congressional History of Railways in the United States* (Madison: University of Wisconsin Press, 1908), 1:16, 20, 23, 24, 43–46, 78.

[51] See, for example, Lee Benson, *Merchants, Farmers and Railroads* (Cambridge, Mass.: Harvard University Press, 1955), pp. 2–6; and E. C. Kirkland, *Industry Comes of Age* (New York: Holt, Rinehart and Winston, 1961), pp. 44, 115.

[52] Henry S. Haines, *Railway Corporations as Public Servants* (New York: Macmillan, 1907), pp. 20, 21.

The Tensions of Public Service Liberalism

The expansion of the American post–Civil War economy put the fundamental ideas of public service liberalism to the test. The development of a large number of new industries and new considerations and technologies in older ones compelled courts and legislatures to ask continuously whether or not the rebuttable presumption in favor of laissez-faire applied in each instance and to seek out justifications for their decisions. But government intervention, especially in the form of valuable franchises, could virtually invite corruption and the facile assertion of sham reasons justifying a monopoly or a favored position. Only the difficult struggle to adopt clear and explicit standards could guard against this unfortunate result. At the very least, government's potential to favor some persons or business groups over others through interventionist policies was at odds with the cherished principle of equality of opportunity. Finally, historian Charles W. McCurdy has persuasively argued that in the third quarter of the nineteenth century, the Supreme Court became sensitive to preventing states from disrupting the nation's emerging free trade network through restrictive regulations that favored local enterprises.[53] These concerns led many policymakers to scrutinize much government intervention carefully, but the fundamental tenets of public service liberalism remained.

Later chapters show that some of the tensions in telephone industry policy reflected some of these concerns. In this respect the young telephone industry was at the center of the struggle to shape specific policies within the framework of public service liberalism that persisted until its erosion. In a word, as the famous 1873 *Slaughter-House* cases illustrate, the central principles of public service liberalism were established, but their specific application was a far more difficult matter in a complex society. In those cases, Louisiana granted exclusive stockyard and slaughterhouse privileges to certain applicants in New Orleans, thereby denying others the right to pursue those occupations in that city. A bare majority of the Court upheld the measure on health grounds and declared it a proper exercise of the state's police power. But the dissenters urged that the monopolistic restrictions went much farther than was necessary as a health measure and that monopolistic privileges should not have been granted.[54]

[53] Charles W. McCurdy, "American Law and the Marketing Structure of the Large American Corporation, 1875–1890," *Journal of Economic History* 38 (September 1978): 648, 649.

[54] *Slaughter-House Cases*, 16 Wall. 36, 83 U.S. 394 (1873); see also Charles W. McCurdy,

The issues of monopoly and the specifics of state intervention would vex the telephone industry from its invention in 1876. Nevertheless, as we will see, public service liberalism worked effectively as government and the regulated industry evolved a relationship that led to one of the most remarkable examples of service provision in the world. Public service liberalism was, in summary, an eminently successful public policy in an increasingly complex society. It was undermined by the ideology of a new clerisy that came to the fore during the Progressive Era.

"Justice Field and the Jurisprudence of Government-Business Relations," *Journal of American History* 61 (March 1975): 970–1005.

2

The Telephone and the Public Service Idea

Public Service and American Society

As we saw, under public service liberalism a rebuttable presumption exists against government restraint. Some activities are, however, regulated in order to attain certain goals, and even more drastically, some industries are comprehensively regulated. One important set of industries comprehensively regulated pursuant to public service liberalism are those called public utilities or public services. These sectors are critical in any economy and include water, transportation, and communications. To understand public service liberalism in these sectors is to understand public service liberalism as a theory of public policy. We will focus specifically on the communications area and thereby evaluate the ability of public service liberalism to serve the public and advance the development and deployment of new technology. In this and later chapters we will see how the institutional structures direct public service companies toward appropriate goals and help guard against their perversion in favor of special interests.

Following the discussion of the public service idea generally, we will see why the telephone was regarded as a public service, even before its enormous economic impact was recognized. The consequences of that status would affect the industry through the dramatic changes that took place from the telephone's invention in 1876 until recent times. To begin, however, we will see how critical the public service sector was becoming in the American economy during the nineteenth and early twentieth century. The entire structure of the economy, and of society more generally, was dependent on that sector and the policies governing it.

At the dawn of the nineteenth century less than 2.5 percent of the American population lived in towns and cities of more than twenty-five thousand people. By 1830 that figure was still only approximately 4 percent. Yet by 1890, whereas the total population had jumped almost twelvefold to approximately 63 million, more than 22 percent of the American people lived in urban areas with populations in excess of twenty-five thousand inhabitants. The urbanization of America was associated with an extraordinary burst of economic energy; per capita GNP rose threefold from the 1869–1878 average to the comparable 1900 figure.[1] Urbanization, industrialization, and the growing complexity and great increases in the number of commercial transactions completely transformed the way most Americans lived and worked, including those in rural areas.

When the nineteenth century began, most Americans supplied themselves with water from wells and such natural sources as springs and lakes. Many of the public services were unnecessary or unknown, including water distribution, gas, electricity, the telegraph, the telephone, the railroad, and the electric-powered streetcar and cable car. But as the century proceeded, these developments occurred, playing a critical role in future industrial and commercial developments as well as the personal and social lives of the urban and rural populations.[2] Water distribution, of course, developed first. Although the first system in Boston was constructed in 1652, at the turn of the nineteenth century only about fifteen existed. By 1896 about thirty-two hundred such systems existed. Gas for public lighting was introduced in 1816 over the objections of those who argued that God intended that it should be dark at night. As the century progressed, incandescent lighting, electric power, the railroad, the electric-powered streetcar, the telegraph, and the telephone made their appearances, developing and improving rapidly.[3]

At one level some of these public service industries were or had become "necessities" in the everyday lives of Americans. Clearly water distribution falls into this category. At another level, reasonable access to these services was considered appropriate to the level of society that America had become or was becoming. Gas for cooking or electricity for illumination are clear examples. These services were conceived in much the way that we now conceive the "right" to primary and secondary education, without which society would be at a lower level. Education is considered necessary to the principle of equality of opportunity.

Progress in the realm of producing and distributing goods was conceived as inextricably linked to these improvements in power, transportation, and communication. The connections between power and manufacturing, and between transportation, communication, and distribution, are too obvious to require exposition. But improvements in transportation and communication are also intimately linked with production. As Alfred D. Chandler has observed:

[1] U.S. Bureau of the Census, *Historical Statistics of the United States, Colonial Times to 1970* (Washington, D.C.: Government Printing Office, 1975), pt. 1, pp. 8, 11, 12, 224.

[2] The classic survey of the invention and economic exploitation of new technologies during the period is Victor S. Clark, *History of Manufactures in the United States*, vol. 2 (1929; rpt., New York: Peter Smith, 1949).

[3] An excellent survey is Eliot Jones and Truman C. Bigham, *Principles of Public Utilities* (New York: Macmillan, 1931), chap. 1.

> But of far more importance to the expansion of the factory system was the reliability and speed of the new transportation and communication. Without a steady, all-weather flow of goods into and out of their establishments, manufacturers would have had difficulty in maintaining a permanent working force and in keeping their expensive machinery and equipment operating profitably. Moreover the marketing revolution based on the railroad and the telegraph, by permitting manufacturers to sell directly to wholesalers, reduced requirements for working capital and the risk of having unsold goods for long periods of time in the hands of commission merchants. Reduced risks and lower credit costs encouraged further investment in plant, machinery and other fixed capital.[4]

In short, the division of labor, as Adam Smith observed, is limited by the extent of the market. And the extent of the market is limited, in turn, by the speed, reliability, and cost of communications. Rapid and extensive communications, thus, can radically transform production as well as distribution.

Although it is customary to think of the period beginning in the 1960s as the era of the "information revolution," the role of information (or what was termed "intelligence" in the nineteenth century) and the means to convey it were as critical to the changes that occurred in that century as those that have occurred in the last quarter century. Less than twenty-five years after Samuel F. B. Morse's invention of a practical telegraph in 1837, there were thirty-two thousand miles of pole line transmitting 5 million messages per year.[5] In 1861 coast-to-coast telegraph communication began, and in 1866 the Second Atlantic Cable opened, permitting rapid communication between the United States and Europe. Among the effects of the telegraph were the expansion and integration of markets, the reduction of information and transaction costs, and the facilitation of capital movements. The time required to enter into transactions dramatically diminished, and with this efficiency gain the number of transactions increased enormously. Moreover, the telegraph made possible the development of institutions that ushered in the modern complex business system: commodity exchanges, a national securities market, wire service, and rapid dissemination of general and business news and credit reporting, which itself encouraged transactions by lowering credit risks.[6]

The telegraph, in short, was not simply another new invention. Virtually every economic activity was significantly affected by the telegraph.

[4] Alfred D. Chandler, Jr., *The Visible Hand* (Cambridge, Mass.: Harvard University Press, 1977), p. 245.

[5] Richard B. DuBoff, "The Telegraph and the Structure of Markets in the United States, 1845–1890," in Paul Uselding, ed., *Research in Economic History* (Greenwich, Conn.: JAI Press, 1982), 8:256, 257.

[6] Ibid., pp. 257–65.

Although its commercial capabilities were not recognized in the nations of Europe (with the exception of Great Britain), the telegraph in the United States was, together with the railroad, critical in the development of national markets.[7] The telegraph in the United States was an important component of the economic infrastructure—those "services which would facilitate subsequent industrial investment and growth of industrial output."[8]

Before the arrival of regulatory agencies, policies for public utilities were made by judges employing an evolving common law and legislators promulgating rules in new situations. They were guided by the maxim set forth by Chief Justice William Howard Taft, speaking for a unanimous Supreme Court: "Freedom is the general rule, and [governmental] restraint the exception."[9] But as we saw in chapter 1, there were many exceptions.

The Background of the Public Service Idea

Sensing differences and precisely defining them so that virtually every case is covered are problems that have frequently vexed judges and other policymakers. Extreme examples begin the analysis. Most of us would agree that a monopolistic firm distributing drinking water in a community should not have the absolute right to refuse service to a resident, but a single firm dealing in precious stones should have the right to effectively exclude customers on economic grounds. (Such refusals on the basis of race or any other civil rights classification are not at issue here.) Now let us assume that there are two or more water companies competing with one another. Clearly, this would not change the rule at which we are grasping, since the competitors collectively could still exclude certain potential customers for economic reasons. Without collusion water distributors could decide that supplying certain areas or customers is uneconomic. In the case of precious gems, we are apt to be indifferent to the fact that all sellers may charge prices so high as to effectively preclude many people who desire gems from purchasing them.

The starting point, then, in distinguishing public service companies from others is that the most important consideration is the kind of service involved and not the number of firms or potential firms in an industry. If

[7] On the slow commercial development of the telegraph in Europe, see A. N. Holcombe, *Public Ownership of Telephones on the Continent of Europe* (Boston: Houghton Mifflin, 1911), pp. 3–15.

[8] Alexander Gerschenkron, *Economic Backwardness in Historical Perspective* (New York: Frederick A. Praeger, 1962), p. 107.

[9] *Wolff Packing Co.* v. *Court of Industrial Relations*, 262 U.S. 522, 534 (1923).

the economic characteristics of making ermine jackets were such that a single firm would be the provider of the product, that business would still not be a public service. Put another way, although the economic characteristics of an industry play important roles in shaping policy (or no policy) toward it, the social characteristics of an industry are primary in determining whether or not a firm is a public service company.

The point is an extremely important one because many contemporary policymakers and commentators employ *only* economic criteria in making their policy recommendations. Under their view, if an industry can be shown not to be a natural monopoly—an industry in which production is done most efficiently by a single firm—it should no longer be subject to economic regulation.[10] For example, a 1975 article by economist Leonard Waverman plausibly argues that because of technological changes, long distance telephony was no longer a natural monopoly. Accordingly, Waverman concludes that competition should supplant monopoly in long distance service.[11] But under public service liberalism the framework for policymaking involves far more than economic criteria.

Monopoly, as we will see later, plays an important role in the policy toward public service companies, but it is not the defining characteristic. Early English public service occupations included chimney sweeps, carters, and auctioneers. References in important cases to monopoly, according to a leading legal historian, "seem to be for the purpose of emphasizing the size and importance of the business and not of delimiting a necessary condition to regulation."[12] Indeed, in the important case of *Brass* v. *Stoeser*, the Supreme Court concluded that the approximately six hundred competitive grain elevators along the line of the Great Northern Railroad in North Dakota were public service companies.[13] Most important, as we will see, telephone systems were considered public service companies even when they were engaged in vigorous competition.

The public service idea has also become enmeshed in important con-

[10] On the definition and characteristics of natural monopolies, see Richard Schmalensee, *The Control of Natural Monopolies* (Lexington, Mass.: Lexington Books, 1979).

[11] Leonard Waverman, "The Regulation of Intercity Telecommunications," in Almarin Phillips, ed., *Promoting Competition in Regulated Markets* (Washington, D.C.: Brookings Institution, 1975), pp. 232, 233.

[12] Breck P. McAllister, "Lord Hale and Business Affected with a Public Interest," *Harvard Law Review* 43 (1930): 769. A contrary, albeit unpersuasive, view is contained in the Reply Comments of International Business Machines Corporation, *In the Matter of Policy and Rules Concerning Rates for Competitive Common Carrier Services and Facilities Authorizations Therefor*, CC Docket no. 72-252 (April 4, 1980). Essentially, IBM was employing the argument to head off the possibility of FCC regulation of computer-related services.

[13] *Brass* v. *Stoeser*, 153 U.S. 391 (1894).

stitutional questions that are apart from its theoretical basis. Behind the passions of the constitutional questions lay important policy issues, analytically separate from the former but, unfortunately, intertwined in practice. In brief, the question of whether a particular business is "clothed with a public interest" and therefore should be heavily regulated because that would be sound public policy is very different from whether the price regulation of a particular business is permissible under the Fourteenth Amendment to the Constitution or whether it constitutes an unconstitutional taking of property without due process of law. Virtually every student of constitutional law is familiar with the leading cases culminating in *Nebbia* v. *New York*, finally rejecting the distinction on constitutional grounds when the Court held five to four that New York could constitutionally fix minimum and maximum retail milk prices.[14] But few are aware of the numerous English and state common law decisions that developed and applied the public service company concept on public policy grounds. We are concerned with the public policy aspect of the distinction between public service companies and other firms, not the now settled constitutional issues that in many ways obscured the fundamental distinction that we are exploring.

Classifying a firm or industry under the heading public service imposed an explicit set of obligations on that firm or industry. In this respect the public service concept differs from other types of regulation and has important policy consequences. For example, a telephone company is obligated to transmit the messages of every person or firm desiring to use the service at a reasonable fee. In contrast, many franchises have no such obligation. For example, cable companies are local monopolies in most communities. Yet they are under no obligation to transmit the signals of many telecasters. Rather, subject to some must-carry rules, they are free to choose which signals they will transmit to consumers. Thus, if we envision a simple three-party flow of information from sender to distributor to receiver, the public service company concept widens opportunity at both the sending and receiving ends, even though the number of firms in distribution is limited. In contrast, the simple franchise, typified by cable regulation, can often restrict opportunity at both the sending and receiving ends. Thus, some cultural services that appeal to small audiences are excluded from many cable systems. But under the public service company idea they would have to be carried, subject only to reasonable prices and regulations.

It is, of course, easier to delineate what a public service company is not than to spell out what it is. Indeed, the U.S. Supreme Court sputtered so badly as it attempted to apply the "clothed with a public inter-

[14] *Nebbia* v. *New York*, 291 U.S. 502 (1934).

est" idea to different services and industries that some observers claimed, toward the end of its constitutional career, the public service concept was worthless. To many observers in the 1930s it was a legalistic relic of the laissez-faire era and made no economic sense. For example, one scholar asserted: "We must approach the future with a frank admission that our . . . public utility concept is not based on an adequate classification of economic, industrial, political or social facts. . . . Public utilities are a part of our great web of industrial activity and . . . their regulation is merely one phase of a necessary broad economic plan."[15] To many legal scholars and jurists in the same period the conception represented nothing more than the misguided values of some Supreme Court justices, unable to provide a workable set of decision rules or even a consistent conception to the increasing complexities of industrial life.[16]

Nevertheless, many commentators have noted the remarkable capacity of common law judges to transform concepts and ideas that originated in feudal and agrarian England into ones that are functional in a capitalist industrial society. The public service company concept can be traced back to the fourteenth-century idea of a "common calling." During an era of limited business activity and internal trade, a common calling was one in which a person was in business—selling to customers—as opposed to being in the exclusive employ or service of another person. Thus, there could be a common carrier, common baker, common distiller, and so on, as well as private counterparts of each occupation. Neither monopoly nor the kind of occupation determined whether one was classified as "common."[17]

One of the original policy purposes of the "common" classification was to impose a duty to serve all who applied at reasonable charge. This requirement, in turn, stemmed from the Black Death, which broke out in England in 1348, drastically reducing the working population. In order to prevent the surviving workers from extracting "unreasonable" wages, the Statute of Labourers (1349) required all common laborers and tradesmen to practice their callings at reasonable rates to whomever applied.[18] Although the original economic reasons for the idea of a "common" calling disappeared, the concept underwent an important transformation. Although exact dates are not known, sometime during the latter part of the seventeenth century most trades began to do business generally with the public. Accordingly, the idea of a common calling began to lose sig-

[15] Irwin S. Rosenbaum, "The Common Carrier-Public Utility Concept: A Legal Industrial View," *Journal of Land and Public Utility Economics* 7 (1931): 167.

[16] Thomas P. Hardman, "Public Utilities," *West Virginia Law Quarterly* 37 (1931): 267.

[17] Edward A. Adler, "Business Jurisprudence," *Harvard Law Review* 28 (1914): 147–55.

[18] Norman F. Arterburn, "The Origin and First Test of Public Callings," *University of Pennsylvania Law Review* 75 (1927): 421–28.

nificance in most kinds of businesses. Certain kinds of businesses, however, most notably common carriers by land and water and innkeepers, were treated differently. This treatment marked the beginning of the idea of the public service company.[19]

The analogy made at that time was to the duties and obligations of public officeholders, who theoretically are held to higher performance standards than others. A 1701 decision comparing the postmaster general with the new conception of common callings explains:

> Whenever any subject takes upon himself a Publick Trust for the Benefit of the rest of his fellow Subjects, he is . . . bound to serve the Subject in all the Things that are within the Reach and Comprehension of such an Office. . . . If on the Road a Shoe fall off my Horse, and I come to a Smith to have one put on and the Smith refuse to do it, an Action will lie against him, because he has made Profession of a Trade which is for the Publick Good. . . . One that has made Profession of a Publick Employment is bound to the utmost Extent of that Employment to serve the Publick.[20]

Thus, certain occupations, because they did things that were public in nature (as yet undefined), were under a special set of obligations that included the duty to serve all impartially and adequately. The central obligation—the duty to serve all—would be of little or no value if firms so obligated could impose any charges that they wished, discriminate between customers in charges or the quality of service, or provide service of a quality that would deter many potential customers from using the service. From these logical corollaries one could readily foresee the vast growth of obligations that were imposed on public service companies, all of which flow from the few defining characteristics of such companies. In brief, since a firm may evade these obligations in many ways, a heavy dose of regulation is required to assure that they fulfill them.

When the public service company conception was devised in the late seventeenth century, there was little need to define the idea sharply. Few businesses were covered, and most important, the number of new businesses that might conceivably be included—namely, those in communications and transportation—did not expand significantly until the major technological breakthroughs of the nineteenth century. Moreover, the sharp intellectual division between what the appropriate roles are for state and free market that began during the time of Adam Smith had not yet taken root. Consequently, there was no great need for the courts or other policymakers to sharpen the conception of the public service

[19] Charles K. Burdick, "The Origin of the Peculiar Duties of Public Service Companies," *Columbia Law Review* 11 (1911): 514–25.

[20] *Lane* v. *Cotton*, 12 Mod. 472, 484, 485 (1701).

company. The short list of industries covered, reasoning by analogy and the common law's mechanism of rule by precedent, provided sufficient guidance.

The Nineteenth-Century Development of the Public Service Idea

By the turn of the nineteenth century public service liberal ideas were in the ascendancy. Coupled with the vast changes wrought by the industrial revolution and the outpouring of new technologies, including many in transportation and communication, the need for refinement and clarity in the public service company concept was manifest.

Thus, a judge or other policymaker considering an exception to the general rule of free competition unrestrained by government intervention was expected to consider whether some "great good" could not be achieved without government intervention. If the "great good" could not be found, the rule of more or less unrestrained competition would prevail. But if laissez-faire could not remedy the situation, the next problem was to design an appropriate solution. In keeping with the common law method of looking at issues only when they are raised by real disputes and incrementally building up sets of principles, there is no elaborated discussion in any single source of what the "great good" that might override unrestrained competition is. Sometimes monopoly in an important service—by definition, the absence of unrestrained competition—prevents the attainment of the "great good" through competitive means. But at other times, such as in the case of warehouses, monopoly was irrelevant.[21]

As nineteenty-century case law developed, several precepts evolved. First, the concerns that led courts to define an activity as a public service applied both to its effects on the economic life of the community and to other aspects of social life as well. Most of the emphasis was on the former, but as we will see, other aspects of social life, such as the ability to be in social communication with other people, were also considered important values by the courts. With these paramount considerations the courts developed three elements that defined a public service. First, the product or service must be requisite to the community's level of civilization or necessary to its economic life. Thus, a court could conclude that a new, undeveloped technology deserved public service status if its potential indicated that it would be appropriate to the community's level of

[21] See, especially, the discussion in Harleigh H. Hartman, *Fair Value* (Boston: Houghton Mifflin, 1920), pp. 20–26.

civilization. Obviously what is requisite to a primitive society is different from what is requisite to modern industrial society. Second, the activity must have current or future widespread external effects on a community. Third, the unrestrained mechanisms of the market will *probably* not provide significant segments of the community with the service or product in sufficient quality and quantity. In the words of the distinguished nineteenth-century economic theorist Henry C. Adams: "The industry of transportation is fundamental in the industrial organization of a community. He who controls the means of communications has it in his power to arbitrarily make or destroy the business of any place or any person."[22]

In the late nineteenth century, railroads had the capacity to make or break businesses and, indeed, whole communities. Individually, collectively, or collusively (and there was a considerable amount of collusion), railroads had the *power* to thwart the canon of free competition by arbitrarily favoring one firm (or its community) over another.[23] Thus, one cornerstone of the theory of a public service company is that some activities are so central to business life, they must be subject to government economic regulation in order to protect the values of free competition and equality of economic opportunity.

Farmers, as well as many businesses and communities, strongly supported the Granger movement in the 1870s and 1880s on the theory that the railroads could virtually nullify the opportunity to compete in price, service, and quality by arbitrarily discriminating among them. They consequently turned to public service regulation.[24]

It is readily apparent that in the last quarter of the nineteenth century railroads were central to business life. They were the sole practical means through which shippers reached distant markets. And in an expanding economy railroads interacted with virtually every other kind of business. Railroad policies could determine the vitality or stagnation of whole communities and, indeed, of the nation. The influential 1886 Cullom Report, which led to the enactment of the Interstate Commerce Act, provides a contemporary view of the centrality of the railroad and why it is a model of a public service industry: "The railroad, as an improved means of communication and transportation, has produced indescribable changes in all the manifold transactions of every-day life which go to make up what is called commerce. Successful commerce brings prosperity, which in turn

[22] Henry C. Adams, "Introduction," in Frank H. Dixon, *State Railroad Control* (New York: Thomas Y. Crowell, 1896), p. 9.

[23] The classic study of railroad collusion and its limitations is Julius Grodinsky, *The Iowa Pool: A Study in Railroad Competition* (Chicago: University of Chicago Press, 1950).

[24] See Solon Justus Buck, *The Granger Movement* (Cambridge, Mass.: Harvard University Press, 1913), chaps. 1, 3-6.

makes possible the cultivation and development of the graces and attributes of the highest civilization."[25]

Although the telephone did not, of course, achieve the commercial importance of the railroad for many years after its invention, its *probable* centrality in business life was grasped almost immediately. Typical of this perception is the 1910 congressional statement in support of including interstate telephone and telegraph under ICC rate regulation: "Now the telegraph line and the telephone line are becoming rapidly as much a part of the instruments of commerce and as much a necessity in commercial life as the railroads."[26] In short, future as well as immediate strategic centrality was important in determining public utility status. For this reason the telephone was considered a public utility almost from its inception.

At this point one may object and suggest that steel or coal were central to business life during the nineteenth century. The refusal of coal companies to supply businesses or communities, then, could have a crippling effect similar to a railroad's refusal to serve or to serve without discrimination. Here we must focus on a critical difference between public service and other industries, no matter how extensive the latter's sales may be. There is no reasonable alternative way of obtaining the service of those engaged in a public service enterprise. In contrast, intraproduct and interproduct competition in other industries usually allows the provision of reasonable substitutes or the same goods or services from distant places. The point is illustrated by an 1858 Wisconsin case in which a gas company refused to supply gas to a resident. Since ordinary businesses can refuse to deal with customers but public service companies cannot, the case turned on the classification issue.[27] The court asked how the provision of gas differed from the acquisition of soap, candles, hats, or carriages. The latter articles are capable of transportation from place to place, the court answered, whereas gas flowing through pipes to fixed connections is not; there was no reasonable alternative.

The formula can be seen in other early cases involving businesses not usually considered public utilities. Ten tobacco warehouses in Louisville, Kentucky, were a necessary conduit between producers and tobacco dealers; there existed no practical alternative to this method of transacting the tobacco business in that region. Accordingly, the Kentucky Supreme Court held in 1884 that tobacco warehouses were public service companies.[28] For similar reasons irrigation companies in Arizona and Col-

[25] The report is reprinted in Bernard Schwartz, ed., *The Economic Regulation of Business and Industry* (New York: Chelsea House, 1973), 1:35.

[26] 45 *Congressional Record* 5534 (1910).

[27] *Shepard* v. *Milwaukee Gas Light Co*., 6 Wis. 539 (1858).

[28] *Nash* v. *Page*, 80 Ky. 539 (1884).

orado were held by their respective state supreme courts to be public service companies.[29] Without irrigation in the arid regions involved in those disputes, farmers could not grow their crops and, obviously, could not obtain water by means other than irrigation.

It follows, then, that industries may at times fall within the public service category, while at other times they do not. Partly for that reason price-fixing statutes for some food products were common in the American colonies at the time of the American Revolution.[30] Primitive transportation and communication at that time coupled with the paucity of suppliers meant that bread (although not grain) and other foodstuffs could practicably only be obtained locally. The inability or ability to obtain something strategic or necessary from distant suppliers is a factor that can vary over time and from place to place. As late as 1841 an Alabama court, after defining the public service concept as an exception to the general rule of competition and state noninterference in business affairs, ruled that baking fell within the exception.[31]

Just as changing conditions could remove certain products and services from public service status, so also conditions could cause other products and services to be placed on the list. An important example is fire insurance. A New Jersey law fixing premium rates was challenged by 121 fire insurance carriers doing business in the state. The New Jersey Court of Errors and Appeals, after observing that neither monopoly status nor state condemnation of land for eminent domain purposes determined public service company status, carefully traced the development of the fire insurance business in the late nineteenth century. Noting the rapid growth of the business, the court observed that other businesses could no longer obtain credit without obtaining fire insurance, which was universally employed as collateral security. Since credit is the lifeblood of industry and commerce from the smallest retail establishment to the largest industrial facility, fire insurance occupied precisely the same strategic position that the railroad did. Fire insurance, in the court's words, "ramifies in its effects from the greatest banking houses, through the homes of the unemployed, or the badly paid, to the smallest retail shops."[32] Each insurance firm is chartered individually by the state because, like banking, insurance occupies a special position of trust of other people's money. Accordingly, like other public services, a consumer's choice is

[29] *Salt River Valley Canal Co.* v. *Nelssen*, 10 Ariz. 9 (1906); and *Wheeler* v. *Northern Colo Irrigating Co*., 17 P. 487 (S. Ct. Colo., 1888).

[30] See the compilation in McAlister, "Lord Hale," p. 767; and Note, "State Regulation of Prices," *Harvard Law Review* 33 (1920): 839.

[31] *Mayor and Alderman of Mobile* v. *Yuille*, 3 Ala. 137 (1841). The general principle favoring competition at this time is nicely stated in *Charles River Bridge* v. *Warren Bridge* 36 U.S. (11 Pet.) 420 (1837).

[32] *McCarter* v. *Firemen's Ins. Co.* 73 A. 80, 84 (N.J.E., 1909).

restricted to those companies permitted by each state to engage in the activity.[33]

Although a substantial dollar volume of business or a large number of customers of a strategically placed business, such as cotton gins in a major cotton-growing state, reinforced a finding of public utility status, neither was a prerequisite. There may be only one customer, yet the service could be classified as a public utility.[34] Only the dedication to private use of an enterprise that otherwise would be classified as a public service company would allow a firm to evade public service obligations.[35] For this reason private telephone companies owned by and serving only its members have usually been exempt from public utility status.[36] The reason for the exclusion of *entirely* private facilities is that they do not serve the greater goals of the public service concept.[37] Just as a set of values governs the purposes of public service companies at the levels of industry, commerce, and community well-being, so also values govern their fundamental purpose at the level of individuals. A decision involving ferries puts the principle as starkly as possible: "Public ferries are established for the accommodation of the public, rather than for the gain and advantage of individuals."[38]

The judges who developed the public service company concept were aware of the linkages between the economy, social life, and individual well-being. The point is articulated in an 1873 Supreme Court decision, one issue of which was whether railroads were public service companies. Concluding that they were, the Court stated:

> It was originally supposed that they [railroads] would add, and . . . they have added, vastly and almost immeasurably, to the general business, the commercial prosperity and the pecuniary resources of the inhabitants of cities, towns, villages and rural districts through which they pass and with which they are connected. It is in view of these results, the public good thus produced, and the *benefits thus conferred upon the persons and property of all the individuals composing the community*, that courts have been able to pronounce them matters of public concern [emphasis supplied].[39]

[33] See the discussion in *German Alliance Ins. Co.* v. *Lewis*, 233 U.S. 389, 411–14 (1914).

[34] *Cawker* v. *Meyer*, 133 N.W. 157 (Wis., 1911). On cotton gins as a public utility, see *Tallassee Oil & Fertilizer Co.* v. *H. S. & J. L. Holloway*, 200 Ala. 492 (1917).

[35] *Barrington* v. *Commercial Dock Co.*, 45 P. 748 (Wash. 1896); and *Thousand Island Steamboat Co.* v. *Visger*, 179 N.Y. 206 (1904).

[36] *Reading Central Teleph. Co.* v. *Fayette Rural Teleph. Co.*, PUR 1915 A. 56.

[37] If the public service aspect of a company was only a small portion of the business, it was still included within the ambit. *Wingrove* v. *Public Service Comm.*, 81 S.E. 734 (W.Va., 1914).

[38] *Montjoy* v. *Pillow*, 64 Miss. 705 (1887).

[39] *Olcott* v. *Supervisors of Fond Du Lac Co.*, 83 U.S. 678, 21 L.Ed. 382, 387 (1873).

Insofar as a public service company has a major impact on commercial, industrial, and community life, it will have an impact on the individuals who compose that community. Conversely, insofar as individuals are affected by the activities of public service companies, there will be an impact on commercial, industrial, and community life.

Because of the nexus between individuals and society, the public service concept applies to economic necessities and services requisite to the community's level of civilization. Street lighting was not so requisite in the eighteenth century, but in the later stages of the nineteenth it was. Commentator Rexford Guy Tugwell observed, "As our consumption standards change, our conception of the goods and services necessary to life and happiness change. . . . A necessity [is] any good or service which contributes to a psychologically full life."[40] Perhaps this definition of necessity is too broad. But it is not easy to state a standard at once flexible yet capable of providing a decision rule. If, however, we examine the cases through the process of exclusion and inclusion, the gradual establishment of a line of distinction is discernible.

At the most primitive level, public services are those necessary for individuals to survive and participate in the larger community. For this reason a 1683 English case concluded that a cause of action was maintainable against an innkeeper for refusing a guest or against a blacksmith on the road for refusing to shoe a horse.[41] He could not turn someone out into the forest. Similarly, an 1849 Massachusetts decision, declaring a water company to be a public service company, noted that the supply of pure water is a public obligation. If government assigns such a task to a particular firm, that company is under public service obligations and cannot refuse service on reasonable request.[42] The interconnected themes of public service, obligation, and a private firm doing tasks guaranteed by government are seen in the limited use of the corporate form of business organization, which until late in the nineteenth century was most often a privilege granted to private firms undertaking "public" purposes.[43]

As the level of civilization in a society advances, more and more services are considered central to the society's well-being in the way that potable water is at the lowest levels. Thus, the Supreme Court of North Carolina in 1898 prohibited rate discrimination by corporations "supplying the great conveniences and necessities of modern city life, as water,

[40] Rexford Guy Tugwell, *The Economic Basis of Public Interest* (1922; rpt., New York: Augustus M. Kelley, 1968), p. 103.

[41] *Jackson* v. *Rogers*, 2 Show. 327 (K.B., 1683).

[42] *Lumbard* v. *Stearns*, 4 Cush. 60 (1849).

[43] Joseph S. Davis, *Essays in the Earlier History of American Corporations* (1917; rpt., New York: Russell & Russell, 1965), 1:103; and E. Merrick Dodd, "Statutory Developments in Business Corporation Law, 1886–1936," *Harvard Law Review* 50 (November 1936): 38.

gas, electric light, street cars and the like. . . . The business interests and the *domestic comfort of every man* would be at their mercy" (emphasis supplied).[44]

The provision of gas illustrates the process by which a service became a public utility. The first introduction of gas for lighting in the United States occurred in 1816. In 1822 the New York legislature granted the New York Gas Light Company the right to incorporate. The company agreed to supply sufficient gas for lighting the homes and public lamps of Broadway. Considerable resistance to the spread of gas lighting developed on many grounds, including that it would frighten horses, encourage thieves, increase drunkenness, and most important, it was contrary to the divine plan of the world.[45] Political resistance, slow technological development, and expensiveness resulted in delayed acceptance of gas illumination. It was not widespread until the 1870s, and it was not commonly seen in many parts of the country during that period as a public utility service. Accordingly, many of the cases during that period did not treat gas as a public utility.[46]

In 1872, however, an important breakthrough occurred. T.S.C. Lowe was granted a patent for making water gas, a considerably less expensive process than the coal gas process used previously. Notwithstanding the introduction of electricity for illumination in 1879 and its rapid development, the new process permitted accelerated growth of the water gas industry and such new uses as heating and cooking, although there was considerable political resistance by the older coal gas firms.[47] Rapid urbanization coupled with the extraordinary possibilities of gas and electricity moved those technologies from ordinary businesses to public utility status. By 1885 the Supreme Court concluded that "there was a public necessity for gas lights on . . . streets and . . . public buildings, almost as urgent as the establishment of the streets themselves."[48] Similarly, the Supreme Court of Wisconsin, exhaustively citing cases, concluded in 1906 that gas and electricity companies were public service enterprises.[49]

Thus, as the nineteenth century progressed, decisions increasingly were made by courts and legislatures on whether to classify new technologies as public services. The threshold question was whether the new

[44] *Griffin* v. *Goldsboro Water Co.*, 30 S.E. 319, 320 (N.C., 1898).

[45] Jones and Bigham, *Principles*, pp. 10, 11; and Louis B. Hartz, *Economic Policy and Democratic Thought* (Chicago: Quadrangle, 1948), p. 108.

[46] See, for example, *Paterson Gaslight Co.* v. *Brady*, 27 N.J. Law 245 (1858); and *McCune* v. *Norwich City Gas Co.*, 30 Conn. 521 (1862).

[47] A study of the politics of the gas industry in Massachusetts during this period is Leonard D. White, "The Origin of Utility Commissions in Massachusetts," *Journal of Political Economy* 29 (March 1921): 189–91.

[48] *Louisville Gas Co.* v. *Citizens Gas Co.*, 115 U.S. 683, 692 (1885).

[49] *City of Madison* v. *Madison Gas & Electric Co.*, 108 N.W. 65 (Wis., 1906).

technology was central to the level of civilization attained by a society. This is, of course, a social and not an economic standard. At the beginning of the nineteenth century the enterprises included were turnpike, canal, bridge, and ferry companies.[50] Later, as we have seen, still other services, including electricity, gas, and telephone, were added to the public service company list.

The Telephone as a Public Service

Alexander Graham Bell claimed to have invented the telephone in 1876—a claim that was the subject of bitter dispute, as we will see in the next chapter. At first the potential of the new technology was not widely appreciated. Bell was an exception, foreseeing in 1878 that "future wires will unite the head offices of the Telephone Company in different cities, and a man in one part of the country may communicate by word of mouth with another in a different place" and that the telephone would become a means of communication among businesses, hospitals, police and fire stations, as well as the general public.[51] Nevertheless, in the beginning most observers were not sure of the telephone's future use or they conceived of it as a means to transmit music, drama, and news—a radio concept.[52] The central switchboard and other devices necessary to a modern telephone system were dimly conceived, if at all, before 1878.

The rate of the telephone's adoption was rapid in its early phases (although density was extremely low by contemporary standards), as Table 2.1 indicates. Thus, by the early 1880s, even those most skeptical at the outset were converted to a belief in the telephone's rapid spread and improvement. Of course, growth alone does not render a new technology a candidate for public service status. The telephone's similarity to the telegraph—notwithstanding important differences—and the developing public policy toward the telegraph provided an important model for government regulation of the newer industry. Legal treatises generally considered the two technologies together during the nineteenth and early twentieth centuries. It is, then, to the telegraph that we should look to understand why the telephone industry was within the public service category early in the industry's history.

[50] M. H. Hunter, "The Early Regulation of Public Service Corporations," *American Economic Review* 7 (1917): 569–81.

[51] Bell's March 25, 1878, talk is reprinted in John E. Kingsbury, *The Telephone and Telephone Exchanges* (1915; rpt., New York: Arno Press, 1972), pp. 89–92.

[52] Sidney H. Aronson, "Bell's Electric Toy: What's the Use? The Sociology of Early Telephone Usage," in Ithiel de Sola Pool, ed., *The Social Impact of the Telephone* (Cambridge, Mass.: MIT Press, 1977), pp. 20–23.

TABLE 2.1
Telephones in the United States (in thousands)

Year	*Number*	*Year*	*Number*
1876	3	1881	71
1877	9	1882	98
1878	26	1883	124
1879	31	1884	148
1880	48	1885	156

Source: Historical Statistics of the United States: Colonial Times to 1970 (Washington, D.C.: Government Printing Office, 1975), pt. 2, p. 784.

The early development of the telegraph in many ways paralleled that of the telephone. Like the latter, many inventors were simultaneously working on the problem of electrically transmitting signals over wires. Again, just as there were many who challenged Bell's invention of the telephone, so did Morse's contemporaries challenge his claim to have invented a practical telegraph in the 1830s. And in part the skepticism in both cases stemmed from the fact that the backgrounds of Morse and Bell were not in electricity; Morse was an artist and Bell was a teacher of the deaf.[53] Both inventions were, for a very short time, treated with derision by many unimaginative persons. The telegraph "was such a complete innovation that at first its possibilities were but dimly realized. So little use was found for the original line between Baltimore and Washington that chess games by telegraph were promoted between experts in the two cities."[54] But within a very short time after the practical application of the telegraph, even the most narrow-minded were aware of the extraordinary possibilities it promised. By 1846 business demand for the new service expanded dramatically, and the telegraph in the United States was rapidly displacing older modes of communication. The telegraph, it should be noted, preceded the railroad in forging interregional and transcontinental links.[55]

By the 1850s the telegraph had revolutionized the ways in which many important kinds of businesses were conducted. Most obviously newspa-

[53] Details of Morse's life are found in Samuel I. Prime, *The Life of Samuel F. B. Morse, Inventor of the Electro-Magnetic Recording Telegraph* (New York: D. Appleton, 1875); and of Bell's life in Robert V. Bruce, *Bell* (Boston: Little, Brown, 1973).

[54] George Rogers Taylor, *The Transportation Revolution* (New York: Holt, Rinehart and Winston, 1951), p. 152.

[55] Richard B. Du Boff, "Business Demand and the Development of the Telegraph in the United States, 1844–1860," *Business History Review* 54 (Winter 1980): 461, 462.

pers and magazines were able to obtain more rapid and accurate information than ever before. The telegraph had made the world smaller and allowed readers the ability to learn more about the world. The Associated Press moved from carrier pigeons to the telegraph, gradually becoming the largest news distributor in the United States. Because of the telegraph, people in remote areas of the country could receive foreign news soon after the event.[56] In short, the telegraph revolutionized the transmission of information. Businesses reduced market uncertainties by securing rapid, accurate information while at the same time expanding the geographical coverage of their markets. Stock, commodity, futures, and money markets developed, on the basis of the telegraph, as never before. The telegraph permitted ultimate buyers and sellers in many industries to make contact without the intervention of costly middlemen, thereby cutting distribution and transaction costs.[57] It soon became apparent that the telegraph was revolutionizing social and economic life.

On every count, then, the telegraph appeared to be a public service industry. Even at the outset it received the attention of government, notwithstanding the initial skepticism of many. In 1843 Congress, in a close vote, appropriated thirty thousand dollars to construct an experimental line between Washington and Baltimore. The line opened in 1844, at first as a free service to encourage use, and in 1845 for a fee. Even before the line's opening, Morse offered to sell his patent rights to the government, arguing that so powerful an invention should not be left in the hands of private interests. While the postmaster general in 1845 concluded that under no tariff schedule could the telegraph's revenues equal its tariffs, he called for close regulation of the new device. Congress, however, failed to act and the telegraph remained in private hands subject to the restraints imposed by courts and state legislatures.[58]

Numerous telegraph companies were organized in the 1840s and 1850s, most of them on a regional basis. Most of the fifty companies operating in 1851 were licensees of the Morse patents, but some were based on competing claims.[59] The basic Morse patents were due to expire

[56] See Alvin F. Harlow, *Old Wires and New Waves* (New York: Arno Press, 1971), chap. 9.

[57] See Du Boff, "Business Demand," pp. 471–78.

[58] The most careful examination of the early political history of the telegraph is Robert Luther Thompson, *Wiring a Continent* (Princeton: Princeton University Press, 1947), pp. 11–34. Important details are provided in James D. Reid, *The Telegraph in America* (New York: Derby Brothers, 1879), pp. 106–11. In cases of conflict I have relied more on the carefully done Thompson account than on Reid's paean to Morse.

[59] James M. Herring and Gerald C. Gross, *Telecommunications: Economics and Regulation* (New York: McGraw-Hill, 1936), p. 1.

in 1860 and 1861, which, of course, *could* have led to yet more competition. In 1851 a movement toward consolidation began, culminating in the formation of Western Union in 1855, its rapid expansion throughout the West and then the remainder of the country, and its merger with its principal rivals by 1866.[60] Though there were always rivals to Western Union, and a major one—Postal Telegraph—was founded after Western Union's dominance was established and endured until the 1940s, the industry that served as the major public policy model for the telephone showed a clear tendency toward concentration.

Several reasons have been advanced for this tendency. First, it is clear that the capital needed to meet the rapidly increasing demand for telegraph service was very high. Firms unable to meets these demands fell behind technologically and were frequently unable to make even routine repairs. Thus, as Western Union invested heavily in improvements and the extension of its lines, the values of its competitors' assets declined, making Western Union's acquisition of them comparatively cheap.[61] Second, in contrast to the local and regional systems, Western Union's unified system allowed it to "send telegrams direct, eliminating transfers from one company's lines to another's. This meant increased speed and reduction of errors. The nation's commercial community naturally desired efficient service."[62] Third, because a telegraph company's fortunes were tied to its ability to maximize the number of points to which it could deliver messages, interconnection with other companies was crucial. Obviously, then, the most widespread system—Western Union—held the whip hand and could exercise considerable leverage on other companies to merge or face limited coverage.[63]

From the perspective of an informed observer in the 1860s and 1870s, then, the services provided by the telegraph industry were becoming increasingly important to commerce and the dispersion of intelligence, while simultaneously the number of service providers was shrinking. To a business firm during that period, a telegraph company's potential for abuse and arbitrary action was enormous. In England, one of the few nations that had a private system, commercial firms triggered a movement for nationalization that bore fruit in 1868. Earlier regulatory statutes had failed to bring rates down, leading to the drive to nationalize a

[60] The details of the consolidation process are spelled out in Reid, *The Telegraph*, chaps. 35–43.

[61] Ibid., pp. 472, 473.

[62] Lester G. Lindley, *The Constitution Faces Technology* (New York: Arno Press, 1975), p. 15.

[63] Du Boff, "Business Demand," p. 463; and Gerald W. Brock, *The Telecommunications Industry* (Cambridge, Mass.: Harvard University Press, 1981), p. 83.

business that was widely conceived as a public service. Indeed, proponents of nationalization in Britain adversely compared the rates and quality of their service to those prevailing in the United States.[64]

The greater progressiveness of Western Union and some other telegraph companies, coupled with the United States' more decentralized public policy decision-making system and constitutional problems that would be raised, prevented nationalization from being a realistic prospect in the nation. Nevertheless, Republican Senator John Sherman (for whom the Sherman Antitrust Law was named) and others clamored for some *federal* government intervention.[65] Although virtually all of these legislative attempts to intervene in telegraph decision making failed, the courts, especially those at the state level, carved out a special set of policy rules for the telegraph. Only one year after the invention of the telephone, the Supreme Court noted the unique and critical contribution of the telegraph in industrial, commercial, and community life.[66]

For these reasons, at the telephone's birth the telegraph was conceived in America as a public service industry that was critical to American economic and social life.[67] In the succinct words of a 1916 Massachusetts case: "Telegraph companies exercise a public employment and are bound to serve all the public without discrimination."[68] Because of the great value of rapid, distant communication in expanding markets, even a single provider of telegraph service (assuming that it behaved in a manner appropriate to a public service company) was perceived as tending to break down local monopolies in goods by encouraging entry of nonlocal firms. Courts and state legislatures, beginning as early as an 1845 New York statute, sought to supervise telegraph companies as public service businesses.[69] But unlike railroads, few saw a need in the nineteenth century for national legislation, in large part because of overall consumer

[64] Jeffrey Kieve, *The Electric Telegraph* (Newton Abbot: David & Charles, 1973), pp. 120–27, 138–53.

[65] A good review of federal government proposals is Harlow, *Old Wires*, chap. 16.

[66] *Pensacola Telegraph Co.* v. *Western Union Telegraph Co.*, 96 U.S. 1 (1877).

[67] The courts divided, however, on whether the telegraph was a "common carrier" in the sense of having higher tort responsibilities than ordinary businesses for goods entrusted to it, such as transportation companies had. See, for example, *Birney* v. *New York & Washington Telegraph Co.*, 18 Md. 341 (1862), holding that the telegraph was not a common carrier for tort liability purposes. However, the telegraph was almost consistently held to be a public service business in the sense I have used the phrase. See, for example, *Western Union Tel. Co.* v. *Call Pub. Co.*, 44 Neb. 326 (1895).

[68] *Western Union Telegraph Co.* v. *Foster*, 113 N.E. 192, 195 (Mass., 1916).

[69] Early telegraph statutes are described in William K. Jones, "The Common Carrier Concept as Applied to Telecommunications: An Historical Perspective," Appendix to IBM Reply Comments in FCC CC Docket no. 72–252, pp. A-17–38.

satisfaction with the price, quality, and progressiveness of telegraph service coupled with the adequacy of state judicial and legislative regulation.[70]

The Telephone's Advantages

The telephone's uses were early seen as virtually the same as the telegraph's. But the telephone had major advantages over the telegraph. First, the telephone could be used by an ordinary person, whereas the telegraph required the employment of an expert familiar with the Morse Code and much else. Obviously, the opportunity for commercial use of the telephone was seen as much greater than that of the telegraph. Second, the telephone permitted discussion because it reproduced speech. The telegraph, in contrast, required the transmission of a complete statement before a reply could be made. Third, the telephone, since it reproduced ordinary speech, allowed the hearer to detect the nuances and emotional content of information—anger, sarcasm, insincerity, and so on. Fourth, because translation of ordinary language into code followed by retranslation into ordinary language was not required in telephone use, it was a much more accurate device than the telegraph, which had given rise to much litigation over the issue of negligent mistranslation. Fifth, for the same reason, the telephone is a much faster device than the telegraph for conveying information. Sixth, because of all of the advantages described above, the telephone was soon seen as a cheaper and considerably more efficient device than the telegraph. Finally, although the initial uses of the telephone were perceived to be largely business related, it did have the capacity to become a social instrument because it could be connected into homes.[71]

Ironically, these advantages were increased even further because of important technological advances that had been made in the telegraph. Because of the demands made on telegraph use, Western Union and other telegraph companies sought to increase the capacity of wires to carry messages. Initially the duplex, allowing separate eastbound and westbound transmissions over the same wire, was deployed in 1872. Next, Thomas A. Edison developed a quadruplex system in 1874. Private telegraph lines entirely within cities were developed as early as 1849. Central office switching, allowing telegraph subscribers to communicate

[70] Lindley, *The Constitution*, pp. 19–24.

[71] There was some residential use of the telephone even at the outset. Aronson, *Bell's Electric Toy*, p. 28.

with each other, was employed in New York and Philadelphia as early as the late 1860s and proved valuable to bankers, stockbrokers, and others. Finally, telegraphically operated burglar and fire alarm boxes as well as district telegraphs—allowing the placement of simple signaling devices in homes connected to central offices—were in place at the time of the telephone's invention.[72] Although the progressiveness of telegraph technology probably initially discouraged the adoption of the telephone, the latter's superior ability to exploit these innovations was manifest only a few years after its invention.

By the 1880s there was, then, universal agreement that telephone firms were public service companies. And this view continued without challenge after 1894, when competitors of the Bell System sprang up after the expiration of the basic Bell patents. An 1885 Nebraska case in which a telephone company refused to allow a lawyer to become a subscriber sums up the sentiment. Forbidding the telephone company to make such an unjust discrimination, the Nebraska Supreme Court said:

> While it is true . . . that it has been organized under the general corporation laws of the state, and in some matters has no higher or greater rights than an ordinary corporation; yet it is also true that it has assumed to act in a capacity which is to a great extent public, and has . . . undertaken to satisfy a public want or necessity.
>
> The demands of the commerce of the present day make the telephone a necessity. All people, upon complying with the reasonable rules and demands of the owners of the commodity . . . should have the benefits of the new commerce.
>
> It has assumed the responsibilities of a common carrier of news. It has and must be held to have taken its place by the side of the telegraph as such common carrier.[73]

In addition to court decisions, states enacted statutes regulating telephone companies as public services or making telegraph laws applicable to telephones. When challenged on the ground that such laws were unconstitutional because the telephone industry was not affected with a public interest, courts almost uniformly upheld their constitutionality. In 1886, the Indiana Supreme Court, upholding a state law regulating telephone charges, said the telephone "has become as much a matter of public convenience and of public necessity as were the stage-coach and sailing vessel a hundred years ago, or as the steam-boat, the railroad, and the telegraph have become in later years. . . . No other known device

[72] Reid, *The Telegraph*, pp. 594–666; Kingsbury, *The Telephone*, pp. 80–88; and Aronson, *Bell's Electric Toy*, pp. 16–19, 26.

[73] *State* v. *Nebraska Telephone Co.*, 22 N.W. 237, 238, 239 (Neb., 1885).

can supply the extraordinary facilities which it affords. It may therefore be regarded, when relatively considered, as an indispensable instrument of commerce."[74] The idea that a telephone company is a public utility received the blessings of the Supreme Court in 1892, ending any possibility of challenging the regulation of such companies.[75]

Thus, the telephone could be heavily regulated as a public service industry. In the Indiana case just discussed, this meant that the monopolistic patent rights granted by the federal government could not override a state's right to prescribe reasonable maximum charges for its use. Once the decision is made that the firms in a certain industry fall within the public service category, these firms must abide by a set of *general* principles that we will examine in the next section. But beyond these are a large number of *particular* principles that are tailored to *each* public service industry. Let us now look at some of the general public service precepts.

Public Service Obligations

When an industry has been declared a public service, the firms in it have four overriding obligations (which will be sketched later). From these four flow a large number of subsidiary rights and duties that are designed to assure fulfillment of these obligations. Responsibility for carrying out the tasks is in the private sphere; supervision is in the public one. Thus, the day-to-day decisions affecting electric, gas, telephone, and so on are left to those who are actually engaged in the business. It is assumed that the combination of their technical experience *and* their obligations to shareholders, creditors, suppliers, and their customers will *tend* to direct them toward achieving such performance goals as efficiency and good service quality. Their incentives are sometimes reinforced by the press and muckrakers, who can obtain information about the activities of public service companies more easily than they can about other enterprises, largely because of the substantial amount of information such companies are required to make public.

Before looking at these obligations, it is appropriate to sketch the frequently voiced view that the public service model is a sham because regulators do little more than follow the dictates of dominant regulated firms. Many of these critics argue that free competition, allowing existing and potential competitors to challenge current industry leaders by producing better, lower-priced, or simply different products and services,

[74] *Hockett* v. *State*, 5 N.E. 178, 182 (1886).

[75] *Budd* v. *New York*, 143 U.S. 517 (1892).

will lead to superior performance than that produced by the public service model. The threat of what Joseph Schumpeter called "creative destruction" is what is supposed to keep firms innovative and responsive to consumer demands. In Schumpeter's words, what really counts is "the competition from the new commodity, the new technology, the new source of supply, the new type of organization . . . competition which commands a decisive cost or quality advantage and which strikes not at the margins of the profits and outputs of the existing firms but at their foundations and their very lives."[76] Thus, if firms fail to so respond, they lose out to those who are attentive to consumer performance demands.

Although public service companies are not necessarily monopolies, as we have observed, they are almost always protected from the threat of creative destruction by some form of entry control that precludes new challengers to existing firms. This protection is often the incentive necessary to assure that public service firms engage in money-losing activities (such as supplying distant areas) in fulfillment of public service goals. Thus, there may have been ten trucking firms with Interstate Commerce Commission operating rights to haul cement between two points. But the fact that the eleventh firm cannot enter the market and challenge existing firms meant that the dynamic of creative destruction is, at least, weakened. But what then impels technological progressiveness in a public utility?

Adherents of public service liberalism argue that government supervision can supply for the public service sectors what creative destruction supplies for other sectors. And government supervision is expected to supply still more—the fulfillment of the four principal obligations of a public service company, which are not necessarily goals of firms in other sectors. But has government supplied the requisite supervision? Much of the criticism leveled at America's economic regulatory agencies at the state and federal levels can be boiled down to two charges. These agencies allegedly do not direct the firms they supervise to achieve the goals attained by creative destruction, or they lead to the attainment of these and the public service goals at too great a cost because public utilities are woefully inefficient.[77]

But as we will see, these charges are misplaced. To what extent has the government-business interface acted as a substitute for creative destruction? The issue has important contemporary relevance, for many of the most rapidly developing economies in the twentieth century, such as Japan, have employed policy models that closely integrate public and pri-

[76] Joseph A. Schumpeter, *Capitalism, Socialism and Democracy*, 3d ed. (New York: Harper & Row, 1950), p. 84.

[77] See the summary in Alan Stone, *Regulation and Its Alternatives* (Washington, D.C.: Congressional Quarterly Press, 1982), pp. 250–54.

vate decision making. Indeed, few capitalist economies have employed a government hands-off policy. Many of the most successful ones have more closely resembled public service liberalism. To be sure, one must avoid sweeping reductionist conclusions. There are obviously factors present other than government intervention that account for performance. For example, technological factors play a critical role in achieving efficiency and other goals. But as we will see in chapter 3, public service liberalism provides institutions for the development and exploitation of novel technologies.

One could clearly argue that in addition to the governmental and technological factors that lead to the attainment or nonattainment of progressiveness and public service goals, there are organizational and economic variables, including the absence or presence of vertical integration and management structure. True! Good public service regulation must be prepared to take cognizance of an industry's technological, organizational, and economic characteristics, as well as of the regulated company's incentives, and harness them to public service goals. Good public service regulation should neither rubber-stamp industry action nor be antagonistic to the supervised firms. For example, in 1940 the Washington Department of Public Service ordered the Pacific Telephone and Telegraph Company (PT&T) to speed up its conversion from manual to dial operation. At the same time the Washington regulators investigated costs, rate of return, and rates to assure that PT&T would earn an adequate rate of return after it made the transition.[78]

There is, of course, no guarantee that public service regulation will assure that the supervised firms respond to their public service obligations and act progressively, any more than there is an ironclad guarantee that a particular firm in a competitive industry will act efficiently or make decent products. Public service decisions can be good or bad, just as market decisions can. Public service decisions may be based on imperfect or incorrect information. Wrong decisions may be based on faulty logic or a confusion of means and ends. Nevertheless, the system of public service decision making is one that has worked well. Decisions must be justified, they usually must be public, they must be based on available facts presented in an orderly fashion, and most important, the decision makers must employ reason in relating facts and rules to the overriding public service obligations and other values. There are grounds for differences in judgment in this process. And of course, these differences can reflect philosophical and political differences. Courts as well as regulatory agencies have majority and dissenting opinions. Nevertheless, the process, as we will see, is constrained to attempt to accomplish public service goals.

[78] *Dept. of Pub. Sev. of Wash.* v. *Pacific Teleph. and Teleg. Co.*, 34 PUR (NS) 193 (1940).

Political preferences enter decisions, but only within the context of furthering public service obligations.

This view is at variance with one that conceives public service decisions as nothing more than shams that guarantee the victory of regulated firms. For example, Samuel Huntington's influential 1952 article claimed that the ICC consistently favored railroads in their conflicts with motor carriers. From this study (sharply disputed by others who have studied surface transportation) the author generalized that large entrenched private interests tend to "capture" regulatory agencies, bending every decision to the will of the powerful regulated firm (or firms).[79] Since the publication of that article, many have come to accept its sweeping generalization as a matter of faith. Yet AT&T and other large telephone companies have lost numerous decisions before courts as well as federal and state regulatory agencies. More important, the "capture theory" fails to grasp the nature of the processes involved in public service proceedings. Most important, it fails to grasp the way in which decisions are related to the four principal obligations of public service firms.

The first and, perhaps, most ancient of the public service obligations is to serve *all* who apply for the service and are willing to abide by the reasonable regulations of the firm. To take a simple example, in an 1839 New Hampshire case the proprietor of a stagecoach refused to carry the plaintiff even though he had room. The court concluded that this was a common law obligation of a stagecoach.[80] For the same reason the Bell Telephone Company was required to install a telephone in the offices of a telegraph company that was a rival of Western Union, even though Bell had entered into a contract that stipulated Bell would not provide connection to any other telegraph company. Notwithstanding that the telephone was still patented, the court held that public service obligations overrode the monopoly rights ordinarily accompanying a patent. "It is not possible to admit the principle that a railroad, telegraph or telephone company may avoid the performance of any part of the paramount duty they owe to the entire public."[81] Like all rules there are exceptions; a telephone company was not required to provide service to an establish-

[79] Samuel Huntington, "The Marasmus of the ICC: The Commission, the Railroads and the Public Interest," *Yale Law Journal* 61 (April 1952): 467–509. But see William Z. Ripley, *Railroads: Rates and Regulation* (New York: Longman, 1913), p. 118; and Gordon P. MacDougall, "Industry Pricing Practices as a Factor in Regulatory Decisions," in *Transportation Research Forum, Proceedings—Sixth Annual Meeting* (Oxford, Ind.: Richard B. Cross, 1965), pp. 53–58.

[80] *Bennett* v. *Dutton*, 10 N.H. 481 (1839).

[81] *State* v. *Bell Tel. Co. of Mo.*, 22 Alb. L.J.363 (1880). A very few cases on this topic held to the contrary, but most are in full accord with this case for the quoted reasons. See, for example, *Commercial Union Tel. Co.* v. *New England Tel. & Tel. Co.*, 61 Vt. 241 (1889); and *Bell Telephone Co.* v. *Commonwealth*, 3 A. 825 (Pa., 1886).

ment that it *knew* had only an illegal purpose.[82] But while the rules became more and more complex as novel cases were presented, the fundamental principle stands today.

Since there are ways other than direct refusal to evade this responsibility, several corollaries follow. Public service companies must extend their facilities so as to meet reasonable demand. For the same reason abandonment of a line will be subject to judicial or administrative scrutiny. But the easiest way to evade responsibility to serve all is to arrange service in such a way that many would-be customers are excluded. This leads to the second crucial obligation: to serve the public adequately. This has both a quantitative and a qualitative aspect. The word *adequate* is, of course, a relative one dependent on the technological and economic capabilities of the firm and industry under consideration. In general—although the rules had become complex—this requirement was translated into a demand for progressiveness, as indicated in the 1940 Washington public utility decision requiring PT&T to move faster in the transition to dial telephones. Similarly, in an 1895 case a telegraph company was required to expand its business. "But it is the duty of the telegraph company to have sufficient facilities to transact all the business offered to it for all points at which it has offices."[83]

Another way in which one may evade the central obligation to serve all is by discriminating between customers in such a way that denies some customers service. This leads to the third critical obligation of public service companies: they must act impartially to all. Indeed, the complaint of "unjust" discrimination in rates and quality of service was the single most important Granger complaint against the railroads and provides the key theme in the most important nineteenth-century federal regulatory statute—the 1887 Interstate Commerce Act—as well as in many state statutes. The canon that rates must be equal for all applicants for similar services under similar conditions can be traced back to the development of English highways and canals; it was readily applied to railroads and then to other public services.[84] Of course, a public service company may discriminate in rate when there is different quality or quantity (that is, distance) of service. Again, the rules in this area have grown complex and detailed as more and more cases have been brought to the courts and commissions.

It is important to point out one item not covered by the nondiscrimination principle. As we observed earlier, one of the obligations of a public service company is to provide service to all applicants who are willing to

[82] *Godwin* v. *Carolina Tel. & Tel. Co.*, 136 N.C. 258 (1904).

[83] *Leavell* v. *Western Union*, 21 S.E. 391, 392 (N.C., 1895).

[84] Isaac Beverly Lake, *Discrimination by Railroads and Other Public Utilities* (Raleigh, N.C.: Edwards & Broughton, 1947), pp. 11–22.

abide by the firm's reasonable regulations. This has sometimes been construed to require public service companies to extend their area of coverage. Obviously, if a public utility was permitted to recover the costs of serving distant or hard to reach customers or customers in sparsely populated areas, these costs could be so high as to preclude use in these areas. Under such circumstances regulators have permitted discrimination through rate averaging, in which a certain class of traffic cross-subsidizes another class: urban may subsidize rural, business may subsidize residential, and long distance may subsidize local traffic. Another way of approaching situations in which the ratios of customer rates to rates of return are unequal is to remember once again the underlying purpose of classifying certain activities as public services. As a matter of public policy it is important to make such services widely available.

The fourth general principle follows, once again, from a possible way that a public service company could evade its obligation to serve all. It could charge rates so high that many would-be customers could not afford the service (which, again, is by definition one that should be widely available). Accordingly, the fourth general obligation is that it should charge only reasonable rates. The phrase "reasonable rates" is one that has kept armies of accountants and attorneys active and affluent since the expansion of public utility regulation in the nineteenth century. The rules, needless to say, are complex and detailed. They involve utility commissions and courts in such collateral issues as the structure of utilities' securities, the costs of replacing existing facilities and modernizing, depreciation schedules, "fair" return, and intangible values. Volumes have been written on these rate principles that differentiate reasonable from unreasonable rates; like most areas of the law, much is settled, but questions remain. And though *exact* rates are always subject to dispute, the rules do usually permit a determination of a zone of reasonableness, below or above which rates are unreasonable.

It is easy to become bogged down in the details of rate making, but like the other principal obligations of public service companies, the underlying conception of a reasonable rate is intended to serve values in addition to economic efficiency—values that embrace a venerable conception of social justice. That there are such other values is the heart of public service liberalism. And while conflicts inevitably arise, the conception operates on the theory that public decision makers can cooperate in achieving the public welfare not by balancing the demands of interest groups, but by making explicit, reasoned decisions designed to attain predetermined goals.

As we will see in the next chapter and later ones, these goals have not been attained at the cost of technological backwardness.

3

Protection of the Newborn

The Lifeblood of Capitalism

THE lifeblood of capitalism is "creative destruction"—the process by which new products, methods of production, transportation, markets, and industrial organization displace the old. Schumpeter observed that the revolutionary force of creative destruction, and not the marginal forms of price, service, and promotional competition, is the essential aspect of capitalist economic progress. "In capitalist reality as distinguished from its textbook picture" what counts is "the competition from the new commodity, the new technology, the new source of supply, the new type of organization . . . competition which commands a decisive cost or quality advantage and which strikes not at the margins of the profits and the outputs of the existing firms but at their foundations and their very lives."[1]

Yet contrary to Schumpeter, many interventionist policies associated with public service liberalism fostered progressiveness. Public service liberalism can provide a strong incentive structure to promote technological innovation. It can do so through regulation and through patent policy, at which we will look in this chapter. At the same time we will see how the technology nurtured by patent policy is an important factor in shaping the structure of enterprises, allowing them to achieve public service goals. Public service liberalism allows enterprises to shape their structures flexibly in response to technological imperatives. Thus, regulation under public service liberalism must be sharply distinguished from antagonistic government intervention, such as the breakup of AT&T, in which public policymakers restructured the industry on the basis of preconceived sweeping generalizations presumably applicable in all situations. In contrast, public service liberalism looks at sectors and asks how well public and private arrangements are achieving the goals of *that* sector.

Public service liberalism may embrace both ease of entry, in order to foster the new, and protection of the new through patent monopolies and other policies designed to nurture the newborn technology, product, dis-

[1] Joseph Schumpeter, *Capitalism, Socialism and Democracy*, 3d ed. (New York: Harper & Row, 1950), p. 84.

tribution system, and so on. Obviously, accommodating both of these goals entails walking a fine line; the potential for abuse is clear. Thus, patent policy itself required creative evolution, as the telephone experience shows. The resolution of patent-related issues is a testament to the problem solving capabilities of public service liberalism.

Few policies are more in conflict with laissez-faire than patent policy, under which the state grants inventors monopolies for limited periods. Modern Anglo-American patent policy can be traced to the closing of the port of Antwerp to English merchants in 1565, resulting in a severe blow to English trade. This led, as so often occurs in periods of severe economic dislocation, to both merchants and policy officials carefully reexamining public policy. The most important new policy adopted was the chartering of monopolistic trading companies, the purpose of which was to exploit and develop new trading patterns in eastern Europe, Africa, and Asia.[2] While the English patent can be traced to 1552, the Antwerp closure led to great strides in its adoption in order to make England as free as possible from foreign dependence. Thus, during the 1560s patents were granted for such diverse products as hard white soap, ovens and furnaces, oil made from rapeseed, and window glass. Other policies short of monopoly patent encouraged the immigration of persons with skills unavailable or in short supply within England.[3] In short, as early as the second half of the sixteenth century, English policymakers had seen the need to adopt interventionist policies intended to spur invention, innovation, and as a result economic progress.

The tension between the legitimate purposes of monopoly patent privileges and their abuse led in 1624 to the enactment of a statute that sought to distinguish legitimate from illegitimate monopolies by carefully carving out the exceptional situations in which monopolistic privileges could be granted. The sixth clause defined the privileges that could be granted to new inventors. It granted the first inventor of a new product the exclusive right to make, use, and sell the product for fourteen years. The invention had to be wholly new, not merely a marginal improvement of an older invention.[4]

In the eighteenth century a marked degree of sophistication had en-

[2] See F. J. Fisher, "Commercial Trends and Policy in Sixteenth Century England," in E. M. Carus-Wilson, ed., *Essays in Economic History* (London: Edward Arnold, 1954), 1:152–72.

[3] William Hyde Price, *The English Patents of Monopoly* (Boston: Houghton Mifflin, 1906), pt. 1; Joan Thirsk, *Economic Policy and Projects* (Oxford: Clarendon Press, 1978), chap. 3; and D. C. Coleman, "An Innovation and Its Diffusion," *Economic History Review* 22 (December 1969): 417–29.

[4] William Holdsworth, *A History of English Law*, 3d ed. (London: Methuen, 1945), 4:352–54.

tered into patent proceedings, which has important bearing on the conception of patentability in the nineteenth and twentieth centuries. This occurred as the courts assumed greater jurisdiction over patent issues at the Privy Council's expense in order to curb abuse. The courts, for the first time, required full disclosure of the details of a patent. Previously applicants for a patent introduced wholly new industries into England, such as the mining of copper or the manufacture of paper or alum. But in the eighteenth century the cutting edge of technological progress was in major improvements of existing processes. Several inventors worked more or less along the same lines to achieve the same purposes. Consequently, inventors who considered their inventions patentable had a strong incentive to describe particularly its unique properties. The courts called on to settle disputes between inventors of similar products and processes inexorably required specification, although substantial differences on the kind and amount of disclosure persisted.[5]

In the eighteenth century a close link developed between the notion of patentability and economic goals. Ideas, no matter how ingenious and unique, are not patentable. Rather, ideas had to be harnessed to practical application so as to produce, in the words of the court in an important 1795 case, "effects in any art, trade, mystery or manual occupation."[6] For the same purpose, economic development, some inventors, such as James Watt, sometimes received special privileges through private acts of Parliament. Thus, patent policy was intended, as economist F. M. Scherer observed, not simply to stimulate invention, but to encourage investment and development as well—the whole process of innovation.[7] Patent policy, in short, was an important component of public service liberalism's goal of economic progress and, contrary to laissez-faire, clearly restrained the free market.

Of course, the idea of patent monopoly privileges generated criticism, as it still does. Issues of the appropriate duration and alternative incentives to innovate continue to be debated.[8] Nevertheless, English public policy on patents rested on the assumption that there is a clear link between patent monopoly, innovation, and economic progress. Certainly the contribution of new technologies, many of them subject to patent, to England's extraordinary economic progress in the seventeenth, eighteenth, and nineteenth centuries made a strong case for that policy. The

[5] William Holdsworth, *A History of English Law* (London: Methuen, 1938), 9:427, 428.

[6] Quoted in ibid., 9:429, 430.

[7] F. M. Scherer, *Industrial Market Structure and Economic Performance*, 2d ed. (Chicago: Rand-McNally, 1980), p. 441.

[8] See the excellent summary in Milton Handler, Harlan N. Blake, Robert Pitofsky, and Harvey Goldschmid, *Patents and Antitrust* (Mineola, N.Y.: Foundation Press, 1983), pp. 3–14.

American Founders, deeply influenced by English economic thought and covetous of England's economic progress, included in article I, section 8 of the Constitution a statement declaring the public interest in new technology and empowering Congress to enact patent laws. Congress obliged in 1790 by enacting the first patent law, and in 1791 the first American patent was granted, covering a process for "making pot and pearl ashes."[9]

Although less than ten thousand patents had been issued in the United States by 1836, the general principles of patentability were firmly established as the great industrial boom occurred after the Civil War. Obviously, issues would arise on whether particular processes were patentable, and there would be disputes over priority of invention, novelty, and the application of language in the patent laws to factual situations. Patent policy had become a valuable tool of public service liberalism. John Stuart Mill typically provided both economic and moral rationales for the patent monopoly. He argued that patents eventually make goods and services cheaper for consumers; the benefit is simply postponed. Further, it would be immoral to allow everyone to appropriate freely the fruits of another person's arduous and costly labor.[10]

But just as the vast outpouring of patent applications and patents granted in post–Civil War America tended to confirm their importance in economic development, so also the value of monopolizing potentially lucrative new technology enhanced the long-standing possibility of abusing the patent privilege. The courts became increasingly skeptical of marginal claims, demanding considerably more skill and ingenuity than a skilled mechanic can manifest. But though the courts provided guidelines, their application was not easy. The story of the telephone's invention and its early history illustrate the importance of patentability, its role in American economic development, and its place in public service liberalism.

The Invention of the Telephone

On January 17, 1876, Alexander Graham Bell moved to new quarters in Boston at 5 Exeter Place. In part he moved because the rent was only sixteen dollars a month, and the board at the restaurant downstairs was but five dollars a week. But more important, one of the two rooms he let could be transformed into a laboratory to carry on his telephonic and telegraphic experiments. Less than a month later Bell filed one of the

[9] Leonard M. Friedman, *A History of American Law* (New York: Simon and Schuster, 1973), p. 225.

[10] John Stuart Mill, *Principles of Political Economy* (London: Longmans, Green, 1929), pp. 932, 933. The discussion in other editions is in book V, chap. 10, sec. 4.

most important patent applications in the nation's history. From these modest beginnings much happened through November 1922, when the lineal descendant of the company that Bell helped to found completed the construction of its imposing new twenty-eight-story headquarters at 195 Broadway in New York City (which it shared with Western Union until 1930, when the latter moved out).

In the forty-six years from Bell's modest move to the completion of its skyscraper, a small partnership for the exploitation of patents had been transformed into one of the largest corporations in the world. In 1922 the Bell System, a labyrinth of companies, leased almost 10 million of the 14.3 million telephones in the United States. Although a miniscule amount by contemporary standards, the company handled an average of approximately 38 million telephone conversations daily in 1922. The book value of its telephone plant was more than $1.7 billion, it employed about 243,000 people, and its 1922 net income was in excess of $86.6 million. Not only were Bell and his associates small at the outset, but they were challenged by the giant Western Union Company and the rival claims of prior invention made by many others, the most important being Elisha Gray. The survival of the Bell interests and the growth of the American telephone system owes much to patent policy.

In the early months of 1874 Elisha Gray made a remarkable discovery.[11] Gray was then one of the foremost inventors of electrical apparatus, holding patents on telegraphic repeaters, printing telegraphs, telegraph station switches, and a host of other devices. Gray's young nephew enjoyed playing with his uncle's electrical apparatus located in the bathroom. For reasons that are unclear, the young man enjoyed "taking shocks" from an induction coil—a type of transformer that transfers electrical energy from one circuit to another or others.

> Having connected one wire to the zinc lining of Gray's bathtub, his nephew held the other wire in one hand. As he moved his other hand along the lining of the tub, Gray noticed that a sound came from under his nephew's hand; the pitch of this sound seemed to be the same pitch as that of the vibrating apparatus of the induction coil. Gray took his nephew's position, changed the pitch of the vibrating apparatus, and found that the pitch of the sound from under his hand had also changed.[12]

Working diligently on the observed connection between electricity and sound, Gray set out to make a "telephone." And according to supporters of Gray's claims, he was the first person to invent the "telephone." Pro-

[11] Most of the details concerning Gray's career and experiments are taken from David A. Hounshell, "Elisha Gray and the Telephone: On the Disadvantages of Being an Expert," *Technology and Culture* 16 (April 1975): 133–61.

[12] Ibid., p. 139.

fessor Lloyd W. Taylor of Oberlin College (which Gray attended) claimed, "But while it is impossible to attribute to Bell the development of the telephone receiver earlier than June 1875, Elisha Gray had constructed and publicly demonstrated, as early as 1874 and before February 1875, not less than four receivers which were prototypes of the modern telephone receiver."[13] Gray went to his death mysteriously claiming that the history of the telephone's invention would never be fully written, obscured as it was by massive amounts of testimony in patent cases and the silence of key participants on critical facts. Certainly the fact that Alexander Graham Bell filed the most valuable patent application in history and Gray filed a caveat (a formal notice of an inventor's idea) within hours of each other on February 14, 1876, lends an aura of mystery to the invention of the telephone.[14]

But the circumstances of the invention of the telephone, and their important implications (which we will examine later), are more complicated than a dispute between Gray and Bell. Even as the Bell interests were scoring legal victory after victory in patent fights, the prestigious *Scientific American* scoffed at Bell's claim, asserting that Philipp Reis, a German schoolteacher, had designed several instruments that transmitted articulate speech as early as 1861.[15] Although a majority of the U.S. Supreme Court found, in one of its most complex decisions, that Bell had invented the telephone, three dissenting members concluded that Daniel Drawbaugh had priority of invention![16] Other students of the question have argued that Antonio Muecci, Edward Farrar, M. Petrina, S. D. Cushman, Amos Dolbear, and many others preceded Bell, or at least deserve a reasonable share of the credit for inventing the telephone. Under this view the Bell patent should not have been granted, for there was no novelty of invention—a fundamental element of patentability.[17]

Yet the evidence that Bell did, in fact, invent the telephone is persua-

[13] Lloyd W. Taylor, "The Untold Story of the Telephone," *American Physics Teacher* 5 (1937): 250.

[14] It is clear that Gray believed *in his later years* that he invented the telephone and that Bell had employed fraud in obtaining his patent. See R. C. Clowry to William H. Forbes, 25 September 1885, AT&T Archives, Box 1288. Clowry, who was general superintendent of Western Union, disclosed the substance of a conversation with Gray to Forbes, president of American Bell. Charges were made that Bell unlawfully saw Gray's caveat, but there is evidence that a Patent Office official unlawfully conveyed information about Bell's 1876 application to Gray. See Fred DeLand, *The Invention of the Electric Speaking Telephone*, pt. 1, p. 23, in AT&T Archives, Box 1108.

[15] Notseep, "The Telephone—Judge Lowell's Opinions Commented upon—Ingenious Transformations," *Scientific American* (October 22, 1881), in AT&T Archives, Box 1098.

[16] *Telephone Cases*, 126 U.S. 863 (1887).

[17] See William Aitken, *Who Invented the Telephone?* (London: Blackie and Son, 1939), chaps. 1–14.

sive and had much to do with the policies adopted by the various Bell companies. Much depends on the conceptions one has of "invention" and "telephone." As early as the 1820s Charles Wheatstone, an English scientist, demonstrated that musical sounds could be transmitted through metallic rods; but he never pursued the discovery. In 1854 the Frenchman Charles Bourseul suggested that a diaphragm making and breaking contact with an electrode could transmit the human voice. But, again, he did not follow through, and Bourseul's contribution remains a suggestion only. Most scientists (and all of the American courts) concluded that the Reis telephone could not articulate speech, although some asserted that with proper adjustment it could.[18] In any event, Reis saw no commercial possibility for his instruments, which could only produce pitch and rhythm but not the intensity and timbre (or quality) of sound.[19] Since the Reis transmitter was a circuit breaker, operating on the principle of breaking and permitting the flow of electricity, it lacked a vital ingredient of a successful speaking telephone—the ability to transmit undulating currents to the receiving instrument.

Indeed, until it became popular several years after the granting of Bell's 1876 patent, the word *telephone* was used to describe any instrument that could convey sound to distant points. This must be kept in mind when one attempts to unravel the complex relationship between Bell and Gray and the nature of invention. As Alfred North Whitehead observed, "The greatest invention of the nineteenth century was the invention of the method of invention."[20] The convergence of three ideas—the conversion of energy, atomicity, and the continuity of matter—unleashed many minds during that century on the road to invention. While the romantic image of the nineteenth-century inventor as a person concerned solely with benefiting humanity is a popular one, it is not realistic. As Jacob Schmookler's careful study of invention has shown, talented would-be inventors were almost always attracted to a particular endeavor because of the prospect of probable large pecuniary gains. Science and prior knowledge, of course, served as bases. Chance played a large role. And of course inventors engage in an activity calculated to serve human wants.[21] But underlying all of these factors, invention during the nine-

[18] Details on Bell's predecessors are found in W. James King, "The Telegraph and the Telephone," in *The Development of Electrical Technology in the Nineteenth Century*, paper 29, bulletin 228 (Washington, D.C.: U.S. National Museum, 1962), pp. 312–18. See also Aitken, *Who Invented*, pp. 1–23.

[19] George B. Prescott, *Bell's Electric Speaking Telephone* (New York: D. Appleton, 1884), pp. 9–14, 400–412.

[20] Alfred North Whitehead, *Science and the Modern World* (1925; rpt., New York: New American Library, 1953), p. 98.

[21] Jacob Schmookler, *Invention and Eocnomic Growth* (Cambridge, Mass.: Harvard University Press, 1966), pp. 197–211.

teenth century was selfish behavior! Inventors competed with one another and often made extravagant claims for their work as they sought to reap the rewards of progress.

Gray and Bell were no exception to this syndrome. Like Gray, Bell's inventive work was initially focused on the problem of multiplex telegraphy. As we saw in chapter 2, the growth of telegraphy in the post–Civil War era was so rapid that it led first to the development of the duplex, in which one wire could simultaneously carry two messages, and then to the quest for the multiplex, which could carry many messages simultaneously. One possible way of solving the multiplex problem that appealed to both Bell and Gray was to vary the pitches of messages so that the number of messages that could be transmitted over a wire would be limited only by the ability of the human ear to distinguish between various tones. Both men began the quest for what became known as the harmonic telegraph in the early 1870s.[22] In 1874 their respective efforts to invent a multiple telegraph had suggested the *possibility* of a speaking telephone.

An important difference in the approaches of the men stemmed perhaps from their different professional backgrounds. Bell was a teacher of the deaf whose primary focus was speech. Gray had gained considerable fame as the inventor of telegraphic devices and as an official of Western Electric, then the principal supplier of telegraphic apparatus to Western Union. Bell clearly paid more, but far from exclusive, attention to the speaking aspects of a "telephone" than Gray did. Indeed, their uses of the word *telephone* differed. Bell wrote to Gray on March 2, 1877: "I have not generally alluded to your name in connection with the invention of the electric 'telephone' for we attach different significations to the word. I apply the term only to an apparatus for transmitting the voice . . . whereas you seem to use the term as expressive of any apparatus for the transmission of musical tones by the electric current."[23] On March 5 Gray responded, conceding: "I gave you full credit for the talking feature of the telephone, as you may have seen in the associated press dispatch that was sent to all the papers in the country—in my lecture in McCormick Hall, February 27th."[24]

Both men received strong financial backing for their efforts. Samuel White, a wealthy Philadelphia manufacturer, agreed to supply capital to

[22] King, "Telegraph and Telephone" pp. 315–18, and Robert V. Bruce, *Bell: Alexander Graham Bell and the Conquest of Solitude* (Boston: Little, Brown, 1973), chap. 9.

[23] Reprinted in *The Deposition of Alexander Graham Bell in the Suit Brought by the United States to Annul the Bell Patents* (Boston: American Bell Telephone Co., 1908), p. 167.

[24] Ibid., p. 168.

Gray in exchange for a share in the profits from his future inventions.[25] Bell's activities as a teacher of the deaf had put him into contact with influential men, including his future partners, Gardiner Hubbard and Thomas Sanders. Hubbard was a patent lawyer with an interest in electrical inventions. He had been involved in public utility business ventures, including water supply and gaslight companies. On a national level he had been involved in a series of campaigns designed to secure national incorporation for a new telegraph company that would act as an agent of the Post Office in transmitting telegraphic messages. Hubbard's proposal to organize such a company assured the enmity of Western Union.[26] Sanders had established a very successful leather business. On February 27, 1875, the three men signed the Bell Patent Association agreement, under which Hubbard and Sanders would supply the capital and all would share equally in any patents Bell might obtain.

Hubbard and Sanders did not, of course, put their capital in Bell's efforts as a matter of blind faith or even friendship. Rather, as the brief sketch of their backgrounds indicates, they were astute businessmen, and what they were investing in was Bell's ability to develop valuable inventions. In the early 1870s Bell had begun his efforts to develop a multiple telegraph, and his talent attracted the attention of his partners. But whereas Hubbard and Sanders viewed multiple telegraphy as potentially the most profitable area of investigation, Bell, as early as 1874, shifted his primary (but hardly exclusive) emphasis to the speaking telephone.[27] According to Bell, while his interests fluctuated, the events that permanently shifted his primary focus were meetings with Joseph Henry on March 1 and 2, 1875. Henry, director of the Smithsonian Institution and the most eminent American physical scientist of his day, strongly encouraged Bell to develop his speaking telephone idea.[28]

At this point the plot becomes sufficiently complex to animate a suspense novel. Indeed, the evidence taken in many matters concerned with the invention of the telephone fills numerous volumes. Bell had first learned of Gray and his experiments on October 22, 1874. It is also clear that Bell was told that Gray was involved in developing a method of transmitting "vocal sounds." A rivalry to develop a harmonic telegraph developed between the two men that led to both secrecy on Bell's part

[25] Hounshell, "Elisha Gray," p. 139.

[26] Lester G. Lindley, *The Constitution Faces Technology* (New York: Arno Press, 1975), chaps. 3–7.

[27] Bell, *Deposition*, pp. 27–40; Frederick Leland Rhodes, *Beginnings of Telephony* (New York: Harper & Bros., 1929), pp. 13–21; and Kenneth P. Todd, Jr., *A Capsule History of the Bell System* (New York: AT&T, n. d.), p. 10. The Todd work is an official AT&T publication and is used here only to verify events, not to interpret them.

[28] Bell, *Deposition*, pp. 46–52; and Bruce, *Bell*, pp. 139, 140.

and a penchant (at the request of his lawyers) to document every inventive step because of a possible patent fight.[29] Nevertheless, by May 24, 1875, Bell could write to his parents: "I think that the transmission of the human voice is much more nearly at hand than I had supposed."[30] The reason for his optimism was his gradual rejection of the intermittent current (or circuit breaker) idea that Reis had devised for the transmission of speech and its replacement by the variable resistance method. According to Thomas A. Watson, Bell's assistant for the construction of apparatus, "Bell told me he had an idea by which he believed it would soon be possible to talk by telegraph. . . . 'If,' he said,'I could make a current of electricity vary in its intensity, precisely as the air varies in density during the production of a sound, I should be able to transmit speech telegraphically.' "[31]

Gray, meanwhile, although moving closer to the invention of the speaking telephone, continued to emphasize the multiple telegraph in his efforts. On more than one occasion Gray emphasized the speaking telephone's impracticality and lack of commercial value. In an October 29, 1875, letter to his patent lawyer Gray wrote: "Bell seems to be spending all his energies in [the] talking telegraph. While this is very interesting scientifically, it has no commercial value at present, for they can do more business over a line by methods already in use than by that system. I don't want at present to spend my time and money for that which will bring no return."[32] The man whose experience, fame, and fortune were based on the booming telegraph industry not surprisingly saw little commercial possibility in an upstart device that would require considerable development before it could become technically and commercially successful. Bell relentlessly pursued the invention of the speaking telephone despite the view of most others that it was of limited practicality. Thus, Bell was fulfilling the intention of patent policy not just to invent, but to exploit commercially new technology.

On February 14, 1876, Bell filed his basic patent application for a telephone transmitter employing a magnetized reed attached to a membrane diaphragm—an apparatus capable of transmitting sounds and changes of pitch, but not speech.[33] Nevertheless, the patent application, like others drafted by skilled patent lawyers, was sufficiently broad to cover other apparatus for transmitting speech. Several hours later that same day Elisha Gray filed a caveat (notice of pending patent application) for a method

[29] Bruce, *Bell*, pp. 128–31; and Bell, *Deposition*, pp. 55, 74.

[30] Bell, *Deposition*, p. 55.

[31] Thomas A. Watson, "How Bell Invented the Telephone," *Proceedings of the American Institute of Electrical Engineers* 34 (1915): 1506.

[32] Quoted in Hounshell, "Elisha Gray," p. 152.

[33] King, "Telegraph and Telephone," p. 320.

of transmitting and receiving speech. But there is serious question as to whether Gray's instruments would have conveyed speech. At the least, the particular instruments he used at the Centennial Exhibition in Philadelphia during the summer of 1876 did not work. But Bell's did at that exhibition, just a few short months after he actually did convey reasonably audible speech to his assistant with the famous words "Mr. Watson, come here; I want you." That statement, however, was heard on March 10, 1876—almost a month after the filing of the patent application.[34]

This fact, along with the suspicion that Bell might have unlawfully seen Gray's caveat, allowing Bell to amend his original application, as well as the claims of many others to have invented the telephone earlier than Bell, led to an intellectual controversy that still rages among historians of technology. But more important, it led to an exceptional amount of patent litigation after the extraordinary worth of Patent Number 174,465 became evident to all.

Let us consider the main charges that have been made against Bell. Patent applications may be amended for the purposes of clarification, which is all that Bell was shown to have done. Second, it is not unusual for the first instruments in a revolutionary breakthrough to be extremely primitive, barely workable. In 1875, well before filing the application, Bell and Watson had dimly heard speech sounds in receiving instruments,

> but it was difficult to make out what was said. . . . As the general result of all the experiments made . . . I was encouraged to believe that I had solved the problem of the transmission of articulate speech by electricity and that carefully constructed instruments, substantially similar to those I had made, would prove to be practically operative speaking telephones.
>
> . . . I was so satisfied in my own mind that I had solved the problem of the transmission of articulate speech that I ventured to describe and claim my method and apparatus in a United States patent without waiting for better results.[35]

Of course, Bell's statements were made in his self-interest. But there are several facts that are incontestable and support his contentions. Gray, although capable of engaging in a protracted and expensive patent interference battle because of Samuel White's financial backing, chose, on the advice of his experienced patent attorney, not to do so.[36] More impor-

[34] Hounshell, "Elisha Gray," pp. 155, 156; and King, "Telegraph and Telephone,"p. 320, 321.

[35] Bell, *Deposition*, pp. 338, 339. For more details, see John E. Kingsbury, *The Telephone and Telephone Exchanges: Their Invention and Development* (New York: Longmans, Green, 1915), chap. 5.

[36] Bruce, *Bell*, pp. 172–76.

tant, the 1876 basic process patent (and the 1877 Bell basic receiver patent) withstood approximately six hundred law suits and patent interference proceedings, including a suit brought by the United States for patent cancellation. While most of the cases brought by the Bell interests for patent infringement were won because the defendants did not file an answer, many were contested in a variety of federal circuit courts, and five cases were considered together by the Supreme Court. The Bell patents were upheld in every contest, although three members of the Supreme Court dissented.

Although questions will always remain, the Bell patents withstood as scrupulous a challenge as could have been mounted. Unless one posits an unlikely conspiracy of such wide proportions as to embrace a large number of patent officials, circuit court judges, and Supreme Court justices, the weight of evidence and authority upholds the Bell patents as *legally* sound.[37] One may, of course, quarrel with the patent laws and, indeed, the underlying rationale of patent protection, but the Bell interests did win the controversy in every forum. If there is fault to assess, it must be laid at the doorstep of the patent system, not the Bell System.

Patents and the Telephone Litigation

Bell and Gray engaged in their race to develop a harmonic telegraph with financial backers interested in exploiting the results of their respective activities. Although motivated by pecuniary gain (as one would expect), the results of their race benefited humanity. Bell ultimately became more successful because he patented the telephone and, in short order, aided in the process of exploiting the invention. He fulfilled the legal requirements, in which there is only one winner. Unlike the Kentucky Derby, patent law does not pay off for place or show. Alexander Graham Bell satisfied the three principal conditions of patentability. His speaking telephone was new and not obvious, it was not invented previously, and it was useful. Certainly, if rubber erasers for wooden pencils, automatic electric scratching machines for dogs, and a crank-operated beach umbrella meet these criteria, so remarkable a breakthrough as the speaking telephone should meet them. And indeed it did.

The Bell patents, as noted earlier, withstood more than six hundred court cases and administrative proceedings. Most of these cases were brought by the Bell interests against alleged infringers who had entered

[37] A comprehensive (although obviously biased) source of information on all of the cases is an internal Bell System report. Charles H. Swan, *Narrative History of the Litigation on the Bell Patents, 1878–1896* (1903), in AT&T Archives, Box 1098.

the telephone business in competition with Bell and its licensees. Notably, from 1876 to 1877 when the commercial viability of the telephone was seen by few persons other than Alexander Graham Bell, no one claimed prior invention. From then until the expiration of the basic patents in 1893 and 1894 the Bell interests were continuously involved in patent litigation against private companies and were involved in defending against a suit brought by the United States charging that the first patent had been obtained by fraud. Not until January 1, 1906—almost ten years after the last steps were taken by government counsel—was the final case arising under the original patents dismissed.

In retrospect, most of the defendants presented claims that courts did not believe. A typical matter involved the Molecular Telephone Company, incorporated in 1880, against which the Bell interests brought infringement suits in the federal district courts in the Southern District of New York and the Northern District of Ohio. Molecular's defense was based on an 1860 patent granted to Alfred G. Holcomb for a phonetic telegraphic relay. But Holcomb admitted during cross-examination that his device was not a talking telephone. Still others, such as the McDonough Telephone and Telegraph Company, based their infringement on telephones employing the circuit breaker principle, which, like the Reis instrument employing the same idea, could not convey timbre. Still other claimants "discovered"—after the success of the speaking telephone—that they had invented the device long before Bell had begun even to experiment with it. Sylvanus D. Cushman, for example, claimed in 1879 that he had invented a speaking telephone in 1851. The Federal District Court for Northern Illinois put it succinctly: Cushman invented nothing.[38] In summary, the Bell interests vigorously challenged every competing telephone claim in many forums and succeeded in demonstrating to a variety of judges that Alexander Graham Bell's patents were validly granted.

The final vindication of Bell's claims occurred in the *Telephone Cases* appeal when a divided Supreme Court upheld the Bell patents in five consolidated appeals in which the respondents claimed, in aggregate, that fifteen people other than Bell had invented the telephone earlier than 1876. Holding in a long and complex opinion that the 1876 patent was valid, the Court stated that the patent was primarily one for a process rather than a product.

> What Bell claims is the art of creating changes of intensity in a continuous current of electricity, exactly corresponding to the changes of density in the air caused by the vibrations which accompany vocal or other sounds, and of using

[38] Ibid., passim. On the Cushman case, see *American Bell Tel. Co.* v. *American Cushman Tel. Co.*, 35 F. 734 (N.D. Ill., 1888).

that electrical condition thus created for sending and receiving articulate speech telegraphically.

It is quite true that when Bell applied for his patent he had never actually transmitted . . . spoken words so that they could be distinctly heard and understood at the receiving end of his line, but in his specification he did describe accurately and with admirable clearness his process . . . and he also described, with sufficient precision to enable one of ordinary skill in such matters to make . . . a form of apparatus which, if used in the way pointed out, would produce the required effort.[39]

Finally, the Court noted that the application described two methods (magneto and variable resistance) to achieve the result. The fact that Bell, at the time of the application, thought that the magneto method was the better one—in fact it turned out to be a dead end—was beside the point. A patent does not require that the process or apparatus be perfect or even good—only that it will work.

Three members of the Court believed that Daniel Drawbaugh had anticipated the invention of the telephone in 1866 or 1867. But not until 1880 did Drawbaugh act commercially on his alleged invention. By this time, of course, the telephone was a success. In the infringement suit brought in the Circuit Court for the Southern District of New York, a record of more than six thousand pages was compiled. The court sustained the Bell patents, rejecting Drawbaugh's vigorous defense.[40] The trial judge noted that Drawbaugh, who had seen Bell's impressive telephone demonstration at the Philadelphia Centennial Exposition in June 1876, was unable to explain any part of the intellectual process by which he allegedly conceived the principle or apparatus. His memory was a blank, and he never exhibited his apparatus outside of his shop. And though many witnesses testified on Drawbraugh's behalf that they had heard "speech" in the shop, this testimony occurred many years after the event. In any event, the trial judge, on the basis of the testimony, concluded that the witnesses probably heard sounds, but not words or sentences. Many witnesses, moreover, were unable to adequately describe the apparatus Drawbaugh was supposed to have used, whereas others described apparatus that could not conceivably have transmitted speech. Notwithstanding the trial judge's opinion and direct contact with the witnesses, the Supreme Court dissenters were sufficiently impressed with the sheer number of witnesses supporting Drawbaugh's claim to conclude that he had invented the telephone.

Even the majority opinion of the Supreme Court did not end the controversy over the 1879 patent. From the perspective of the three original

[39] *Telephone Cases*, p. 989.

[40] *American Bell Tel. Co.* v. *People's Tel. Co.*, 22 F. 309 (S.D.N.Y., 1884).

partners—Hubbard, Sanders, and Bell—as well as their successors, the continual raging conflict had an important impact on the company's behavior. If a solitary loss before a court or patent tribunal occurred in this most bitterly contested patent struggle, all would have been lost. Their primary strategy was to battle every alleged infringement vigorously in the courts.

A second and complementary strategy was to effect compromises in some cases without giving up its fundamental claims. The Bell interests adopted this strategy in the critical fight in 1877-1879 with the Goliath—Western Union. Hubbard had early offered to sell the 1876 patent to Western Union for one hundred thousand dollars and was turned down. Western Union's president, William Orton, then viewed the telephone as an impractical invention.[41] By November 1877 Western Union recognized its error and organized a subsidiary called the American Speaking Telephone Company. It acquired the Gray, Amos Dolbear, and other telephonic patents. Further, Western Union employed Thomas A. Edison, already recognized as a brilliant inventor, to develop telephonic inventions. By 1879 Western Union, far superior in resources to the Bell interests, was actively engaged in the manufacture and rental of telephones. Armed with a deep pocket, it could easily undercut the Bell rates. Moreover, much of Western Union's equipment and technology was at times superior to Bell's. The Bell interests' principal weapons were to develop or buy superior technology and to bring an infringement suit in September 1878.

Things looked bleak for the Bell interests at the beginning of 1879. But on November 10 of that year, the Bell and Western Union interests settled their controversy. National Bell agreed to pay Western Union a royalty of 20 percent on all telephones used in the United States, and Western Union agreed to withdraw from the telephone business until the expiration of the basic patents, to assign exclusive licenses to National Bell under Western Union's telephonic patents, and to assign certain future telephonic inventions to the Bell company. Western Union clearly was not hoodwinked, for four days after the settlement, National Bell stock was sold at $955.50 per share—an increase of $505 above the September market price.[42]

Four reasons have been advanced for Western Union's agreeing to the settlement, which even then was perceived as a Bell victory. First, Western Union was being frontally attacked in its major business by a new telegraph rival founded by Jay Gould in May 1879 with $10 million cap-

[41] The best source of details on the initial negotiation is Rosario Joseph Tosiello, *The Birth and Early Years of the Bell Telephone System, 1876–1880* (New York: Arno Press, 1979), pp. 81–83.

[42] See ibid., pp. 484–91.

ital. Moreover, Gould threatened an alliance with the Bell interests, promising to commit his vast resources to their prosecution of the infringement suit. The president of Gould's American Union Telegraph had already begun to acquire Bell franchises in various areas. Rather than fight on two fronts, according to this view, Western Union made peace on the telephone front so as to wage war more effectively on the telegraph front.[43] Although this view may carry some weight, one must put it in perspective. Western Union's revenues during this period were in the $9 million range, which completely overshadowed the infant Bell interests, just as the telegraph's commercial importance was then far greater than the telephone's. In short, the "second front" theory must be viewed with some skepticism because Western Union's fight with the Bell interests was more of a skirmish than a battle.

Second, some have claimed that Western Union, already a bloated bureaucracy, lacked entrepreneurial vision. Third, it has been suggested that Western Union's desire to monopolize the telegraph industry overcame its commitment in a related, yet different, field. These second and third explanations are highly speculative. But the final one is not. Western Union agreed to a settlement because its attorneys and other officials became convinced that the Bell claims would prevail and that the Bell interests would persist in the prosecution of their claim.[44] George Gifford, Western Union's lawyer, believed that Bell had invented the telephone and that Western Union was an infringer. Accordingly, he urged his client to settle.[45] While the conspiratorial mind will be dissatisfied with this explanation, it is not uncommon for lawyers who feel that their clients claims are weak to attempt a compromise in which the other side gains more. Further, we must remember that the Bell interests agreed not to use its telephones in competition with Western Union in the lucrative business message, market quotation, or news-for-publication markets.

Ultimately, then, the Bell interests' principal weapon was a strong patent position. Yet at the same time it was under attack from a variety of quarters on this very point. If the Bell interests lost in *any* of the numerous cases, they would lose everything—or at least a great deal. The litigation and compromise strategies were, of course, designed to prevent defeat, or even sharing the basic patents. Yet because Bell could lose,

[43] Alvin F. Harlow, *Old Wires and New Waves* (1936; rpt., New York: Arno Press, 1971), pp. 409–11; and Federal Communications Commission, Special Investigation Docket 1, Exhibit 2096F, *Financial Control of the Telephone Industry* (1937), pp. 13–29.

[44] Robert W. Garnet, *The Telephone Enterprise* (Baltimore: Johns Hopkins University Press, 1985), chap. 4.

[45] Affidavit of George Gifford, 19 September 1882, in AT&T Archives, Box 1006; and FCC, *Financial Control*, pp. 23, 24.

still another strategy was required—obtaining more and increasingly advanced patents so that even if the basic ones were held to be invalid, Bell would still be ahead of its adversaries. Thus, patent awards do not lead to laxity, but frequently to a strong commitment to research and development (not necessarily done in-house). The Bell interests developed this attitude long before it became common in large American enterprise—indeed, long before the phrase "research and development" was devised.

A strong commitment to research and development at the outset and other factors played a role in its continuation. Among these factors were the desire to head off or control alternative technologies (such as radio), the force of competition, and perhaps what economists have termed the Averch-Johnson (AJ) effect during periods of regulated monopoly.[46] Under the AJ effect—a topic of controversy among economists—firms subject to rate-of-return regulation have a positive incentive to add capital costs, such as elaborate research facilities, to their rate bases. By choosing more capital-intensive input combinations than would prevail under competitive conditions, regulated firms enlarge their revenues and, presumably, their profits. There are, additionally, distinctive characteristics of the telephone industry that provide strong pecuniary incentives for technological progressiveness. Finally, public service liberalism combines the regulatory aspects of public utility regulation and patent protection—two forms of government intervention—to spur technological innovation.

Patents Beget Patents; Research Begets Research

It would be foolhardy indeed to make sweeping generalizations about the rate and pace of inventions. As Nathan Rosenberg, one of the leading experts on the subject, observed, differentials over time depend on a large number of institutional factors, values, and incentive structures—the very functioning of social systems.[47] During periods of rapid innovation, different industries advance technologically at different rates, and some industries fail to survive. Nevertheless, it is safe to say that several factors present in the telephone industry clearly provided strong incen-

[46] On the Bell companies' early commitment to research and development, see Leonard S. Reich, "Industrial Research and the Pursuit of Corporate Security: The Early Years of Bell Labs," *Business History Review* 54 (Winter 1980): 504–29. On the AJ effect, see Harvey Averch and L. L. Johnson, "Behavior of the Firm under Regulatory Constraint," *American Economic Review* 52 (December 1962): 1052–69.

[47] Nathan Rosenberg, *Inside the Black Box: Technology and Economics* (Cambridge: Cambridge University Press, 1987), pp. 8–14.

tives to undertake the development and exploration of innovation through and outside the patent process. First, the network aspect of telephony encouraged the development of processes and products that would enlarge the network. Second, system reliability and quality improvement would also make telephony more commercially viable. Similarly, new services (based to some extent on what the telegraph already offered) would enhance the telephone's value. Conversely, a weak link in any part of the network degrades the entire network, making it less valuable. For all of these reasons, the telephone industry was one committed to technological progressiveness and in which patents begat other patents and research begat research.

Of course, the social organization of research was different in the nineteenth century from what it is now. Large companies then primarily relied on the efforts of individual inventors acting independently, who would assign inventions for a fee or royalty. Second, wealthy individuals or companies would financially back talented people before they invented a product or process in exchange for the right to profit from it. Both Gray and Bell illustrate this style. Third, firms would frequently rely on the ingenuity of workshops that would supply machinery or apparatus to order. Such shops would employ people who could design or invent products that met the requirements of customers. Western Electric, before it became associated with the Bell interests, enjoyed such a relationship with Western Union, meeting the latter's equipment requirements. One of the first versions of the typewriter was an offshoot of Western Electric's efforts for Western Union.[48] Not until the 1890s did it become common for major firms to have their own specialized research departments.[49]

Yet even earlier the Bell interests had established research and development departments. These were, of course, rudimentary by contemporary standards. From the time of the formation of the first Bell Telephone Company in July 1877, modest efforts were instituted along these lines. In June 1878 Western Union, then vigorously competing in the telephone business with the Bell interests, began employing a transmitter based on a Thomas Edison patent (which the Bell company had previously refused to purchase) that was technologically far superior to the comparable Bell instrument. Although Bell soon recovered when it purchased an even better transmitter from the inventor Francis J. Blake and

[48] See Frank H. Lovette, "Western Electric's First 75 Years: A Chronology," *Bell Telephone Magazine* 23 (Winter 1944–45): 271–76; and A. R. Thompson, "History of the Western Electric Company, 1869–1924," pp. 1–10, AT&T Archives, Box 2061.

[49] Alfred D. Chandler, Jr., *The Visible Hand* (Cambridge, Mass.: Harvard University Press, 1977), pp. 374, 375; and Leonard S. Reich, *The Making of American Industrial Research* (Cambridge: Cambridge University Press, 1985), pp. 39–41.

put it into service in December 1878, an important lesson had been learned. As George David Smith succinctly stated: "The Bell Company could no longer rely on itself or on a small group of easily identifiable friends for its technical needs."[50] An electrical and patent department under the leadership of the distinguished technical expert Thomas Lockwood had by 1885 been joined by a mechanic and testing department directed by Dr. Hammond V. Hayes, trained in physics and engineering at Harvard and MIT. The very fact that a highly educated scientist held an important position in a business firm was unusual enough. Even more remarkable was that one of the divisions of the new department was devoted to general experimental work.[51]

Equally important, the Edison transmitter experience emphasized the interactive network aspects of telecommunications. Thomas Lockwood, recounting early company strategy, testified: "It has been the practice of the company . . . to select everything in the way of conductors, apparatus or telephones that it has seemed would be of benefit to the *system at large*, and either to buy them outright . . . or else to buy sufficient rights under them to enable the American Company and its system to carry on the telephone business in high efficiency everywhere" (emphasis supplied).[52] In an interactive network such as that involved in a telephone system, a weakness in any part of the system reduces the ability of other parts of that system to operate up to full potential. A bad transmitter, for example, will be reflected in inferior sound emanating from the receiver. Early in the telephone's history Bell policymakers realized that such weaknesses, in turn, can retard the expansion of the network, lower quality, or otherwise impede a more widespread acceptance and use of the telephone system, with a resultant negative impact on revenues.

For this reason Bell's research efforts and patent acquisition policies had to be very broad and not limited to particular products or technologies. Bell's early research program was based on a host of diverse problems that had to be solved to increase revenues, "including: attenuation (a loss of energy of telephone signals during transmission), cross-talk (transfer of speech energy from one circuit to a parallel circuit) and inter-

[50] George David Smith, *The Anatomy of a Business Strategy* (Baltimore: Johns Hopkins University Press, 1985), p. 47; see also pp. 52, 106. See also Tosiello, *Birth and Early Years*, pp. 343–48.

[51] M. D. Fagen ed., *A History of Engineering and Science in the Bell System: The Early Years (1875–1925)* (New York: Bell Telephone Laboratories, 1975), pp. 37, 38.

[52] Quoted from Lockwood's testimony in *William A. Read et al.* v. *Central Union Telephone Company*, reprinted in Federal Communications Commission, Special Investigation Docket 1, Exhibit 1989, *Report on Patent Structure of the Bell System, Its History and Policies and Practices Relative Thereto* (1937), p. 23.

ference with other electrical systems."[53] The problems to be solved included not only those associated with electricity, but more traditional ones as well, such as the best woods to use for telephone poles, tensile strength of wire, how to place cables under rivers, the best materials for insulation, and how to minimize the number of overhead wires.[54] In many ways telephone technology paralleled electric technology in lighting, driven by the goals of lowering costs and applying new technology so that the service could come into more general use while at the same time devising new uses for the service. In turn each improvement in a part of the system called for compatible improvement in other parts of the system. "Because electrical equipment typically operates as part of a system, certain features of all apparatus on one system must be the same."[55]

The result of this dynamic was, of course, to impel the Bell interests toward standardization and close control over each phase of telephony so as to assure compatibility and coordination. As telecommunications became more and more complicated—and strides were rapid almost from the outset—the problems of control and coordination became similarly complex. The costs of not exercising organizing authority within a central firm (as opposed to relying on other firms as suppliers or contractors) consequently grew with the increasing complexity of telephony. For example, as the network grew and became more complex, the costs associated with the installation of even a minor defective part from an independent supplier would grow commensurately.[56]

Largely for this reason events led to the gradual enlargement of the Bell System so that it embraced local transmission (or local loops), long distance transmission, the manufacture and supply of customer premises equipment, and the manufacture and supply of switching gear and transmission equipment, as well as the requisite research. This structure, virtually unique in American business, came about, as we will see in the next chapter, not because of any grand plan—indeed, the Bell interests sometimes resisted enlargement—but gradually as particular problems arose. But an interactive technology drove the telecommunications industry, even at the outset of the new business, to emphasize technological progressiveness and the accumulation of more and better patents. The

[53] Lillian Hoddeson, "The Emergence of Basic Research in the Bell Telephone System, 1875–1915," *Technology and Culture* 22 (July 1981): 515.

[54] Rhodes, *Beginnings of Telephony*, chaps. 6–14, provide much detail on these problems and their solutions.

[55] Harold C. Passer, *The Electrical Manufacturers, 1875–1900* (Cambridge, Mass.: Harvard University Press, 1953), p. 363.

[56] See the argument on the enlargement of firms in R. H. Coase, "The Nature of the Firm," *Economica*, n.s. 4 (November 1937): 386–405.

ability to add new services, higher standards of reliability, and a larger and larger network—all critical to greater profits for the industry—depended on scientific and engineering advances. Public service liberalism was sufficiently flexible to allow technology to shape the Bell System, which became unraveled later because of modern liberalism's eventual breakup of that system.

The Essential Technology of Telecommunications

We can best understand Alexander Graham Bell's achievement when we consider the number of components and operations that were required to make the telephone commercially viable. (For the same reason we can understand why some of Bell's ingenious rivals did not grasp the telephone's commercial possibilities.) At its simplest level, telephony is a system requiring close coordination and compatibility between parts. The electric transmission of speech from a sender to a listener on a simple private line requires, first, a "transmitter" (a transducer that converts sound energy into electric energy); second, a transmission medium (such as a wire); and third, a "receiver" that reconverts electric energy into sound energy. Fourth, a source of current (such as a battery) is required. Obviously, the aural characteristics emanating from the receiver must allow the listener to understand with reasonable precision the information provided by the speaker. Note that this primitive model does not include amplification, signaling that a call is about to be transmitted, switching, or a host of other functions that were taken for granted only a few years after the telephone's invention. Further, this model is a one-way rather than a two-way circuit, in which there are transmitters and receivers at both ends.

While the four-part system (transmitter, medium, receiver, source of current) is *theoretically* the minimum required for a telephone system, even the most elementary practical system requires amplification. Because of leakage and the resistance of materials, transmission loss inevitably occurs in a telephone circuit. Accordingly, amplifiers, the function of which is to increase the level of power of signals to compensate for loss, must be placed in the system. Of course, the longer the line, the more difficult the problem of amplification becomes. Indeed, attenuation increases exponentially so that "if the output/input ratio is X for one section of line, it will be X^2 for 2, X^3 for 3 and so on."[57] Thus, by 1900 because of the attenuation problem the *practical* limit of long distance telephony by wire was believed to be about twelve hundred miles—the

[57] Fagen, *Engineering and Science*, p. 62.

circuit between Boston and Chicago. But during that period George A. Campbell, an in-house Bell scientist, and Michael I. Pupin, a free-lance inventor (who sold his invention to the Bell interests), independently invented the loading coil, which not only dramatically reduced attenuation, but overcame much distortion as well.[58]

The feat was not only a technological one (with great economic benefit), but it required the development of scientific principles concerning loading. In many cases over the years, the solution of technological problems required the development of underlying scientific principles. One should consider that it was only in 1882 that MIT offered the first electrical engineering degree, and while other schools soon followed suit, a company requiring underlying theory often had to do the work itself. It is not surprising, therefore, that the Bell technological facilities should also have included a major scientific component.[59] And as the attenuation example illustrates, problems tend to become ever more complex as the system expands. Consider that line problems include not only attenuation, but also cross-talk, noise, echo, delay, and interference. These problems were relatively simple when local telephone systems were small and the only other wires that could cause interference through external electrical waves were telegraph wires. Soon power lines, electrical railways, and even other telephone lines compounded the problems to be solved. Consider, too, the problems of electrical and fire protection. Even materials that could provide better transmission required development. For example, iron wire, which was less satisfactory than copper wire, had to be used in the early years because copper wire of sufficient tensile strength to be strung on poles had not yet been developed.[60]

Signaling, too, was a difficult problem in the early years of telephony. We now take for granted that a bell or some other sound signals an incoming call. But before Watson invented the "Thumper," the first primitive signaling device, in June 1877, one signaled by tapping on the front of the telephone's diaphragm with a pencil. The "Thumper" was a button that electrically caused the hammer within the sending set to beat on the diaphragm. The sounds produced were transmitted over the line and reproduced telephonically in the receiving set. Then in 1878 Watson developed the first call bell.[61] But like so much else in telephony, solutions

[58] Ibid., pp. 244–47; and James E. Brittain, "The Introduction of the Loading Coil: George A. Campbell and Michael I. Pupin," *Technology and Culture* 2 (January 1970): 36–57; and Neil H. Wasserman, *From Invention to Innovation* (Baltimore: Johns Hopkins University Press, 1985), chap. 4.

[59] See especially Reich, *The Making of American Industrial Research*, pp. 22–25.

[60] J. O. Perrine, "The Development of the Transmission Circuit in Communication," *Bell Telephone Quarterly* 4 (April 1925): 115.

[61] Rhodes, *Beginnings of Telephony*, pp. 176, 177; Fagen, *Engineering and Science*, pp.

raised new problems, such as transmission loss. Consider the enormous range of problems that have to be addressed in contemporary signaling, which permits calls to be established, supervised, charged for, and disconnected.

The speed with which phone calls currently take place masks the fact that even the most basic signaling involves a large number of discrete operations, each with a separate set of signals. These include, first, an off-hook signal when one lifts the receiver. Second, the serving office sends a dial tone. Third, the customer dials. Fourth, the called telephone, if not engaged, is alerted by a ringing current that activates a bell or some other sound source. Fifth, a tone is sent to the originating telephone indicating that the receiving instrument is ringing, busy, or cannot be reached in the system. Sixth, if the receiving phone is answered, an off-hook signal is returned to the sending instrument. Seventh, the receipt of the answer causes the ringing at the receiving instrument to stop. Eighth, a signal is then relayed from the sending telephone to the billing office initiating charges for the call. Ninth, at the end of the call an on-hook signal is sent indicating that the connection between the two instruments should be severed.[62] As one can readily surmise, a great deal of effort and resources had to be expended before the system worked as simply, effectively, and rapidly as it now does.

Other products and services have components, of course, but there is one critical difference between telephone communication and other businesses that requires reemphasis here—the interactive nature of the telephone system. "If a toaster is defective, it may blow a fuse, but it cannot affect the electrical service of others. An improperly designed or maintained telephone, when put into operation, can distort the electric current flowing to it from the switching center in such a way as to cause return signals that interfere with the quality of service not only for that customer but for his neighbors as well, causing wrong numbers, busy signals, incorrect billings, transmission difficulties, and similar problems."[63] Clearly, too, since receiving and sending are part of the same telephonic process to the customer, impairment at one end will have a deleterious effect on the other end. Similarly, improper design, manufacture, installation, maintenance, or use of any equipment can have an adverse effect on the performance or efficiency of every other link in a telephone system.

Thus far we have only considered a two-person network (without switching), akin to a private line, which indeed the earliest telephone use

---

113–19; and W.C.F. Farrell, "Thomas A. Watson's Contributions to Telephony," *Bell Telephone Quarterly* 14 (April 1935): 88–102.

[62] See Stipulation/Contention Package, episode 1, par. 58, *United States* v. *AT&T*.

[63] Ibid., episode 1, par. 89.

was. A typical private line in the early days was that between a physician's home and office. A May 1877 circular advertising the telephone suggested private line service as the principal use. At that stage a lessor was expected to arrange for the construction of the line, although the Bell interests would do it for $100 to $150 a mile. The circular proclaimed, however, "Any good mechanic can construct a line; No. 9 wire costs 8.5 cents a pound, 320 pounds to the mile."[64] Before the advent of switching and networking, then, the Bell interests were primarily concerned with leasing telephone instruments for private line service. The first line was built for Charles Williams, an electrical manufacturer who supplied apparatus for Bell, between his home in Somerville, Massachusetts, and his office in Court Street, Boston, and was completed on April 4, 1877.

Switching

The advantages of switching were soon seen. In a telephone system connected entirely by private lines, the number of lines needed to connect all nodes (points at which telephones are located) is governed by the formula $\frac{N^2-N}{2}$ where N is the number of nodes. Thus, six nodes would require fifteen lines, ten would require forty-five, and ten thousand would require almost 500 million lines. It is obvious, then, that switching—the interconnection of transmission paths—would have come into existence simply on physical grounds, without considering the benefit of each subscriber conversing with ever increasing numbers of other subscribers. But as the telephone gained in popularity, its entrepreneurs could look to the telegraph, which had rudimentary exchange systems as early as 1865.[65] The economies of a switching system are evident when we consider the ten-node example. Instead of forty-five lines, only ten are required.

The first telephone switchboard was placed at the central offices of the Holmes Burglar Alarm Company in May 1877. Holmes installed telephones in four banks that subscribed to his service and connected the phones to a switchboard constructed to his order. Soon conversations were being held between the four subscribers and the Charles Williams shop, to which a fifth line ran.[66] But it was in January 1878 that the first telephone company, acting on Alexander Graham Bell's conception that

[64] The circular is reprinted in Kingsbury, *Telephone and Telephone Exchanges*, p. 67.

[65] Ibid., pp. 82, 83.

[66] G.K. Thompson and R. B. Hill, "The First Telephone Switchboard," *Bell Telephone Quarterly* 9 (July 1930): 205–11.

the telephone was potentially an instrument of mass communication, installed a manual switchboard. Eight lines with a total of twenty-one subscribers were connected to a switchboard in New Haven, Connecticut, by the District Telephone Company of New Haven. Less than one month later the company issued a list of subscribers (without addresses or telephone numbers, however)—the first telephone directory. George W. Coy, who devised the New Haven switching system, imitated the local district telegraph system and Western Union.[67]

Within two years major technological developments had occurred in switching that, combined with the obvious economic benefits of interconnection, changed the telephone business. As one history of switching summarized: "Development in the design of manual exchanges followed fast, with a flood of inventions which, although they may appear very simple to us, nevertheless ensured steady progress in an entirely new technology."[68] In 1888 the Cortland switchboard in New York City, capable of handling ten thousand lines, opened. As one might gather, the size of this manual switchboard raised important considerations for the system at this stage. The first was a physical one—how to have the cords in so vast a system connect with the large number of other subscribers. The second problem was how to make switching more efficient by substituting machine switching for the cumbersome manual system as it gradually became economically feasible to do so.

The first problem was solved by the use of "trunks"—initially lines connecting distant areas within switching offices. From this idea grew the elaborated conception of trunking, in which switching centers reasonably distant from one another are connected by trunk lines. Centralized switching is an efficient means for interconnecting in small areas, but is less efficient than trunking in large areas, because in centralized switching each line has to extend to the central switching point, whereas in trunking the lines of individual subscribers only have to run to a nearby switching office. In trunking, communication to a subscriber connected to a distant switching office is undertaken by a loop between the two offices and then to the party called, who is connected to the latter switching office.[69]

From the basic concept of trunking a variety of other devices evolved as telephone usage increased, including tandem offices (which switched

[67] J. Leigh Walsh, *Connecticut Pioneers in Telephony* (New Haven, Conn.: Morris F. Tyler chapter, Telephone Pioneers of America, 1950), chap. 4.

[68] Robert J. Chapius, *100 Years of Telephone Switching* (Amsterdam: North Holland, 1982), pt. 1, p. 49. For the early development of switching, see also Rhodes, *Beginnings of Telephony*, chap. 11.

[69] Fagen, *Engineering and Science*, p. 470.

only trunks) and a hierarchy of switching centers.[70] As in the case of other aspects of the evolving telephone system, it was the economic drive to minimize costs that led to intensive technological effort in switching technology. But the problems involved in this effort as the telephone network expanded became increasingly more complex, requiring not just technological activity but abstract scientific work on the theory of networks as well. Thus, the Nobel Prize–winning effort that led to the invention of the transistor at Bell Laboratories in 1948 was stimulated by the economic goal of switching telephone connections more rapidly and efficiently through electronic means rather than electromechanically through relays.

Automatic Switching

The growth of the network required not only trunking, but also the *gradual* substitution of machines for manual switching. There is no question that the Bell company was a follower not a leader in the the early phases of automatic switching. As a Federal Communications Commission (FCC) report observed: "The automatic exchange patents were the one outstanding exception to the Bell System's record of securing all telephone patents of major importance during its early history."[71] The Bell interests did not seek to suppress automatic switching, which, in any event, was beyond their power; patent policy allows technological progress by outsiders as well as insiders. The Bell interests were merely tardy, but as we will see, their caution may have been justified. In any event, they would have been forced to come to automatic switching on the grounds of increasing marginal labor costs. Consider that in 1902, with virtually an all-manual system, the ratio of Bell operators to telephones was twenty-two per one thousand. If the same ratio had been maintained in 1970, more than 2.3 million operators would have been required at the low 1902 utilization level! Thus, the incentives to move to automatic switching were strong. By 1970, when the Bell switchboard system was fully automated, with each telephone used on average for many more calls, the ratio was down to 1.7.[72]

Although applications for patents covering automatic switching were made as early as 1879, the first such system to be widely used in the United States was the "step-by-step" system, for which Almon B.

[70] For details on the theory and technology of switching, see Hiroshi Inose, "Communications Networks," *Scientific American* (September 1972): 117–28.

[71] Federal Communications Commission, *Report on Telephone Investigation* (Washington, D.C.: Government Printing Office, 1939), p. 219.

[72] Fagen, *Engineering and Science*, p. 550.

Strowger, a Kansas City undertaker, filed a patent application in 1889, which was granted in 1891. According to one (possibly apocryphal) story, Strowger felt that Kansas City telephone operators were deliberately steering potential customers to a competitor. Strowger was so irate that he was determined to invent a switching system that would bypass operators. In the step-by-step system, dial pulses emanating from the transmitting telephone directly control switches within the system as they establish a communications path step by step. A central office built under the Strowger system could only serve ninety-nine telephones and required each telephone to run five wires to the central office. Although *initially* of limited commercial value (like the telephone), rapid progress was made so that a successful exchange was established in 1895 in La Porte, Indiana, and—more ominous for the Bell interests—an automatic system was installed in Augusta, Georgia, in 1897 for a company directly competing with the local Bell company.[73]

The Bell interests had begun working on their own automatic system earlier than the Strowger application and had conducted a laboratory trial of an automatic system for small communities in May 1889. In 1902 AT&T authorized the development of a ten-thousand-line automatic exchange.[74] But the policy thinking within the Bell System had to undergo a substantial change before a commitment to automatic switching for any but small communities would be approved. T. D. Lockwood, who was responsible for much of Bell's patent policy during the pertinent period, wrote in 1891, "Experience and observation have united to show us that an operation so complex, as is that of uniting two telephone subscribers lines, and bringing the substation at the outer end of the two lines into communication, can never efficiently or satisfactorily be performed by automatic apparatus dependent upon the volition and intelligent action of the subscriber."[75] Nevertheless, after the 1895 La Porte trial experience, Lockwood's attitude began to change in favor of considering such equipment for small towns and villages. Accordingly, the Bell interests undertook a threefold strategy of acquiring useful automatic exchange patents and licenses, undertaking internal investigation and development, and devising semiautomatic systems involving both manual and automatic operations.

Gradually the economic benefits of manual systems compared to automatic systems in larger communities were reduced and service reliability increased so that a semiautomatic Bell installation was made in Newark,

[73] *FCC, Report on Telephone Investigation*, p. 219.

[74] "Automatic Exchange Switching Development, 1899–1940," AT&T Archives, Box 1014.

[75] T. D. Lockwood to John E. Hudson, 4 November 1891, quoted in FCC, *Report on Patent Structure*, p. 94.

New Jersey, in 1914 and a fully automatic installation was made in Omaha, Nebraska, in 1921.[76] Moreover, in 1916 Western Electric was manufacturing Strowger-type equipment under license for smaller communities. But the problems of serving larger communities with automatic equipment had to await the solution to the economic-technological difficulties inherent in the Strowger system, including capacity limitations, the difficulty of selecting alternative routing when one of the steps in step-by-step is occupied, and the need for considerable spare switching capacity in the system.

During the period 1905–1907 AT&T made important progress toward devising an alternative automatic switching system that would suit the telephone systems of larger cities. In 1905 AT&T scientist E. A. Gray invented the "graded multiple," an early system of alternative routing that rank ordered (or graded) trunks and switches so that if the preferred path from transmitter to receiver was occupied or in heavy use, alternative paths could be chosen. But the graded multiple in turn raised the problem of controlling the added burdens placed on switches through the dialing system with which subscribers were familiar—another example of the interactive nature of the system. E. C. Molina, another Bell researcher and mathematician (whose formal education extended only through high school), devised in the same year a new central office pulse translation system adaptable to large switches and other equipment.

Based on these and other developments, company development policy in 1906 shifted to a new automatic switching system for large cities—the "common control" system, in which selectors hunted for idle switches and trunks, choosing different routes from among a hierarchy of alternatives. Although laboratory models involving common control took place in 1921, not until the 1930s were the combined equipment, technological, and economic problems sufficiently solved to institute full common control.[77]

Complementarity and Integration

Public service liberalism is not planning in which government administrators dictate the structure of an industry. Rather, as telephony illustrates, firms are permitted to choose the structure to which their technology and economic characteristics point, subject to the attainment of public service goals. As the examination of switching shows, the devel-

[76] Frank B. Jewett, "The Telephone Switchboard—Fifty Years of History," *Bell Telephone Quarterly* 7 (July 1928): 164, 165.

[77] Fagen, *Engineering and Science*, pp. 569–76; and Stipulation/Contention Package, episode 1, pars. 47, 48, *United States* v. *AT&T*.

opment of one component of the telephone system has a major impact on the other parts. Interactivity is critical in understanding the drive to innovate and obtain new patents. The capabilities of a switching system, for example, make demands on the design and capabilities of transmitting and receiving instruments and the transmission lines. The instruments and lines, similarly, impose requirements and limit the capabilities of switching equipment and so on. Changes in one part of the telephone system entail changes in its other parts.

As noted earlier, a telephone system is not like a toaster, in which a defect does not injure the whole electrical system. Because of the need for complementarity and close coordination between the parts of a telephone network, there is a strong management incentive to engage in centralized business direction, capital promotion, and technological planning, thereby reducing the risks and possible losses that could occur if any part of the system lags in development. Similarly, the need for close coordination of the various components of the telephone business gives rise to one of the classic incentives for backward vertical integration—assurance of the supply of items needed in a company's major business focus.[78] As we will see, the Bell interests' acquisition of Western Electric as a principal supplier stemmed from precisely these considerations. And in general, technology to a large extent determined the structure of the industry.

The great need for complementarity, coupled with the continual progressiveness of telecommunications technology, implies the need for continuous supervision over the various parts of telephony. The more complex and changing a product or service is, the greater the costs of instructing and supervising others who are responsible for component parts. The less the tolerance for deviation, the larger the costs will be for the failure to fulfill one's part in an overall interactive system. When complementarity is critical yet at the same time the components of an industry progress at different rates, the costs of discoordination can be very high. Accordingly, as the economic historian James P. Baughman, testifying for AT&T, concluded, a firm requires "systematic, anticipatory kinds of management and then in terms of implementation, the need for a greater control over the plant equipment."[79] Again, insofar as licensees had an incentive not to cooperate with the Bell interests in implementing new technology or providing a target (or rising) standard of service, the effects would be felt throughout the network—and would be costly. All of these considerations, which ultimately stem from the system's comple-

[78] On supply and vertical integration, see Edith T. Penrose, *The Theory of the Growth of the Firm*, 2d ed. (Oxford: Basil Blackwell, 1980), pp. 147–48.

[79] Record, *United States* v. *AT&T*, pp. 23, 559.

mentarity and integration, help explain the Bell System's structure. And as the nineteenth century progressed technological developments made this even more apparent to Bell management. Nowhere can this be more clearly seen than in the example of the common battery.

The Common Battery

Almost everyone who has seen films that take place in the 1880s or early 1890s has noticed that in order to begin the process of signaling the other party to a conversation, the person originating the call turned a crank. At the same time a battery (or batteries) associated with the subscriber's equipment provided the power for the telephonic transmission. Similar apparatus was located in the central office. Although obviously cumbersome, the system (known as the local battery system) provided reasonable service because it assured a fairly loud output relative to line attenuation. Yet the local batteries were one of the weakest links in telephonic communication and were widely disliked by customers as well. Until the 1890s the battery cells were of the wet variety, which were extremely corrosive and could cause substantial damage if they leaked or spilled. Additionally, they were bulky and often required the subscriber to mix chemicals for their proper use.[80] Dry cell batteries, first produced in the United States in 1890, improved the situation, but they also were bulky and of uncertain durability. Moreover, the early dry cells developed holes through which the corrosive sal ammoniac leaked. There was wide variation in their durability, although typically they would last only five or six months with light service. The companion signaling instrument for both types of battery, the magneto, was also bulky and required special instructions for beginning and ending a call. The latter was sometimes forgotten, resulting in unnecessary tie-ups.

The maintenance and inspection of both types of battery and the magneto were very costly to the company. E. J. Hall, AT&T vice-president, noted at a technical conference in 1892, "We all realise what an enormous part of our operating expense is due to the maintenance of batteries at subscribers' stations. If we can get rid of that, we get rid practically of ninety percent of our inspection and a very large proportion , of course, of all of our difficulties."[81]

The system that solved these problems was the common battery, in which a battery located in the central office was used to power both signaling and transmitting.

[80] Details are provided in Fagen, *Engineering and Science*, pp. 81, 108–12.

[81] Quoted in Kingbury, *Telephone and Telephone Exchanges*, p. 365.

As early as 1880 Bell technicians and others were working on the common battery system. But it was during the 1890s that the convergence of conditions made the common battery system feasible. By then line construction had improved, loops had been shortened because of trunking, and transmitters had been developed that could operate on extremely low amperage. An 1892 report by Hammond Hayes outlined some of the advantages that would be realized through the installation of the new system. Most obviously, substantial savings would be realized through reduced repair and inspection of equipment on the customer's premises. Second, the reduction in bulkiness would permit the design and production of new customer equipment, including desk-type telephones. Third, the location of a defective local circuit could be more easily found. Fourth, a poor local circuit would not unbalance a longer circuit.[82] Thus, the installation of the common battery system would have major implications for equipment, repair, and long distance, so any changes in these aspects of the telephone system would have to be closely coordinated with the installation of the common battery system.

But this system implied more than simply closer coordination between the components of the telephone system. In the words of Henry Boettinger, a former AT&T executive, before the common battery system "end-to-end responsibility was somewhat slapdash. . . . If you wanted the great advantages of the common battery system, you had to seize control of every bit and piece within that system in order to make it work."[83] Thus, new circuits had to be designed and installed that would reduce the cross-talk that could result from the large number of lines supplied by the common battery. This entailed the design and installation of new circuits and transformers. Because of the obviously greater impact of a power failure, central office power systems had to be designed that would be not only larger, but more reliable as well. New transmitters would have to be designed to compensate for the fact that the transmitter current from common batteries was lower than that obtainable with local batteries. New high-capacity switchboards had to be designed for the radically new system. And each of these components (such as switchboards) consisted of many complex interrelated parts, which had to be integrated with the other segments of the whole system.[84]

The common battery, then, made top Bell officials more aware that the system required centralized direction and control and close integration of each of the parts. Top Bell scientist Hammond Hayes reported in the

[82] Ibid., p. 366; and Fagen, *Engineering and Science*, pp. 497, 498.

[83] Record, *United States* v. *AT&T*, p. 17,397. See also testimony of James P. Baughman, on p. 23,553, and Defendant's Exhibit D-T-243.

[84] Fagen, *Engineering and Science*, pp. 81, 489, 498, 499, 697, 698; Rhodes, *Beginnings of Telephony*, pp. 163–66; and Jewitt, *Telephone Switchboard*, pp. 160, 161.

1898 Annual Report of the Engineers Department: "With the growth of the common battery system, an exact knowledge of the actions of the various circuits has become of such great importance that I determined to devote sufficient time to the matter."[85] In part because of the common battery the company management was compelled to change the system of manufacture as well. Under the older system, parts were made in as close conformity as possible to physical samples, the foreman determining whether the copies were sufficiently close to the original. The large number of required parts and the greater precision needed—a factor of even more importance as telephony advanced over the years—led to: (1) the abandonment of subcontracting, (2) closer coordination between the operating and manufacturing components of the industry, (3) the use of engineering drawings rather than samples whenever possible, and (4) the improvement of inspection procedures.[86] In summary, the common battery system had the inexorable centripetal tendency to locate all parts of the telephone system in one firm, rather than to employ buying and selling arrangements with other companies.

The theme provided by the common battery system could be duplicated by examining other major breakthroughs, such as the loading coil. As telephone technology advanced in the late nineteenth and the twentieth century, the degree of tolerance that could be permitted in any component of the interactive system correspondingly diminished. Close coordination had to be maintained between the various functions involved in telephony: design, manufacture, installation, maintenance, and use. Any improvement or change in one part of the network—local loops, customer premises equipment, switching and transmission equipment, and long distance—required consideration of the probable impact on other parts of the network and the ways in which these other parts of the network could (or could not) change in complement with the proposed changes in the first part.

An examination of the Bell companies' documents shows this concern for central integrated company planning of each part and subpart of the system. Since problems arise during the implementation stages of each significant change, the process of central integrated management planning became an ongoing one entailing frequent feedback in which the information developed in one department was transmitted to others in the interdependent parts of the network. The growth of the network, too, required not just continuing coordination, but also drastic redesigning at times. Telephone cables, for example, were capable of containing only

[85] AT&T, 1898 Annual Report of the Engineers Department, in AT&T Archives, Box 2021.

[86] Wasserman, *Invention to Innovation*, pp. 107, 108.

one hundred pairs of no. 19 wires in 1892. By 1928 cables had been substantially redesigned nine times so that the largest cable was then able to contain 1,818 pairs of no. 26 wires.[87] New devices designed to encourage use, such as the coin box, also had to be integrated into the network carefully in connection with the various parts of the system.[88]

Public Service Liberalism and Industrial Structure

Under the philosophy of public service liberalism, the state had an obligation to actively promote economic progress; it should not idly sit back and expect the competitive market to achieve such progress unassisted by state intervention. Like some of the most dynamic political economies, such as Japan and South Korea, the state was expected to lend a helping hand to the development of economic enterprise. Patent policy, as we saw, applicable to all industries, was one of the most important tools. Land grants, under the Morrill Act for the establishment of state colleges, and various subsidies to encourage railroad development were other policies designed to improve American technology. Tariff policies, both those favoring protection and those lifting barriers, were in their different ways designed to promote domestic technology, either through giving laggard domestic industries an opportunity to catch up or by stimulating native companies through foreign competition.

Public service liberalism did not automatically provide ready-made simple answers to the political-economic issues faced. Rather, it provided a framework for looking at such issues and guidelines for addressing when government intervention was appropriate and when it was not. Public service liberalism was, then, like the common law, an ongoing, dynamic way of approaching the issues it was expected to treat. Starting with a rebuttable presumption in favor of the free market, it did not dictate a single industrial structure—monopoly, oligopoly, or many units. Rather, as the telephone industry illustrates, such factors as technology and the *probable* performance of an industry under different industrial structures would shape public policy. Mistakes could be made and policy would move in new directions until the best (or, perhaps more accurately, least worse) structure was devised. Again, the telephone industry ably illustrates the approach of public service liberalism to the critical problems of industrial structure.

[87] F. L. Rhodes, "1800-Pair Cable Becomes a Bell System Standard," *Bell Telephone Magazine* 8 (January 1929): 25–29.

[88] On the coin box, see Thomas D. Lockwood to John E. Hudson, 5 September 1890, in AT&T Archives, Box 1231.

4

Structural Liberalism: The Issues of Economic Structure

The Political Economy of Business Organization

TODAY enterprises from the smallest mom-and-pop store to the largest industrial firms and service companies style themselves corporations. The corporate form dominates in virtually every economic activity except certain professional ones, such as the practice of medicine or law. Yet only a century ago corporations were far less common, and considerable antipathy existed toward them. Both the hostility toward corporations and their gradual acceptance as the dominant business form in nearly all branches of enterprise are best understood within the dynamic of public service liberalism. The gradual transition from hostility to acceptance in both England and the United States was, like patent law, designed to encourage the flow of capital to industry and to encourage the investor.[1]

Under public service liberalism corporations were stewards of the public interest. Corporate charters were meaningful. Since the decline of public service liberalism, obtaining corporate charters has become pro forma; lawyers simply fill in applications for clients who seek to use the corporate form for their business activities. Their current concerns are almost always tax liability or other practical effects of alternative business forms, and certainly not the public obligations corporations were once expected to discharge. Thus, the corporate form in contemporary society may superficially be the same as that prevailing under public service liberalism, but the consequences of incorporation are vastly different.

When individual entrepreneurs and partnerships were sufficient to operate businesses, corporate privileges were granted under public service liberalism to serve only important public purposes. Later, when the scale of enterprise increased dramatically, there was a clear need for more widespread use of the corporate form to raise large amounts of capital. Public service liberalism was flexible enough to facilitate this transition, although not without resistance. The early history of AT&T illustrates the changes in the critical field of business organization and how public ser-

[1] See William Holdsworth, *A History of English Law* (London: Methuen, 1965), 15:4, 44–62.

vice liberalism coped with the rapid transformation of business in the late nineteenth and early twentieth century. Similarly, we will see how public service liberalism is flexible with respect not only to economic structure, but also to business practices, such as telephone companies' historical resistance to selling instruments—their lease-only policy. There is an important policy consequence of this history with potential contemporary relevance: if a corporation fails to achieve its public service goals as embodied in the charter or does so at too great a cost, its charter *can* be lifted.

The idea of the corporation as a fictitious legal entity was known as early as classical Greek and Roman times. After the spread of Christianity, its various churches found many uses for the corporate idea; the Roman Catholic church, for example, was conceived as a corporate entity independent of the pope, clerics, and members at any particular time. Thus the church endured although the persons occupying roles within the institution changed over time. As the church developed, various orders and monasteries were also conceived as corporations. Based on these same ideas, peace guilds, consisting of persons living near one another, underwent a transition to the form of municipal corporations.[2]

Difficulties began to arise when the idea of corporateness moved from the religious and governmental realms to the economic one. For the corporation became intertwined with the ideas of monopoly and restrictive entry into trades and occupations. When English foreign trade expanded in the sixteenth century, the joint stock company with an exclusive right to trade in certain areas was devised to reduce risk and accumulate capital for vessels and voyages. Thus, the Russia Company, formed in 1553, was granted a monopoly of trade with Russia. But in exchange for the monopoly, the joint stock company was expected to confer a public benefit. That is, unless the company agreed to confer the public benefit, it would not be granted a charter.[3] The contrast between that idea and the contemporary corporation, which exists primarily for its own profit, could hardly be greater. In the past the overriding public purpose of the company's charter was taken seriously.

One variant of the joint stock company, the colonial company, created an immediate predisposition to the corporate form in the New World. In 1609 the Virginia Company was chartered for the public purpose of providing an English settlement in that colony. It was followed in 1628 by the chartering of the Massachusetts Bay Company, and others followed later. In each case the basic charter became the fundamental document of self-governance. Religious, educational, and charitable corporations

[2] John P. Davis, *Corporations* (New York: Capricorn Books, 1961), chaps. 3, and 4.

[3] See, for example, ibid., p. 34.

followed settlement, and in 1732 Connecticut chartered the first business corporation. Shortly thereafter other business corporations came into existence, notably in the fields of constructing and maintaining such public works as wharfs, piers, and water supply facilities or providing insurance against fire. Again, the key to incorporation during this period, in which charters were granted individually rather than through general laws, was whether the activity would promote the general welfare. This was as true of water supply and fire insurance companies as it was of religious institutions.[4]

Thus, if we focus on the public purpose element of the corporation, it is apparent that the state's activist role in corporate chartering fits clearly into the conception of public service liberalism. The state used corporate privileges to achieve certain public purposes that the free market, unaided by such state intervention, would not have achieved or that would have been attained at a considerably slower pace. Therefore, corporations were used to develop the country's infrastructure during the last part of the eighteenth and in the nineteenth century. Monopolistic privileges were often granted—but also resented. The justification for the exclusive privileges granted such infrastructural corporations as those involved in canals or railroads was that such high-risk activities required enormous funds. Returns on such investments were not only uncertain but very slow, since construction time was necessarily lengthy. The principal ways to make such investments more attractive were to reduce risk through various forms of state subsidy, tax concessions, state purchase of stock, and the promise of monopoly.[5] Thus, during the canal and early railroad era, the success of the corporate form for achieving public purposes, including economic development, had been aptly demonstrated and, with it, the collaborative public-private form of policymaking that characterized public service liberalism.

Nevertheless, the potential for abuse of monopolistic privileges remained. Indeed, as the stakes, scale, and variety of economic activity burgeoned in the nineteenth century, the potential for abuse correspondingly increased. For this reason many policymakers sought to divorce the corporation from monopolistic privilege. A corporation might still be granted a monopolistic privilege if it could be shown that the public would benefit more from this arrangement than one in which several eco-

[4] Joseph S. Davis, *Essays in the Earlier History of American Corporations* (New York: Russell & Russell, 1965), 1:75, 87–89, 103–6.

[5] See G. S. Callender, "The Early Transportation and Banking Enterprises of the States in Relation to the Growth of Corporations," *Quarterly Journal of Economics* 17 (1902): 111–62; George R. Taylor, *The Transportation Revolution* (New York: Harper & Row, 1968), pp. 86–90; and Frederick A. Cleveland and Fred Powell, *Railroad Promotion and Capitalization in the United States* (New York: Johnson Reprint, 1966), chaps. 5, 6 and 13.

nomic units undertook an activity. But chartering a corporation should not *necessarily* result in a monopoly grant. This divorce was an achievement of Jacksonianism. The judicial opportunity to divorce the corporate form from an inherent link with monopoly came in 1837, when Andrew Jackson's close associate Roger Taney was chief justice of the Supreme Court. In a suit to enjoin the construction of a second bridge across the Charles River between Cambridge and Boston, the Court, speaking through Taney, held that unless a monopoly grant is clear and unambiguous, the presumption is that none was intended.[6]

As the link between corporate status and monopoly was severed and the American economy expanded rapidly, the flood of many different kinds of enterprise seeking to incorporate inevitably led state legislatures to enact general incorporation laws instead of chartering each one through an individual statute. But while one legacy of the *Charles River Bridge* case was to make incorporation easier, another was that state legislatures began to take a careful look at not only monopolistic privileges, but even the right of a company to attain a greater size than originally allowed. Fear of market power lurked behind the latter attitude. Not only populists and other forces antipathetic to big business cast jaundiced eyes on large companies with monopolistic or near monopolistic privileges, but so did some moderate guardians of public service liberalism. The issue of whether a monopolistic or a competitive structure will best serve public purposes in any particular industry can be vigorously debated. While public service liberalism provides a structure for asking questions and determining when government should act, it does not mechanically guide one beyond doubt to the precise configuration of a specific industry.

Probably no other company was as caught up in the conflicts between the forces shaping the complex new business enterprise and the varying responses of public policymakers than AT&T and its predecessor companies. Public service liberalism faced a difficult task in responding to the dramatic changes in industrial structure and market conduct that began after the invention of the telephone.

New Year's Day 1900

The Bell System began 1900 with a new structure created only two days before. AT&T took over the assets and ongoing business of its former parent, the American Bell Telephone Company, which had gradually as-

[6] *Proprietors of the Charles River Bridge* v. *Proprietors of the Warren Bridge*, 11 Pet. 420 (1837).

sumed control over a vast communications empire embracing virtually every aspect of the telephone business. Many things had occurred in the nearly quarter century since the Bell interests had begun business, and much more would happen in the next three-quarters of a century before the fundamental changes resulting from divestiture and certain FCC decisions. Nevertheless, it is remarkable how much of the Bell System's basic structure and most fundamental policies were in place in 1900.

Structurally the centerpiece of the system was AT&T, whose genesis we will look at in the next sections. At the time of the 1984 divestiture, AT&T was a holding company with controlling interest in a vast network of subsidiaries. But it also contained the Long Lines Department, which operated the Bell System's long distance service, and most important, AT&T operated as a high command, making decisions for the system as a whole. Among its subsidiaries at the time of the 1984 breakup were twenty-one operating companies that provided local loop service in localities within designated territories and toll service between points within their respective territories. Thus, the Bell Telephone Company of Pennsylvania operated within Pennsylvania, while Southwestern Bell provided similar service in a five-state area. Some of these operating companies were wholly owned by AT&T, whereas in other cases AT&T owned most of the shares. In addition, AT&T owned a minority interest in Southern New England Telephone Company (SNET) and Cincinnati Bell.[7]

Thus the horizontal structure of AT&T that emerged in 1900 was one that included a wholly owned long distance division and varying degrees of control over operating companies that provided local and some toll service. We saw in the last chapter that the technology of telephony influenced the way this horizontal structure evolved. But we must also look at two of the early Bell System policies—leasing rather than selling telephones and the license contract with operating companies—to understand AT&T's horizontal structure. Similarly, both technological and business reasons must be considered in explaining AT&T's vertical structure. The principal components of that vertical structure were Western Electric Company and Bell Telephone Laboratories. Western Electric was acquired by the Bell interests during 1881-1883 to become the manufacturing arm of the Bell System. Later it took on other functions, including the supply of equipment not of its manufacture to Long Lines and the operating companies.

In the course of Western Electric's history it has had a number of subsidiaries and partial subsidiaries, the most important of which has been

[7] Details are spelled out in Plaintiff's Third Statement of Contentions and Proof, vol. 1, pp. 38–62, *United States* v. *AT&T*.

Bell Labs, incorporated in 1924. Bell Labs is one of the best-known and most prestigious research organizations in the world and had been jointly owned by Western Electric and AT&T.

Thus, the Bell System gradually developed a structure based on the nature of the telephone industry and early in the twentieth century had adopted such policies as retaining ownership and leasing telephones rather than selling them; licensing others to operate local loops and gradually absorbing most of the licensees; controlling its own manufacturer; supplying long distance connections and service to licensees; and engaging not only in applied research, but in pure scientific and mathematical research as well.[8]

The Genesis of American Bell

In 1965 a *Fortune* article described the Bell System's structure as "federal."[9] The description, clearly intended to be laudatory, employed the constitutional model of the United States, in which powers are divided between central and local authorities. But while the American Founders conceived a two-part division of power between the national and state governments, the structure that evolved in the Bell System was in many ways even more complex. That evolution resulted from the combined effect of government decisions, telephone technology, the profit incentive, and the managerial decisions of the Bell System's founders and its later business leaders, notably Theodore Vail and William H. Forbes, who became associated with the Bell interests in 1878 and 1879, respectively.

Because the styles and personalities of these figures (and others) played roles in the business decisions that led to the shape and structure of the Bell System, one must be careful not to view the company's history mechanistically as one that inexorably led to a particular organization and structure of power. But on the other hand, the men who made the critical decisions did so with a keen awareness of the technological, political, and economic considerations. That is not to say that they did not make mistakes. As we will see, they engaged in considerable experimentation, often changing directions until they found what they believed to be the best course. And, of course, changed industry conditions (such as the new competition that began in 1893 and 1894) called for new responses, some of which were inefficacious. Yet in a fundamental sense the early

[8] "Some Important Decisions in the History of the Bell System," 4 December 1953, in AT&T Archives, Box 2028.

[9] See Robert Sheehan, "AT&T: A Study in Federalism," *Fortune* (February 1965): 143 et seq.

leaders of the Bell interests ultimately made the decisions that led to the complex structure that lasted until altered by the AT&T breakup and the FCC's *Computer II* decision.

This complex structure came about after considerable experimentation and tension. It can be traced to the formation on February 27, 1875, of what had become known as the Bell Patent Association, an unincorporated partnership. Under the agreement, Thomas Sanders, the leather merchant, and Gardiner G. Hubbard, the attorney and businessman, agreed to finance Bell's *telegraph* work in exchange for which the three partners would share equally in whatever patents would result. (Later Thomas A. Watson was admitted to a one-tenth share in the partnership in exchange for his full-time commitment to telephone work.) The 1875 partnership agreement was, in fact, a written restatement of an October 1874 oral agreement and was committed to writing when Bell's harmonic telegraph invention was ready for submission to the Patent Office.[10] Other patent applications followed, but even the 1876 basic telephone patent application did not immediately lead the partners to devise a more formal agreement that would permit raising additional capital for expansion. The inference is, then, that *at first* Bell's businessmen partners did not share his more extensive vision of the telephone's future. That Hubbard offered to sell the telephone invention to Western Union for a hundred thousand dollars in 1876—and was turned down—tends to confirm this.

Almost immediately after the 1876 patent filing, Bell and Watson began to undertake a series of public demonstration-lectures, the most important of which were at MIT in May and at the Philadelphia Centennial Exposition in June. Later, in November and December 1876, successful demonstrations were made, first between Boston and Salem (16 miles) and then between Boston and Conway, New Hampshire (143 miles). Although the second demonstration was not satisfactory, the improvements that had already been made and the prospects of still further advances led to wider interest in the commercial possibilities of the telephone.

In May 1877 the Bell Patent Association published its first advertisement, and in that same month the first telephones were rented for business use on a private line between Boston and Somerville, Massachusetts. Later in May E. T. Holmes opened the first switchboard in Boston, hinting at the commercial possibilities of this type of arrangement in place of private lines. During this period Bell and Watson demonstrated the new means of communication in New York, drawing widespread at-

[10] Robert V. Bruce, *Bell: Alexander Graham Bell and the Conquest of Solitude* (Boston: Little, Brown, 1973), pp. 130, 203; and Rosario Joseph Tosiello, *The Birth and Early Years of the Bell Telephone System* (New York: Arno Press, 1979), pp. 7–10.

tention. Perhaps most important, in May 1877 the company was paid its first telephone bill by one James J. Emery, Jr. who had rented two telephones for a private line.[11]

The time had clearly arrived for the institution of a more formal arrangement. The result was a Massachusetts Trust (a business form then popular) that came into existence on July 9, 1877, under the name "Bell Telephone Company, Gardiner G. Hubbard, Trustee." Superseding the original patent association, the Bell Telephone Company issued five thousand shares of stock. The officers other than Hubbard were Sanders (treasurer), Bell (electrician), and Watson (superintendent). Bell and Watson had little interest in the business aspects of the enterprise, so the critical early business decisions were shared by Sanders and Hubbard. Among these was the crucial decision to lease rather than sell telephones. The original Declaration of Trust (a fundamental business document akin to a corporation's charter) stated, in part, that the company was in "the business of manufacturing telephones and *licensing parties to use the same for a royalty*" (emphasis supplied).[12]

By early 1878 Sanders and Hubbard had invested substantial amounts in the new business with very little return. The prospects at this early stage looked very attractive, but the time had clearly arrived to interest other potential investors in the new business to finance further expansion. Thomas Sanders succeeded in interesting a group of Massachusetts and Rhode Island businessmen who agreed to invest but wished to limit their activities to New England. Accordingly, the New England Telephone Company, capitalized at $200,000—2,000 shares of $100 each—was incorporated on February 12, 1878. The corporation's purpose was "carrying on the business of manufacturing and renting telephones and constructing lines . . . in the New England states."[13] Under the agreement setting up the new company, Hubbard, as trustee of the Bell Telephone Company, assigned all rights under the basic patents to the New England Telephone Company in exchange for stock in the new company issued to Sanders, Hubbard, and Watson. The agreement required the new corporation to buy all of its telephones and call bells from the Bell Telephone Company and forbade the new corporation from selling instruments.

[11] Bruce, *Bell*, chap. 20; and AT&T, *Events in Telecommunications History* (New York: AT&T, 1979), pp. 3–5.

[12] Quoted in M. D. Fagen, ed., *A History of Engineering and Science in the Bell System: The Early Years (1875–1925)* (New York: Bell Telephone Laboratories, 1975), p. 28.

[13] William Chauncey Langdon, "The Early Corporate Development of the Telephone," *Bell Telephone Quarterly* 2 (July 1923): 141. This article provides a good summary of the subject. Details on the negotiations and the formation of the New England Telephone Company are in Tosiello, *Birth and Early Years*, chap. 8

The New England Telephone Company, like the Bell Telephone Company, appointed agents to develop exclusive territories, compensating them by commissions on telephone and call bell rentals. By the end of February 1878 it had sent agents 342 telephones and 54 call bells, and though a dip occurred in subsequent months, it was clear by June that the new corporation was going to do well. Switched central office systems, it will be recalled, had begun, thus creating a market that would supplement the private line business. But this development, in turn, required a "high fixed investment represented by the central office plant, a condition that changed the economics of telephone service dramatically."[14] Watson began working on the switchboard in early 1878.

At this point the financial strain on the Bell Telephone Company was extraordinary, due to high demand for the new switched service as well as a hybrid messenger–private line service known as the district system; the intrusion of Western Union into the telephone business and the high costs of fighting the battle that would be waged on the legal and economic fronts; development costs for the switchboard and other devices; and acquisition costs for new patents. These and other factors placed a great strain on the Bell Telephone Company, which financial flow from the New England Telephone Company only partly alleviated. Accordingly, in March 1878 Sanders wrote to Hubbard proposing the formation of a new company with a capital base for future as well as present needs.[15]

The result of their efforts was a new Bell Telephone Company—this time a corporation and not a Massachusetts Trust—incorporated in July 1878. The new investors, as well as Theodore Vail, who accepted the position of general manager, apparently demanded a new organizational structure. The new company was capitalized at $450,000 consisting of 4,500 shares at $100 each. Interestingly, the articles of association allowed the possibility of selling as well as, or in place of, leasing telephones: "The purpose for which the Corporation is constituted is to manufacture and sell telephones and their appurtenances, and to construct, maintain and operate telephone lines and rent telephones throughout the United States outside of the New England states."[16] While Hubbard, Sanders, and Bell (who played no role in management decisions) received three-quarters of the new corporation's stock, the new investors demanded and got a management structure that limited Hubbard's authority and placed day-to-day management responsibilities in Vail, one of the earliest American professional managers.[17]

[14] Robert Garnet, *The Telephone Enterprise* (Baltimore: Johns Hopkins University Press, 1985), p. 23.

[15] Ibid., p. 25.

[16] Quoted in Langdon, "Corporate Development," p. 143.

[17] Garnet, *Telephone Enterprise*, p. 27; Langdon, "Corporate Development," p. 143; and

Vail, who had established a system of fast mail delivery for the Post Office, began his career in the late 1860s as a lowly mail clerk for the Union Pacific Railroad. Even at this stage of his career he devised a system that made it possible for the mails to arrive earlier than they previously had. A meteoric rise in transportation services led to his appointment in 1876, at the age of thirty, as general superintendent of the Railway Mail Service. Dissatisfied with Congress's economizing measures and at the same time searching for new opportunities, Vail was approached by Gardiner Hubbard, whom Vail had known in Washington. Part of the inducement offered to Vail to join the Bell interests was the promise of the New York City telephone franchise. With characteristic energy Vail obtained the financing for this highly important franchise. Vail thus became associated both with the franchisor and a franchisee, an early portend of the Bell System's horizontal structure.[18]

Vail began as general manager on July 1, 1878, when there were 10,755 Bell telephones in service. His experience, which combined professional management, a familiarity with communications and transportation networks, and a sophistication and knowledge of government, was certainly rare at the time, and combined with his energy and commitment, led to the development of a strategy for the Bell Telephone Company. Paramount was a resolve to fight Western Union. Consequently the major patent infringement suit against the American Speaking Telephone Company (Western Union's affiliate) was filed on September 12. At the same time, Vail encouraged the Bell franchisees to file suits against their competing Western Union franchisees. Since the control of patents was clearly critical to Vail's strategy, he also embarked on a campaign of internal development as well as the acquisition of promising patents developed by outsiders, the earliest important one of which was the Blake transmitter acquired in November 1878.[19]

The second prong of Vail's strategy was to increase the number of uses for the telephone. By the end of 1878 switchboard exchanges had opened in a large number of cities and towns, including Portland (Oregon), Detroit, Keokuk (Iowa), and Philadelphia. The first Bell Telephone fire alarm system used solely for that purpose was installed in Burlington, Iowa, in October. Most important, while long distance covering substan-

---

Albert Bigelow Paine, *In One Man's Life* (New York: Harper & Bros., 1921), chaps. 20, 21. Although the Paine book is an uncritical hero worship of Vail, it contains much information and is the only published biography of Vail.

[18] Paine, *One Man's Life*, passim.

[19] Fagen, *Engineering and Science*, pp. 69–74; AT&T, *Events in Telecommunications History*, pp. 6, 7; Paine, *One Man's Life*, chap. 22; Garnet, *Telephone Enterprise*, pp. 29–34; and H. A. Frederick, "The Development of the Microphone," *Bell Telephone Quarterly* 10 (July 1931):169–74.

tial distances had to await subsequent technological developments, Vail was determined to expand toll lines, a few miles of which were in operation as early as January 1878. Whether it was the success of these modest toll lines or Western Union's ability to transmit telegraphic messages over great distances, the possibility of an extensive long distance system, although not necessarily operated by the Bell interests, was in Vail's mind very early in his career with the Bell Telephone Company. For this reason, Vail refused to accept Western Union's proposal to take toll service and leave the local business to the Bell interests, one of the early proposals that Western Union made during the settlement negotiations.[20]

Vail's offensive and defensive commitments, of course, led to further financial strain on the Bell Telephone Company. Inexorably this led to the next step in the genesis of AT&T—the creation of the National Bell Telephone Company, a Massachusetts corporation created in March 1879 as a successor to a combination of the Bell Telephone Company and the New England Telephone Company. At this juncture William H. Forbes, another important figure in the evolution of the Bell System, entered the picture. Forbes, a Civil War hero and a partner in his father's vast enterprises in railroads, plantations, land sales, and the shipment of goods, was the major provider of the additional cash needed to advance the Bell interests. The new corporation was capitalized at $850,000 and issued 8,500 shares at $100 each. A by-law of the new company effectively placed control in Forbes and the other people supplying the new money to the company. Forbes became the president and Vail the general manager of National Bell, the purposes of which were "to manufacture, sell and rent telephones and their appurtenances, and to build, maintain and operate lines for the transmission of messages by electricity or otherwise."[21] The transformation to National Bell thus marked the point at which the telephone's fortunes were no longer within the control of the three original partners.

The most important event of National Bell's brief life—little more than one year—was the settlement of the patent infringement suit against Western Union on November 10, 1879. Part of the settlement called for the elimination of competition between the two companies: Western Union and its American Speaking Telephone subsidiary were to turn over the rights to their telephone inventions, and National Bell was to buy all of American Speaking Telephone's exchanges and equipment. American Speaking also was to receive 20 percent of the rental fees from instruments leased by National Bell during the life of the two basic patents. At

[20] AT&T, *Events in Telecommunications History*, pp. 6, 7; Paine, *One Man's Life*, p. 137.

[21] Quoted in Langdon, "Corporate Development," p. 145. On Forbes's role in the new company, see Arthur S. Pier, *Forbes: Telephone Pioneer* (New York: Dodd, Mead, 1953), pp. 116–22. Details on negotiations are in Garnet, *Telephone Enterprise*, pp. 38–43.

the heart of the agreement was the understanding that the Bell interests would not enter the telegraph business and Western Union would not enter the telephone business during the life of the basic patents. Further, National Bell (and its successors) agreed not to connect with or do business with any competitor of Western Union in the telegraph business, thus depriving Western Union's competitors of an important service to speed transmission of messages.[22]

The leaders of National Bell, no longer faced with the prospect of competition from Western Union, could breathe a sigh of relief. Indeed, although in subsequent years there would be much litigation over the terms of the agreement, the former rivals had in some respects become allies. The Bell interests, however, needed still greater funds for expansion, and its stock was being sold for extremely high prices. Additionally, the Bell interests were obligated to purchase Western Union and American Speaking Telephone's telephone equipment. Further, as we will see, a new Bell strategy was to hold stock in its licensee companies (requiring a special act of the Massachusetts legislature). National Bell sought to obtain loans for these purposes from investment bankers but was unsuccessful. For these reasons National Bell passed out of existence and was succeeded by a new company—the American Bell Telephone Company. The new company was incorporated on April 17, 1880, and was capitalized at $7,350,000, consisting of 73,500 shares with par value of $100 each. Forbes became president of the new enterprise and Vail the general manager. Bell (with whom Forbes apparently did not enjoy a good relationship) was retained as a consulting engineer and Watson became general inspector, but Sanders and Hubbard were no longer part of the management structure.[23]

The corporate charter was not bestowed on American Bell by the Massachusetts legislature without an extensive struggle. As we have noted, it was incumbent on a firm to petition the state legislature for the right to incorporate. Although still requiring a showing of public purpose, not until the revolutionary 1875 New Jersey General Incorporation Law did states become less restrictive in permitting incorporation.[24] Nevertheless, opposition persisted. It was not uncommon for some business interests to attempt to show state legislatures that incorporation or a change in a charter served no public interest. While National Bell was already

[22] Details of the agreement and negotiations, together with questionable interpretations of them, are in Federal Communications Commission, Special Investigation Docket 1, Exhibit 2096F, *Financial Control of the Telephone Industry* (1937), chap. 2.

[23] Langdon, "Corporate Development," pp. 148–49. Pier, *Forbes*, pp. 123–28; Garnet, *Telephone Enterprise*, pp. 58, 59; and Tosiello, *Birth and Early Years*, pp. 446, 447.

[24] See John W. Cadman, Jr., *The Corporation in New Jersey* (Cambridge, Mass.: Harvard University Press, 1949), passim.

incorporated, the reason for the transition to American Bell was to enlarge the company's capitalization, since the former company's limits were too small. But, a group of influential newspaper publishers wrapped themselves in an antimonopoly, public interest mantle to prevent the Massachusetts legislature from chartering American Bell.

Underlying the newspaper sentiment was a clause in the agreement with Western Union requiring the Bell interests to withdraw from the business of transmitting news messages. Because the transformation of National Bell into American Bell was important for the Bell interests to carry out their part of the agreement, including the purchase of Western Union equipment, Western Union relaxed the restrictions on telephone transmission of news accounts. Accordingly, the newspaper publishers' opposition to the incorporation of American Bell evaporated. The last hurdle to incorporation was based on the trepidations of some legislators concerning the new corporation extending control over Massachusetts licensees. American Bell, therefore, agreed to a prohibition on acquiring more than 30 percent of a Massachusetts licensee's stock.[25]

In the years following the organization of American Bell the company grew rapidly. According to company data, American Bell and its licensees operated about 30,000 miles of wire in 1880. The comparable figures in 1885 and 1890 were, respectively, 156,000 and 332,000. Operating revenues rose from $3,098,000 in 1880 to $16,405,000 in 1890 according to government figures.[26] Equally impressive was the new company's geographical spread. As early as 1881 there were only nine cities with a population of more than ten thousand that did not have a telephone exchange operated under a Bell franchise.[27] Technological progress was rapid, too, especially in the areas of modernizing subscriber equipment, advancing the art of switching, producing better circuits, and increasing the distances that intelligible speech could be transmitted and heard. All of these changes led to the formation, and then the dominance, of the American Telephone and Telegraph Company within the Bell System.

Long Distance and the Foundation of AT&T

In 1877 a Bell Telephone pamphlet set twenty miles as the limit for satisfactory conversation. Only two years later a twenty-eight-mile line had been strung between Boston and Lowell, Massachusetts. The develop-

[25] See Garnet, *Telephone Enterprise*, pp. 59, 60, and materials cited therein; and Tosiello, *Birth and Early Years*, pp. 447–49.

[26] Bureau of the Census, *Historical Statistics of the United States: Colonial Times to the Present*, pt. 2, pp. 785, 786.

[27] See Stipulation/Contention Package, episode 2, par. 16, *United States* v. *AT&T*.

ment of toll business paralleled that of local business. Initially the Bell company's conception was that it would franchise others to actually provide service to customers, while Bell would lease the instruments to them and provide some of the know-how. The Pioneer Telephone Company, financed by Lowell businessmen who were associated with the local telephone business, began commercial toll operation in October 1879. But the line was plagued with transmission difficulties and could only be used for telephone conversation when conditions were exactly right, even though Pioneer received considerable technical help from Thomas Lockwood of American Bell. Additionally, disputes arose between Pioneer and the Boston local licensee over the allocation of repair responsibilities. The Bell interests' attempts to settle the disputes failed. Similar ventures had similar experiences.[28]

The problem of dispute settlement alone might have suggested to American Bell's executives that it should control the toll business. In an expanding business resources could be better employed than in arbitrating differences between licensees, providing technical and business advice to often resistant licensees, and coordinating the equipment and technical interfaces between licensees so as to avoid deteriorated service in which the lowest-quality link in a system governed the overall performance of the system. Nevertheless, most of American Bell's management (with the apparent exception of Vail) were still reluctant to enter the toll business and, accordingly, in 1880 voted to license a company that would build a line from Boston to New York. The Inter State Telephone Company was formally organized in July 1880 with a capital of $100,000. Its incorporators, drawn from several states, included ex-governors of Massachusetts, Rhode Island, Connecticut, and Kentucky and a former secretary of the treasury. Its charter called for the establishment of a vast long distance network, initially throughout eastern points and then beyond.[29]

Inter State's first project was considerably more modest—the construction of a forty-five-mile line between Boston and Providence. Although the plans were drawn up by Inter State officials, they were thoroughly revised by Watson and Vail of American Bell. Thus, even at this early stage the Bell interests were being drawn into the role of network manager. Inter State hoped to complete the line around October 31, 1880, but difficulties intervened, delaying the opening until January 12, 1881,

[28] Garnet, *Telephone Enterprise*, pp. 63–65.

[29] J. Leigh Walsh, *Connecticut Pioneers in Telephony* (New Haven, Conn.: Morris F. Tyler chapter, Telephone Pioneers of America, 1950), p. 110; Garnet, *Telephone Enterprise*, pp. 65, 66, and the materials cited in these works. On the divisions about the toll business within the American Bell leadership, see Herbert N. Casson, *The History of the Telephone* (1910; rpt., Freeport, N.Y.: Books for Libraries, 1971), pp. 171, 172.

when the first call was made. Public response was slight, however, largely because of very low transmission quality and extensive cross-talk in the line. Despite some improvement in transmission quality due to the adoption of metallic circuits, Inter State did not prosper and the enthusiasm of some of its founders waned. Losses and low revenue prospects led to the directors' decision on July 12, 1881, to dissolve the company within ninety days unless remedies for the technical and financial problems could be found. Notwithstanding American Bell's aid, conditions did not improve. The subsequent reorganizations, which merged Inter State and other companies into the Southern New England Telephone Company (SNET) strongly suggested to American Bell's management the need for a single firm—a quarterback—that would make the technical and financial decisions and coordinate the disparate parts of the growing system.[30]

This conclusion could only have become apparent as the development of the means to solve the problems of toll service proceeded. The metallic circuit, first successfully employed in the Boston-Providence line, illustrates the point. The first telephone circuits were based on that prevailing in the telegraph industry, consisting of a single iron wire between points with ground return. Because the frequencies transmitted in telegraphy were relatively low, attenuation (the difference between transmitted and received power) was low. Telephony, however, required the transmission of frequencies as much as one hundred times as high as those used in telegraphy. The attenuation problem was consequently much greater, and the problem increased as distances between points increased. Consider, then, only some of the solutions that had to be offered in order to provide even rudimentary toll service. The metallic circuit consisting of two wires on a line (instead of a single wire with ground return) was found to reduce interference significantly. But to take full advantage of the ability to reduce interference the wire pairs had to be redesigned, a process of such complexity that it resulted in patent awards. The "twisted pair," however, doubled the cost of wire between points and called for the development of new switchboards at both ends that could handle the metallic circuit. Much effort was also employed in developing a method for drawing copper—a much better conductor than iron—so that it would not easily break. And there was a great deal of experimentation on insulator development and even on the best design of telephone poles.[31]

The interrelated needs for coordination, uniformity, and technological

[30] Walsh, *Connecticut Pioneers*, pp. 111–20.

[31] Fagen, *Engineering and Science*, pp. 196–205.

development, coupled with the disappointing performances of the first licensees, led American Bell to look more favorably on advancing toll telephony itself. Continual disputes, such as that between the Wisconsin company and the Chicago Telephone Company (including the use of different gauge wires for the construction of a Chicago-Milwaukee link), inexorably pushed American Bell toward the best solution—developing and managing its own toll service.[32] Only its own long distance operation could reduce the unacceptably high transaction costs in the forms of interconnection disputes, incompatible plant and operating procedures, traffic priority problems, questions involving the division of costs and revenues, routing difficulties, and the burden of development expenses. It was considerably cheaper to incorporate long distance within the single firm than to engage in market transactions for it.

During this period SNET, building on the efforts of Inter State, had completed the connections between Boston and New York in early 1884. On March 27 it opened the line for business. But by May 1885 the president of SNET announced to his shareholders that the Connecticut–New York link (which it had retained after selling the remainder of the line to American Bell) was a commercial failure. "We are still in very serious doubt whether it can be profitably conducted. Our efforts to develop it have not met with the results and success which we anticipated. The burden of expenses and the amount of capital required for plant appear to be very much larger than was at first thought."[33] Accordingly, the remaining links between Boston and New York were sold to American Bell. In its 1884 Annual Report American Bell concluded: "It therefore becomes necessary for one company to take the lead in building lines of connection between the large cities, and a beginning will be made this summer between New York and Philadelphia."[34]

Vail, the prime mover in long distance experimentation, took advantage of the opportunities presented by resigning his post with American Bell in order to organize a long distance subsidiary. To finance long distance ventures, American Bell applied to the Massachusetts legislature to increase its capital stock from $10 to $30 million. And as in the case of the American Bell incorporation, the Massachusetts legislature balked. But this time American Bell decided to incorporate its new subsidiary, which would be named American Telephone and Telegraph Company, in the far more hospitable climate of New York. Its certificate of incorpora-

[32] Defendants' Third Statement of Contentions and Proof, vol. 1, pp. 105–8, *United States* v. *AT&T*.

[33] Quoted in Walsh, *Connecticut Pioneers*, p. 146.

[34] See Fred De Land, "Notes on the Development of Telephone Service," chap. 21: "The Beginning of Long Distance Lines," p. 10, in AT&T Archives, Box 1109.

tion, executed February 28, 1885, contemplated a network that would extend throughout the United States, Canada, and Mexico.[35]

The creation of AT&T was intimately linked with the next important step in long distance telephony—the building of an experimental line between New York and Philadelphia. Considerable thought had already gone into the economic and technological problems. E. T. Gilliland, superintendent of American Bell's mechanical department, had in July 1884 pointed out that the commercial development of long distance would generally require superior apparatus and wiring in every part of the system.[36] The concept of an interactive network in which each part affects the other parts was becoming clear to American Bell policymakers. And though American Bell planners still attempted to interest their licensees in jointly financing long distance to and from their territories, they met with very limited success. In short, technological and economic exigencies led AT&T reluctantly to assume complete control over long distance.[37]

Construction on the first AT&T long distance line (expenses for which were shared with Metropolitan Company, the New York licensee), from New York to Philadelphia, began on August 18, 1885, and was completed the next year, after the company experienced considerable difficulty in getting terminal facilities in Philadelphia. Whereas the 1887 American Bell Annual Report treated long distance as an experiment, the 1888 Annual Report confidently asserted:

> The success electrically as well as commercially has been beyond our expectations.
>
> The income from the long lines is now more than sufficient to meet the current expenses, and there is every reason to expect that before next year it will pay a moderate profit. The great cost of the plant is to be borne in mind in considering the direct profit from the long line service, but the importance of the system as a safeguard to our business cannot be overestimated.
>
> It is intended to complete this year the lines between New York and Boston, to extend from Albany towards Buffalo, and to build a line from Chicago to Milwaukee.[38]

The statement provides a glimmer of the strategy that AT&T and its then parent would take. The reference to long distance "as a safeguard to

[35] The certificate of incorporation is reprinted in Arthur W. Page, *The Bell Telephone System* (New York: Harper & Bros., 1941), pp. 214–16.

[36] E. T. Gilliland to Theodore N. Vail, 24 July 1884, in AT&T Archives, Box 1223.

[37] Garnet, *Telephone Enterprise*, pp. 76–80; and George H. Bliss to Theodore N. Vail, 8 July 1884, in AT&T Archives, Box 1223.

[38] Quoted in Frederick Leland Rhodes, *Beginnings of Telephony* (New York: Harper & Bros., 1929), p. 199.

our business" is a clear allusion to the concern of American Bell's top management about the principal sources of revenue when the basic Bell patents ran out in 1893 and 1894. At that time rivals could lease or sell competing telephone instruments. One Bell answer, already provided by the statement, would be the creation of a long distance network linking the various Bell licensees. Such a network, if well under way before 1894, would provide a competitive advantage compared to new rivals, which could not offer potential subscribers such extensive coverage. Moreover, a vigorously pursued head start would place a rival long distance system at a serious disadvantage.[39]

But such a strategy required, in turn, two critical underpinnings: new creative financial management and a redoubled commitment to research and development in order to solve the three major problems of long distance telephony in the mid-1880s: better and faster communication, more powerful transmitting instruments, and improving the lines. We have already looked at two of the solutions to the third of these technological problems—replacing iron wires with copper ones and replacing single-wire grounded lines with metallic circuits. It is obvious that these were costly activities. Consider, then, expanding the long distance network while simultaneously attempting to solve the other technological problems of long distance and replacing primitive telephone instruments and switching and transmission gear. Consider, too, that unlike telegraph lines, which could turn corners to avoid obstacles, for technical reasons telephone lines in the 1880s had to be comparatively straight. Consider, finally, that AT&T officials foresaw that a simple system consisting of private lines was economically infeasible. It had to be a full-scale system involving the local exchanges.[40]

In view of the substantial risks, of which AT&T officials were aware, the only reasonable way to finance such an undertaking was through the sale of bonds. Although the Bell companies from their inception had adopted a conservative financial strategy, events once again conspired to change that tradition. Typical of large-scale projects in their early stages, the costs of creating a long distance network were greater than initially foreseen. In 1888 AT&T was compelled to float a $2 million, ten-year, 7 percent debenture bond issue to help finance what the 1889 Annual Report said was now 26,038 miles of long distance wire.[41] This bond issue and the later decision to raise even more money by recapitalizing American Bell to $20 million in 1889 and then to $50 million in 1894 (in the

[39] For details, see John V. Langdale, "The Growth of Long Distance Telephony in the Bell System," *Journal of Historical Geography* 4 (1978): 145–59.

[40] These considerations and others are detailed in E. J. Hall to Theodore N. Vail, 12 May 1885, in AT&T Archives, Box 2028.

[41] Cited in Rhodes, *Beginnings of Telephony*, p. 200.

face of considerable resistance from the Massachusetts legislature) had important consequences for the firm's long-term incentives to operate efficiently.[42]

In its frequent resort to capital markets, especially in its sale of debentures, notes, and bonds, AT&T has had to appeal not to potential customers of its telephones, but to investors who are ordinarily indifferent to the product or service in which they are investing. To them return matters; telephones or potato futures are equally good. Investors consider risk, yield, and other financial variables. Any company seeking to sell debt instruments in financial markets is reasonably expected to know that it must deal with sophisticated investors who will compare the seller's performance (and probable future performance) to those of other prospective sellers of financial securities, including those in highly competitive industries. If a company such as AT&T wishes, then, to sell its debt on favorable terms, its performance and probable performance must measure up to that of its rivals in competitive industries. This, of course, is doubly true if it plans to make repeat sales of financial instruments, for it is unlikely that potential investors will be twice fooled.

Thus, the expansion of a company such as AT&T through frequent resort to financial markets (instead of almost entirely through retained earnings) has an important consequence. The discipline imposed by the financial markets acts on the firm as a surrogate for competition. Of course, this does not necessarily guarantee that such a firm will be efficient any more than competition necessarily assures that a firm will be efficient. But in both cases a strong incentive to be efficient is provided. Coincidentally, government regulatory institutions were being developed to prod the Bell System toward the same goal. In this way the complementary public-private nature of public service liberalism compels subject firms to be attentive to efficiency as well as social goals.

The Bell interests' spur toward increasing their capitalization limits and the continued resistance toward it shown by the Massachusetts legislature led the heads of the two companies to reach an important decision in 1899. In that year the capitalization of AT&T had been increased from the original $100,000 to $20 million under New York's more liberal corporation laws. For these reasons on the last day of 1899 AT&T became the parent company of the Bell System, and corporate headquarters were moved from Boston to New York. American Bell transferred its assets to AT&T but continued for a few years as a patent-holding company before terminating its existence. Long distance services were placed in the hands of the newly established Long Lines Department within AT&T.[43] Thus, AT&T had become the holding company that it was at the time of

[42] On American Bell's recapitalizations, see Langdon, "Corporate Development," p. 151.

[43] See Stipulation/Contention Package, episode 2, pars. 29, 30, *United States* v. *AT&T*.

the breakup. In the next three sections we will look at how leasing, licensing, and vertical integration furthered this structure.

Leasing or Selling Telephones?

The practice of only leasing and not selling telephones was, too, consistent with public service liberalism's impulse toward accommodating social and economic goals. Leasing a product can be an effective marketing tool to expand use, or it can be a device to deter entry into an industry. In either case the practice can have a major effect on an industry's structure. The telephone industry provides an illuminating look at the underlying reasons for leasing and the impact of the practice.

From the beginning of the telephone business in the United States (and most other countries) until many years after the end of World War II, telephones were routinely leased and not sold to subscribers. Only recently have subscribers had the option of leasing or buying telephones. Remarkably, the practice continued unabated and without a significant deviation from 1876 until comparatively recent times. Even while it was a struggling firm in 1877 the Bell interests adamantly refused to sell telephones. An August 28, 1877, letter to a prospective Chicago purchaser flatly stated that the telephones are not for sale, only lease, and it was recommended that the potential customer contact the Chicago agent. A September 11, 1877, letter from Gardiner Hubbard to William Hovey, a prospective Grand Rapids, Michigan, agent, plainly indicates that agents were expected to respect the rental-only policy or lose their franchises.[44]

This consistent policy, adopted when the Bell Patent Association was a small company struggling to sell a novel product and maintained through its transformation into a giant corporation, raises two separate but clearly related questions. Why did the Bell interests adopt this policy when it was introducing a novel, patented product? Why did it continue the policy in the era of competition after the expiration of the basic patents and then into the era of regulated monopoly? Frequently the business strategies that a company uses when it introduces a new product or service vary considerably from the strategies it employs when the product or service is successfully established. Looked at another way, a small company with tenuous financing cannot afford the haughtiness of a highly successful one. Yet the Bell interests refused to deviate from the lease only policy, although in its later transformations the Bell company corporate charters allowed for this option as a possibility.

One of the more common explanations of the origins of the lease-only

[44] See the materials in folder marked "Sale vs. Lease of Telephone—Bell Tel. Co., G. G. Hubbard, Tr," in AT&T Archives, Box 1001.

policy is based on Hubbard's experience as attorney for the Gordon MacKay Shoe Machinery Company, which engaged in a lease only policy. According to this view, even Hubbard's wife begged and pleaded to have phones sold, not rented. But Hubbard stood firm, and thus the Bell companies were oriented to selling service rather than products, to their ultimate good fortune.[45] Of course, on its face this argument proves nothing, because most products with which Hubbard was familiar (including highly profitable ones) were sold, not leased. Why should Hubbard have imitated shoe machinery, certainly not an industry that resembles telephony in most ways? Interestingly, although earlier Bell materials advanced the shoe machinery explanation of the lease-only policy, the defendant soft-pedaled it in *United States* v. *AT&T*, whereas the Justice Department advanced the theory.[46] The reason for this twist and the lawyers' motives for making leasing into a point of contention are not hard to see. Leasing arrangements were held to have contributed to the exercise of monopoly power in one of the most important antitrust cases of the twentieth century—a case against the dominant firm in the shoe machinery business.[47]

Yet it is clear that the shoe machinery business had some influence on the decision to lease rather than sell telephones, even if at the outset of the business it was not as crass as a desire to dominate all aspects of telephony. Indeed, one should recall that the early private line subscribers were encouraged to construct and control their own lines. The short-term reasons for leasing instruments included the greater ability to keep track of potential infringers of the instruments that leasing afforded. Ingenious, skilled mechanics could easily take apart and copy these primitive instruments—an early example of reverse engineering. Then as the need for equipment standardization and system compatibility became clearer to the company's founders, the need for the greater control that leasing provided became even more manifest.[48] Similarly, if any of the early franchisees left the business—and there were many shaky ones—leasing afforded greater control over what could become a chaotic situation. Leasing allowed the Bell interests to step in and at least exercise temporary control over the affairs of troubled franchises.

Like most business firms, Bell's underlying concern has been profit maximization. If we put ourselves in the shoes of the partners of the Bell Patent Association and its successors, a leasing policy was superior to selling in virtually every respect. First, the lower outlays required for leas-

[45] W. C. Langdon, "Two Founders of the Bell System," *Bell Telephone Quarterly* 2 (October 1923): 278, 279.

[46] See Stipulation/Contention Package, episode 2, pars. 51, 52, *United States* v. *AT&T*.

[47] *United States* v. *United Shoe Machinery Corp*., 110 F. Supp. 295 (D. Mass, 1953).

[48] See the materials cited in Tosiello, *Birth and Early Years*, pp. 77–81.

ing telephones would certainly encourage a larger number of potential subscribers to try the revolutionary new instrument. Clearly, a lower outlay would provide a greater stimulus to wider acceptance than a policy of selling telephones. Second, the early instruments were problem ridden and required frequent technical improvement. Accordingly, the Bell System's commitment to repair defective equipment or replace it with technologically superior telephones, which is concomitant with leasing, provided important incentives to prospective subscribers. Similarly, as early advertisements show, the guarantee of free maintenance—another concomitant of leasing—was an important incentive offered to prospective subscribers.[49] All of these advantages of leasing, would better expand the market for a new but imperfect instrument and are consistent with a rational business firm's desire to maximize profit in both the short and long run. Thus, leasing appeared to be the sounder marketing strategy.

To some extent warranties and service agreements, coupled with a selling strategy, could accomplish the same result. But such a strategy would be inferior to leasing. First, the frequency with which transmitters and receivers were improved from the earliest days of telephony would make ownership with warranties highly risky from a subscriber's perspective. A costly purchase could quickly become obsolete. In many ways this factor makes early telephony similar to mainframe computers, for which leasing was, for a long while, the prevalent marketing strategy. Under the circumstances of a technologically progressive industry with high obsolescence risks, sellers are in a better position to evaluate the risks of obsolescence. They know what is being developed in the laboratories, whereas users ordinarily do not. Thus, the Bell interests were much more likely to bear the risks associated with rapidly changing technology than their customers. Further, there are risks associated with the suitability of a product to a customer's needs. In the case of so novel an instrument as the telephone, customers could not be sure of how much use they would make of it relative to the telegraph, mail, or messenger. Even after the telephone became more commonplace, subscribers were frequently unsure how extensively they would use it. Here too, then, the Bell interests would be much more likely to assume the risk than subscribers. This factor points to a leasing arrangement rather than a sale.[50]

Leasing offered other advantages to a prospective customer that selling did not. Customers were assured that the equipment in their custody was technologically advanced. As James P. Baughman, an economic historian

[49] See Defendants' Third Statement of Contentions and Proof, vol. 1, p. 98, *United States v. AT&T*.

[50] The best discussion of the computer industry's leasing policy is Franklin M. Fisher, John J. McGowan, and Joen E. Greenwood, *Folded, Spindled and Mutilated* (Cambridge, Mass.: MIT Press, 1983), pp. 191–93.

testifying for AT&T, put it in connection with early leasing policy, "When it's installed in the user's premises, they replace it, which they frequently did, with a better quality instrument when they take it out to service it. They felt those were very important features."[51] This feature, too, must be considered in light of the fact that during this period there were few repair facilities or trained technicians for electrical products, so that it would be extremely difficult for customers to have their equipment independently repaired or upgraded. In summary, then, from the potential subscriber's perspective, the risks and benefits weighed very heavily in favor of leasing rather than buying telephones. Since the Bell interests' policy was to encourage use, its marketing strategy was to lease phones. This policy is one reason, too, that the residential rental fee was lower than the business fee. Since the social use of the telephone was not immediately apparent to early subscribers, an added incentive to leasing had to be provided. The risks associated with a potentially wasteful expense were thus higher for residential than for business subscribers.

All of this is not to suggest that the shoe machinery experience played *no* role in the decision to lease telephones rather than sell them. At best it played a minor role. But few of the reasons favoring leasing shoe machinery applied to telephones. For example, the leasing of a shoe machine allows a manufacturer to keep close control over it so that a competitor does not obtain and then reverse engineer it. Although such close control would work in the case of bulky machines, that consideration would be futile in the case of telephones, a small item subject to a public utility obligation to serve all customers who abide by reasonable rules. Restraining the copying of telephones under these circumstances would have been a hopeless gesture since they could be easily transported and copied. Economist Carl Kaysen has argued that tying service to a leased device makes entry into an industry more difficult. But this certainly was not a reason for leasing in the early phases of the business and, moreover, did not prevent an avalanche of new entrants after the expiration of the basic patents. Unlike the shoe machinery business, in which lease terms were very long, presumably to tie up customers and prevent entry, this was not the case in telephony. About the only significant parallel was that leasing in both cases prevented the development of a secondhand market. But even here a secondhand market for telephones was largely a market for obsolete equipment.[52]

Hubbard may have gotten the idea of leasing from the shoe machinery business, but there is nothing to suggest that this astute man slavishly

[51] Record, *United States* v. *AT&T*, p. 23,581.

[52] Carl Kaysen, *United States v. United Shoe Machinery Corporation* (Cambridge, Mass.: Harvard University Press, 1956), pp. 64–73.

modeled telephone marketing on a business that was so very different. Rather, leasing was undertaken for the sound business reasons discussed above.[53] And as the telephone business developed, it was continued for technological reasons, even as some of the reasons for leasing in the introductory phase of the business no longer applied. As we saw in the last chapter, the Bell System management gradually came to appreciate the interactive nature of telephony. They realized that the weakest link in the system would govern overall quality. And they further realized that substandard customer premises equipment or equipment in a bad state of repair could adversely affect the use of other parties to conversations, fellow party line subscribers, and others not privy to a conversation. As we saw, the common battery system virtually mandated integrated control over the network.

Leasing afforded much greater technical control over the network and avoided the potential high transaction costs that would result from the difficult job of policing equipment that subscribers could purchase themselves. For these reasons during and after the competitive era utility commissions throughout the country were hostile to subscriber ownership of telephones. In their view subscriber ownership would inexorably lower the quality of service. Thus, the New Jersey Board of Public Utility Commissioners declared in 1920 that the 1911 public utility law forbidding unsafe, improper, or inadequate service required telephone company "ownership and control of the instrumentalities used in furnishing service." The board approvingly quoted a New York decision:

> "The individual subscriber is not the only one interested in his telephone equipment. Every subscriber is interested in the entire service, including the equipment of every other subscriber with whom he has occasion to communicate. . . . The public has a right to generally efficient telephone service and it has a right to look to the telephone company for such service and to hold it responsible therefor. Generally efficient telephone service would, we believe, be impossible if subscribers generally owned their own station equipment, buying such as they saw fit and maintaining it as best they could."[54]

The business and technological reasons for the lease-only policy were, thus, joined by the public service demands of many utility commissions. Commission public policy even extended to the requirement that independent telephone companies (after the basic patent expiration) discontinue the practice of selling equipment and reinstitute leasing. A 1917

[53] See George David Smith, *The Anatomy of a Business Strategy* (Baltimore: Johns Hopkins University Press, 1985), pp. 163–65.

[54] *Quick Action Collection Company* v. *New York Telephone Company*, PUR 1920D 137, 141 (N.J. Bd. of P.U. Comm., 1920).

Nebraska decision, for example, rejected the selling practice of a local independent, declaring, "The present practice of permitting a majority of the subscribers to own their own telephone instruments should be discouraged and ultimately discontinued. It makes it impossible for the company to standardize its equipment, and without a standardization of the equipment the service will never be as good as it should be."[55] To take two further (from among many) examples, the California Railroad Commission (which had jurisdiction over telephones) complained in 1921 that "much trouble is experienced in furnishing service to the seven subscribers who own their own telephone instruments," and the Wisconsin Railroad Commission in 1916 concluded that "the practice of subscribers buying and installing their own equipment is poor practice, detrimental to good service and should be discontinued."[56]

Unless one accepts the paranoid view that utility commissioners throughout the country were mere puppets of the telephone industry, their decisions, based on experience supervising the telephone industry, tend to confirm the reasons for the lease-only policy. In this respect we can contrast public service liberalism's flexibility in permitting firms or industries to shape their practice as long as they meet their public service goals with modern liberalism's deeper government intervention in shaping business practices.

The practice of leasing, however, had other consequences, one of the most important of which concerned licensing. Leasing, because of its installation, maintenance, repair, and replacement obligations, brought the central company into more intimate contact with subscribers than a selling policy would have.[57] In brief, leasing was an important factor on the road to end-to-end responsibility, which the public utility commission statements above imply. These obligations, in turn, imply a larger degree of control over licensees of particular territories than a situation in which a seller is relatively indifferent to the uses of a product or service made by a buyer. In view of the high transaction costs of policing licensees to assure that obligations of AT&T and its predecessors were being upheld, it is not surprising that it was drawn into a policy of acquiring licensees rather than engaging in arm's-length market arrangements. Network integration, advancing technology, and the development of long distance provided further incentives to incorporate licensees within the Bell System.

[55] *In Re Palisade Teleph. Co.*, 10 Ann. Rep. Neb. S.R.C. 277 (1917).

[56] *Tognini, Ghezzi & Dalidio Teleph Co.*, PUR 1921C 72, 75 (Cal. R.R. Comm., 1920); and *Franksville Teleph. Co.*, PUR 1917A 270, 276 (Wis. R.R. Comm., 1916).

[57] See Stipulation/Contention Package, Defendants' Contentions, pars. 52, 53, 54, *United States* v. *AT&T*.

The License Contract and Horizontal Integration

As in the case of leasing, several questions are raised by the Bell policy of entering into licensing agreements with operating companies that were granted rights to operate within a territory in exchange for a consideration paid to the licensor. Why did the policy originate? Did this structure, once again sharply criticized by AT&T's opponents, further public service liberalism? That is, why did the Bell interests engage in franchising instead of developing territories themselves? Second, why and how did the Bell interests gradually absorb most of their licensees? But before looking into these questions, it will be useful to examine in some detail the early license agreements.

A company holding a valuable patent may license others to make and/or sell the product or service based on the patent, it may engage in these activities itself, or it may undertake some combination of the two. Clearly, one of the critical factors in devising a strategy is access to capital. Licensing obviously entails a much lower capital commitment for a company than undertaking the activity itself. Hubbard sought to interest potential investors in telephony during 1877, traveling through the eastern states, but was then unable to do so. That failure led to a strategy begun in June 1877 of enlisting agents whose principal tasks were to set up private line systems in assigned (and often overlapping) territories and to lease the Bell instruments. The agents derived their incomes from setting up the lines and a commission on the lease of instruments. In this way the Bell Patent Association was able to gain considerable coverage with a minimum capital outlay.[58]

The beginning of the exchange business in early 1878 changed the considerations surrounding licensing considerably, for licensees (or someone else) had to incur higher costs (and risks) in constructing a switching service. Consequently, the licensees were granted a major incentive in the form of exclusive territories. Pursuant to Form 107, the then prevalent contract, the Bell interests were required to supply the instruments, and the licensees constructed and equipped the lines. The contracts were effective for short periods—usually five years. Most of these license contracts contained a clause requiring the licensor to assume all the expenses of defending the licensees in infringement actions under the Bell patents. Finally, while the licensor agreed that it would extend the agreement if the licensee fulfilled its obligations, the Bell interests also reserved the

[58] Garnet, *Telephone Enterprise*, pp. 14–17; and Federal Communications Commission, Special Investigation Docket 1, Exhibit 130, *Origin and Development of the License Contract* (1936), pp. 1, 2.

right to purchase the licensee's property "at a fair valuation to be determined by arbitration or otherwise."[59]

No one, of course, forced the licensees to accept these terms, and many early contracts were negotiated with terms different from those found in Form 107. Before the Western Union settlement, 185 license contracts were executed, covering most of the principal cities. From the perspective of the Bell interests, the short term of most of these contracts, coupled with their ability to purchase the franchise if necessary, was intended to protect the long-range prospects of the company. In the words of AT&T's assistant comptroller in 1935, the Bell interests used the leverage of a short-term agreement so that no arrangement would be made "which would make it possible for persons to exploit the business for financial advantage to the detriment of good service or the proper extension of the service."[60] Although uttered a long time after the fact, the statement makes considerable sense. The Bell interests wished to preserve the long-term prospects of the telephone against short-term exploitation. And one must recall that until the Western Union settlement the Bell interests' prospects were questionable, thus providing the licensees with an incentive to engage in short-term exploitation or be half-heartedly committed because the licensee had a primary business interest in some other field—a common occurrence. To guard against these contingencies, the Bell interests initially entered into short-term contracts with the right to acquire the franchisee's property, a condition to which the franchisee voluntarily consented.

After the Western Union settlement two factors played key roles in the revision of the license contract. On the one hand, the value of a franchise was substantially enhanced, and the risks associated with operating it correspondingly diminished. But on the other hand, the Bell management, especially Vail, was keenly aware that the basic patents—their most valuable resource—would come to an end in 1893 and 1894. Therefore, they had to supplant and eventually replace income based on the patents with something else. The "something else" was not then specified. Thus, in the renegotiation of license contracts, the new basic form (109) was little changed from Form 107, only incorporating conditions required by the Western Union settlement.[61] In short, Bell management had not yet considered the prospect of engaging in local loop operations (except in very few areas in which it had so engaged from the beginning) either alone or jointly with local interests.

A major change occurred in 1881 when a permanent license was issued

[59] FCC, *Origin and Development*, appendix 2.

[60] Ibid., p. 5.

[61] Ibid., p. 13.

to the Bell Telephone Company of Missouri, which served as the prototype for most of the new license contracts. In exchange for a permanent license, the licensee granted American Bell a certain percentage of its stock. Thus, American Bell effectively became a partner in local loops, while at the same time limiting its capital involvement. This new arrangement provided one answer to what the Bell interests would do after the expiration of the basic patents. But the promise of long distance, the manufacture of customer premises equipment and switching and transmission equipment, as well as the control of other patents and perhaps other ventures, afforded still other opportunities. In view of this, we must still ask why American Bell opted for an ownership interest rather than better royalty terms and another term license. Why did American Bell not enter the operating business itself and refuse to renegotiate with the licensees, as it had a contractual right to do under the short-term agreements?

The solution chosen was a compromise. On the one hand, the joint ownership arrangement limited the amount of capital that American Bell would have had to invest if it decided to go it alone. The licensees, on the other hand, were happy to obtain permanent licenses and would now have an incentive to make a greater investment than they would have under short-term licenses. And in turn, such growth would benefit the licensor.[62] Joint ownership would also solve the vexing problem of assuring that licensees would take a long-run, fully committed interest in the telephone. By giving American Bell an ownership interest in the licensees, the licensor could try to assure that the local companies would be well run and that each local system's interaction with the overall system would meet American Bell's technological standards. The ownership interest in local companies would presumably help to assure that American Bell would be able to enforce its will as network manager of the expanding system. The "control" over the industry that Vail and other American Bell executives sought during this period was not an end in itself; it was subservient to the goals of profit maximization, expansion of the network, and technological integration.[63]

A further important reason that American Bell opted for some degree of participation in ownership and management was its experience with its special agents. Although most of these agents were clearly not technological troubleshooters of the type common today, neither were they simply salesmen. They were expected to offer advice to the licensees and report back to American Bell on how well the franchisees were operating their systems. There were few such agents; in 1881, for example, there were

[62] Ibid., p. 17.
[63] Ibid., pp. 6, 7.

five, only two of whom had practical technical experience. All of them, however, were capable of investigating whether a local company was being run reasonably well—and some companies were not. In some cases licensees, for example, refused to make the transition to newer technologies.[64] At the same time the agent system was insufficient as a means of technical advice or assuring the compatibility of licensee behavior with developing technical requirements. In an April 18, 1881, letter to Vail, Bell patent attorney Thomas Lockwood stated: "It is a fact that no electrical expert of this company is now on the road or visiting exchanges. . . . Only two of our special agents . . . have any pretension to ability of a technical character."[65]

Some of the surviving material pertaining to the agents' work shows clear dissatisfaction with the way the local licensees were operating. For example, in April 1882 an agent visiting the Saint Joseph, Missouri, licensee reported: "The lines are sadly in need of overhauling all over town. . . . The old switch in use here . . . probably contributes its part towards the general dissatisfaction. . . . The St. Joe man was not attending to business." The same agent reported about the Iowa and Minnesota Telephone Company: "In every office except one the manager is a person selected with respect to his local influence . . . and can give very little personal attention to the business. Not one of them is possessed of a particle of technical knowledge or could fix a broken line or a disabled telephone."[66] Lockwood himself, traveling through the East in 1880, was asked to give technical advice to the president of the Litchfield, Connecticut, exchange and reported from Fishkill-on-Hudson, New York, that he "was shocked to find that beyond having their lines built they had nothing done. . . . I cannot persuade them of the absolute necessity of making good joints and ground connections."[67]

In summary, then, the system combining arms-length local ownership and a few traveling agents appeared to be insufficient to assure proper service, sound use of equipment, or technological progressiveness; nor were rentals and growth reasonable. The short-term license coupled with the field agent supervisory system, therefore, was not a success. Some franchisees apparently performed well, demonstrating their long-term value to the Bell interests. But in other cases the short term acted as a disincentive to install more modern equipment, such as the duplicate switchboards developed in 1878 and 1879.[68] In any event, as the above

[64] See the material and testimony of AT&T witness James P. Baughman in Defendants' Exhibit D-T-243 and Record, *United States* v. *AT&T*, pp. 23,556–559.

[65] Quoted in Stipulation/Contention Package, episode 2, par. 69, *United States* v. *AT&T*.

[66] Ibid., episode 2, par. 71.

[67] Ibid.

[68] On the duplicate switchboard, see Fagen, *Engineering and Science*, p. 490.

correspondence and other materials indicate, the Bell interests expressed considerable dissatisfaction with the existing system and could do little more than cajole or threaten the licensees. They could not control licensees to assure technological progressiveness, market expansion, or system integration.

For these reasons the Bell interests began issuing permanent licenses in 1881. In a typical agreement the permanent licensee granted American Bell stock in the enterprise—usually 35 percent. In some cases, however, the percentage was as high as fifty, while in others (such as that with Cincinnati and Suburban Bell) less. In some cases hard bargaining preceded the signing of the permanent license.[69] Once concluded, the licensees received a permanent license and a closer coordination with the Bell interests. Because the latter were now directly interested in the profits of the operating company, there was a joint interest in Bell sharing information and providing advice, technical assistance, and other services on a regular, rather than sporadic, basis. These included annual switchboard and cable conferences attended by the licensees, at which engineering and other technical information was provided. The licensor also tested equipment for many of the licensees.[70] American Bell received, in addition, an important source of income that would continue after royalties from the basic patents could no longer be collected.[71]

In its 1884 Annual Report, American Bell reported: "The tendency toward consolidation of telephone companies noticed in our last report has continued, and is for the most part in the interest of economical and convenient handling of the business. The connection of many towns together, causing large territories to assume the character of great telephone exchanges, made it of importance to bring as large areas as possible under one management to insure simple and convenient arrangements for furnishing rapid intercommunications."[72] The movement toward the consolidation of operating companies into larger units, thus, stemmed at first from the need to coordinate the anticipated growth of intercity communication. Although modest in comparison to contemporary long distance, the traffic between nearby points was then seen as a major source of growth. Forbes, influenced by past experience in reorganizing small railroads into larger units, and Vail, based on his experience in the Railway Mail Service, viewed larger, more centrally inte-

[69] FCC, *Origin and Development*, pp. 19–22; and Walsh, *Connecticut Pioneers*, pp. 145, 146.

[70] FCC, *Origin and Development*, p. 37 and appendix 15, sheet 2, and Plaintiff's Third Statement, vol. 1, pp. 102, 103, *United States* v. *AT&T*.

[71] American Bell Telephone Co., *Annual Report of Directors to Stockholders*, March 29, 1881, p. 7.

[72] Quoted in Stipulation/Contention Package, episode 2, par. 80, *United States* v. *AT&T*.

grated and managed systems as more efficient.[73] As the network expanded and required the implementation of new technologies, the capital and coordination demands imposed on operating companies increased, providing further impetus to their geographical enlargement. By 1910 the structure consisting of AT&T as network manager of large regional operating companies, in turn divided into districts and subsidiaries, was largely in place.[74]

During this period American Bell (and then AT&T) gradually increased its equity holdings in most of the operating companies. The revised contracts called for approximately 35 percent of the licensee's stock to be owned by American Bell. According to FCC data, the outstanding voting stock of the combined associated operating companies held by AT&T rose to 42.13 percent in 1900, 61.61 percent in 1910, and 69.90 percent in 1915. In 1935 the percentage was 92.89. But even before the turn of the century, American Bell had significantly increased its stock holdings in certain key licensees such as the Central New York Telephone and Telegraph Company.[75] The reasons for this policy of increasing share ownership are clear. Foremost, of course, is the fact that it was a profitable business. But of equal importance was the apparent failure of the policy of acquiring minority interests in the licensees to assure that their policies closely fit into AT&T's complementary network. Minority stock ownership, an inability to dominate boards of directors, and a power to veto only some financial policies were found to be insufficient. As E. J. Hall wrote to Vail in 1909: "When we acquire the ownership of all the stock of any company, we are in a position for the first time to say just how [the business] should be handled. Up to this time, we have not been in that position with relation to any company, although we have approximated it in a few."[76]

Even earlier Hall forcefully advocated full-scale integration of the licensees. In 1886 he urged that AT&T was "the natural link between the scattered local companies and could readily bring them into close and systematic working relations." He claimed that the licensees were not cooperating in developing the growing long distance business and that the local managers were performing inadequately, requiring "keeping

[73] Garnet, *Telephone Enterprise*, pp. 67–69; and Federal Communications Commission, *Investigation of the Telephone Industry in the United States* (Washington, D.C.: Government Printing Office, 1939), pp. 21–26.

[74] Alfred D. Chandler, Jr., *The Visible Hand* (Cambridge, Mass.: Harvard University Press, 1977), p. 202; and FCC, *Investigation of the Telephone Industry*, p. 24.

[75] FCC, *Investigation of the Telephone Industry*, pp. 20, 21. See also Stipulation/Contention Package, episode 2, par. 97, *United States* v. *AT&T*.

[76] Quoted in Garnet, *Telephone Enterprise*, p. 138. See also FCC, *Investigation of the Telephone Industry*, pp. 18, 19.

the pressure on day by day to raise their standards." Further, he saw the need for much greater uniformity of practice than that then prevailing. For these reasons Hall advocated the absorption of the operating companies.[77]

The growth of party line service, the implications of automatic switching raised by the Strowger step-by-step system and Bell automatic switching research, the common battery system, the spread of long distance, new transmitters, the replacement of grounded circuits by metallic circuits, dial telephones, new gauge cables, and a host of other inventions and innovations furthered the need for complementarity and the ability of American Bell and AT&T to assure that operating companies implemented the new technologies—and did so in a timely fashion. Full control over these companies was, therefore, sought. Because of the same coordination issues, the Bell interests were also drawn into manufacturing and research.

Vertical Integration and Western Electric

The last major pieces in the structure that composed AT&T before the breakup were Western Electric and Bell Telephone Laboratories. The categorization of industries into natural monopolies and competitive structures underpins contemporary thinking about which enterprises should be subject to economic regulation and which should not. Under public service liberalism, industrial structure issues were framed differently. The primary question was whether a given structure would better promote the attainment of public service goals. The integration of Western Electric into the Bell structure illustrates the point.

Bell Telephone Laboratories, as we noted, was formed by consolidating existing facilities in Western Electric (Western Electric Research Laboratories) and AT&T (part of the engineering department) in a new facility in early 1925; Bell Labs, thus, was born because of an internal reorganization. Western Electric, however, is another matter. It was acquired by the Bell interests in a gradual process that began on July 5, 1881, with American Bell's purchase of 33.3 percent of the stock of the Western Electric Manufacturing Company (a predecessor of the Western Electric Company of Illinois, which was incorporated on November 26, 1881). Not until August 3, 1883, did American Bell obtain controlling interest in Western Electric, when its holdings rose to 52.5 percent of the stock. From that time on, American Bell, and then AT&T, held con-

[77] See the documents quoted in Stipulation/Contention Package, episode 2, par. 98, *United States* v. *AT&T*.

trolling interest in Western Electric, closely integrating it into the parent company's overall operations through its reorganization as the Western Electric Company of New York in 1915.[78]

The path that led to the convergence of American Bell and Western Electric began before the invention of the telephone. Again, the story involves the complex relationships of Elisha Gray, Western Union, and the three original members of the Bell Patent Association. Like the Bell triumvirate, Western Electric's origins can be traced to a partnership of three men. In April 1869 Elisha Gray, whose record of developing telegraphic inventions was already established, General Anson Stager, Western Union's general superintendent, and Enos Barton, Western Union's chief telegrapher in Rochester, reorganized the partnership of Gray and Barton, a company that made telegraphic instruments for Western Union and working models for inventors of electrical apparatus. Stager's participation was premised on the firm moving from Cleveland to Chicago, which occurred in 1870. In 1872 Stager, who had become a vice-president of Western Union, interested that company in purchasing an interest in Gray and Barton. The new arrangements led to the incorporation in Illinois of the Western Electric Manufacturing Company in 1872. Under the new arrangements Stager owned about one-third of the stock, Western Union another third, and the rest was held by various other people.[79]

Western Electric thus became intimately associated with Western Union, but because Stager's shares were held personally, Western Union's control would last only as long as Stager's and their interests coincided. At first this was the case and the firm grew rapidly, manufacturing apparatus that had been invented by Gray (including the printing telegraph), fire alarm systems, burglar alarm calls, and various apparatus required by Western Union. After the entry of Western Union into the phone business, Western Electric began to manufacture telephone equipment for Western Union's subsidiary, American Speaking Telephone Company. By 1881 Western Electric had become the largest manufacturer of telephone and telegraphic apparatus in the United States. That year it doubled its capital stock to $300,000, issuing the new stock

[78] The chronology is taken from H. W. Forster memo, 3 January 1935, *Re: Western Electric Co.*, AT&T Archives, Box 1046; and FCC, *Investigation of the Telephone Industry*, pp. 27–31.

[79] Details on Western Electric's origins are taken from A. R. Thompson, *History of the Western Electric Company: 1869–1924*, AT&T Archives, Box 2061; Frank H. Lovette, "Western Electric's First 75 Years: A Chronology," *Bell Telephone Magazine* 23 (Winter 1944–45): 271–75; Charles G. DuBois, "A Half Century of Western Electric Achievement," *Western Electric News* 8 (November 1919): 1–6; and FCC, *Investigation of the Telephone Industry*, pp. 26–28.

to its shareholders (which no longer included Gray, since he had disposed of his shares in 1875) as a 100 percent stock dividend, and purchased a controlling interest in the Gilliland Electric Manufacturing Company of Indianapolis. Despite this great success, deep dissatisfaction with events led Western Electric to rupture its close relationship with Western Union and form a new one with the Bell interests.

The critical policy of leasing rather than selling telephones was the germ that led the Bell interests to acquire Western Electric. In 1877 Bell was obliged to supply its licensees with the hardware that embodied its patents. The three Bell partners naturally first turned to Charles Williams, Jr., the proprietor of the electrical machinery shop that employed Watson, and in which many of Bell's early experiments occurred. Williams was engaged to manufacture both the basic equipment (receiver and transmitter) and what was termed "apparatus" (the equipment used with the receiver and transmitter). Even in 1877 the Williams shop was hard pressed to keep up with demand. By February 1879 Vail complained to Watson, "Williams is 200 behind on magneto bells and way behind on other work."[80] One solution that the Bell interests tried was to license four other companies to manufacture apparatus, but still retain the Williams shop (which was very generous in its credit arrangements with the Bell interests) as the sole manufacturer of the basic instruments. Thus, although the vertical relationship with Williams was a loose-knit one based on friendly ties and proximity, it is apparent that even at this early stage Bell interests sought to keep tight control over the manufacture of the key instruments based on their patents. The compromise on apparatus was based on the limitations of the Williams shop.

Aside from the short-run coordination problems that the Bell interests faced in this period, several long-term considerations were moving them toward a policy of integrating an equipment manufacturer into the fold, either through acquisition or otherwise. The market for licensing telephones depended on the profitable operation of the licensees. In turn this implied that the equipment licensed would not only be reliable, but progressively improving as well. Further, the expansion of telephone use and an increase in the number of subscribers necessitated (just as in the case of the electrical manufacturers during this period) that equipment would become cheaper, and that, therefore, production processes would become more efficient. Thus, the Bell interests had a strong incentive to assure that an equipment manufacturer would be efficient, progressive, and able to supply their products at low prices.[81]

[80] Quoted in Stipulation/Contention Package, episode 2, par. 140, *United States* v. *AT&T*.

[81] See the discussion in Harold C. Passer, *The Electrical Manufacturers* (Cambridge, Mass.: Harvard University Press, 1953), p. 351.

When one adds to these considerations the need to innovate and control as many patents as possible in an industry characterized by rapid technological advance and the equally great need for standardization and compatibility in a changing and expanding interactive network, it is clear why the Bell interests soon placed a high priority on controlling a manufacturer of telephonic equipment. Because the Bell interests' concerns embraced not only customer premises equipment when it was primarily a private line business, but switching and transmission equipment as well, the need for a tight-knit combination with a manufacturer became even more manifest.[82] And as the business expanded and production moved from the Williams shop, the quality of equipment became a greater problem. Charles E. Scribner, for a time chief engineer of Western Electric, wrote in 1906 that the Williams shop products were excellent, but its equipment and methods "old fashioned" and that the other manufacturers of apparatus produced inferior products or treated telephone manufacture as a mere sideline compared to more important products.[83]

On July 14, 1880, a Bell internal memorandum (unsigned but probably prepared by Vail) to Forbes recommended that "negotiations be opened looking to the consolidation of our Manufacturing interests into one Co. to this Co. a perpetual license to manufacture under any and all of our Patents. . . . And for this perpetual license to give us a consideration in paid up stock and a voice in control of the Co."[84] In February 1881 negotiations between American Bell and Anson Stager looking toward a connection between Western Electric and American Bell began. Western Union, at the same time, sought to obtain complete control of Western Electric or to sell out. After lengthy negotiations among not only Western Union and American Bell, but also the Williams shop and the Gilliland Manufacturing Company, a basis for an agreement was found.[85] On July 5, 1881, American Bell acquired 33.3 percent of the Western Electric stock. On November 26 of that year Western Electric Company—consolidating Western Electric Manufacturing Company, the Gilliland Electrical Manufacturing Company, and the Charles Williams shop—was incorporated in Illinois.

From the perspective of Western Electric the key to the deal lay in the shifting loyalties of Stager, who *personally* controlled one-third of the

[82] Smith, *Business Strategy*, pp. 31–33.

[83] See Stipulation/Contention Package, episode 2, par. 145, *United States* v. *AT&T*.

[84] Quoted in Defendants' Third Statement of Contentions and Proof, vol. 1, p. 120, *United States* v. *AT&T*.

[85] Documents showing details of the negotiations are found in a folder marked "Western Electric Company—Acquisition of—By American Bell Telephone Company—1881–1883," AT&T Archives, Box 1046.

stock. The withdrawal of Western Union from the telephone business pursuant to the November 10, 1879, patent settlement suit had clearly deprived Western Electric of a major potential product market for which the printing telegraph market could not adequately compensate. Further, the Gilliland company had become a significant apparatus supplier to Bell before Western Electric acquired controlling interest in it. This fact alone required Western Electric and the Bell interests to reach an agreement relating to future supply. But the most important factor from Stager's perspective was his estrangement from Western Union. In January 1881 financier Jay Gould announced that he had wrested control of Western Union from William H. Vanderbilt. Gould engaged in a wholesale purge of Western Union managers and directors, including Anson Stager. The hostility of Stager and other Western Electric shareholders to the new Western Union management assured the latter that it would be a permanent minority shareholder. Western Union was, therefore, ready—perhaps even eager—to unload its Western Electric stock. Stager and his new allies in American Bell were thus able to bring Western Electric under Bell control after a series of complex negotiations and contracts.[86] On August 3, 1883, American Bell's stock interest rose to 52.5 percent and its control of Western Electric was complete.

One of the first steps undertaken after American Bell's stock purchase was the establishment of a formal agreement between it and Western Electric. Western Electric was granted the exclusive right to manufacture all telephones and apparatus under the Bell patents. Consequently, American Bell's other contracts with such manufacturers would be terminated. Western Electric was required to devote its energies to supplying telephones and "appliances" for American Bell and its licensees. "Appliances" were defined to include "calls, switches, switchboards, annunciators, exchange furniture and other apparatus and devices adapted for use on or for the telephone lines." American Bell was granted a general supervisory power over Western Electric's workmanship, material, and so on. American Bell was obligated to purchase all of its telephone needs from Western Electric insofar as the latter was capable of supplying them, but was under no similar obligation with respect to "appliances." The licensees (and later the operating companies) were under no obligation whatsoever to purchase anything from Western Electric, although, of course, they carried out the policy of leasing Bell telephones only as a condition of their licenses.[87] Nevertheless, over the years much of the licensees' "appliance" needs were met by Western Electric.

Western Electric, too, assumed still other roles under the contract. As

[86] See Smith, *Business Strategy*, chap. 4, for details.

[87] The contract is reprinted in ibid., appendix E.

long as it met American Bell's needs, it was free to engage in other activities as well. At the very time that it entered into the contract, Western Electric was engaged in work on such things as arc lamps and dynamo machines. At a more mundane level, Western Electric began in 1901, initially on an experimental basis, to act as a jobber for the Bell Telephone Company of Philadelphia, buying, warehousing, and distributing supplies not made within the Bell System. The cost savings were apparently significant, for soon the other operating companies followed suit, establishing another Western Electric function that continued until the divestiture of the operating companies.[88]

The acquisition of Western Electric appears initially to have been primarily directed at the problem of assuring a steady supply of high-quality standardized products required in a growing business. In this respect the acquisition was consistent with the empirical findings of Harold Livesay and Patrick Porter (covering a somewhat later period) that firms generally engage in backward vertical mergers "by a desire to rationalize flows by . . . assuring needed raw materials rather than from a desire to add the profits of the manufacturer to that of those downstream in the business."[89] Of course, there were other short-term gains to be made, such as the control of patents held by Western Electric. Further, American Bell was concerned about maintaining manufacturing quality. The continual back-and-forth communication when a firm is vertically integrated can lower the costs of improving products and processes more than they would be if the firms at each level dealt at arm's length. In Vail's words, "Nearly all the valuable auxiliary apparatus comes from suggestions that are made in the course of doing business."[90]

As time went on the advantages of vertical integration became even clearer to the Bell System. Vertical integration occurs to some extent in virtually every firm and is usually chosen because administrative direction is seen as a more efficient method of directing flows than market transactions. The decision can vary from industry to industry and within industries for a variety of reasons. Some of the important factors that would seem to point toward administrative direction rather than market transactions would lead one to expect telephony to have chosen the former path early on—precisely what happened. For example, when the needs for coordination and precision are great in an interactive network and technological change is rapid, administrative direction is more prob-

[88] Material on cost savings is found in Stipulation/Contention Package, episode 2, pars. 154–59, *United States* v. *AT&T*.

[89] Harold C. Livesay and Patrick G. Porter, "Vertical Integration in American Manufacturing, 1899–1948," *Journal of Economic History* 29 (September 1969): 495–96.

[90] Quoted in Stipulation/Contention Package, episode 2, par. 148, *United States* v. *AT&T*.

able than market transactions. Under these circumstances centralized planning can significantly reduce transaction costs. Further, in a capital-intensive industry with a high degree of technological complexity communication between different levels must be very frequent, and the vertically integrated firm will be better able to shift resources to a unit for long-term profitability, even if the short-run profit prospects are nonexistent or dim at best. The history of many Bell Labs innovations and the persistence with which they were pursued over a long period of time bear this out.[91]

Industrial Structure and Public Service Liberalism

The development of AT&T's structure and conduct was, of course, unique. Each large corporation whose roots can be traced to the last quarter of the nineteenth century has a unique history. Each of them considered various structures and practices that were rejected and instituted. Some instituted later had to be abandoned. Although each large company's history is unique, each shares a common fate as well. Each was the beneficiary of the cooperative attitude that eventually prevailed under public service liberalism. As we have seen, there was considerable fear of the dramatic changes that were reshaping the American industrial landscape. Yet ultimately, corporation laws and other policies were shaped in such a way that each industrial company could experiment and find the practices and structures that best suited it to attain public service goals while attempting to operate efficiently. And the antitrust laws, before they became vehicles for restructuring firms and reshaping practices by administrators and judges, could be used to curtail clear cases of abuse.[92]

[91] See generally the statement of Malcolm Schwartz on behalf of AT&T. Defendants' Exhibit D-T-125, *United States* v. *AT&T*.

[92] On the background of the antitrust laws and early enforcement, see Hans B. Thorelli, *The Federal Antitrust Policy* (Baltimore: Johns Hopkins University Press, 1955), chaps. 1–4, 7, 8.

5

The Progressive Impulse and the Telephone

Progressivism and Public Service Liberalism

FEW TOPICS in American history have led to more heated debate—and extremely high levels of scholarship—than the Progressive Era. As we saw in chapter 1, progressivism was associated with the rise of the professional classes and their impulse toward planning, often through government intervention as the initial response to a problem. As advocates for change usually do, they provided a justification for their ideology in the form of progressivism and characterized their proposals as progress. In Richard Hofstadter's words, progressivism was "that broader impulse toward criticism and change that was everywhere so conspicuous after 1900 when the already forceful stream of agrarian discontent was enlarged and redirected by the growing enthusiasm of middle class people for social and economic reform. . . . Progressivism in this larger sense . . . affected in a striking way all the major and minor parties and the whole tone of American political life."[1]

Under this view the impulse for reform came from a diversity of groups, including engineers, social workers, the clergy, journalists, professional muckrakers, and political reformers, each with its own agenda of grievances and reforms. An interpretive counterrevolution set in under the influence of economic historian Gabriel Kolko, who argued that Progressive Era "Federal economic regulation was generally designed by the regulated interest to meet its own end, and not those of the public or the commonweal."[2]

Both perspectives recognize that the era was one of conflict ultimately stemming from the dramatic changes taking place in the American political-economic landscape. Massive immigration, large-scale enterprises, the rise of self-conscious professionalism, rapid urbanization, and other changes about which historians of the period have made us familiar inevitably triggered intellectuals and others to propose solutions to the new problems. During the Progressive Era there were several competing phi-

[1] Richard Hofstadter, *The Age of Reform* (New York: Alfred A. Knopf, 1955), p. 5.

[2] Gabriel Kolko, *The Triumph of Conservatism* (New York: Free Press, 1963), p. 59.

losophies of public policy, as well as variants of each. In addition, different conceptions had bases of support in different institutions. For example, socialism had a powerful base of support in municipal government, whereas laissez-faire had one in the court system. Other institutions, such as many of the independent regulatory commissions, were dominated by public service liberalism. And as we shall see in succeeding chapters, modern liberalism, led by the new professionals, was beginning its ascendancy.

Thus, if one examines the whole of American public policy during the Progressive Era, it follows that any reductionist theory that purports to explain all or most of it is apt to be wrong. It was a period in which different conceptions of public policy were dominant in varying institutions and locations or in which a policy eventually adopted represented compromise between different conceptions. Overall, however, it was a period of government expansion in response to the new problems. All of the major perspectives except laissez-faire inexorably led to increased government involvement, and the resistance to that increased involvement was centered in the courts—the bulwark of laissez-faire strength.

When we take all of this into account, we can understand why state legislatures and local governing bodies were, under their police powers, legislating on behalf of the public in the very broad areas of public health, safety, morals, and welfare and why they had to withstand judicial scrutiny by showing that restrictive rules were reasonably related to the objectives of the police power. We can appreciate why cities became involved in water, sewage, and disease control in the face of rapid urban development. Again, even though extended and stable families and private and religious charities were still the institutions primarily expected to fulfill the welfare function, states increasingly became involved in it. New York enacted a weak tenement house law in 1867. But in 1901 a comprehensive statute was passed. It was imitated in other states.[3] State governments were increasingly involved in the field of public education, an area of great importance to new professionals. At the same time the Supreme Court and lesser tribunals struck down many statutes.

But even those closest to the laissez-faire position, such as Justice Stephen J. Field, upheld maximum hours laws, compensation statutes for employees in hazardous occupations, and other exercises of the police power.[4]

The rise of the independent regulatory commission—relatively inde-

[3] Edward Chase Kirkland, *Industry Comes of Age* (New York: Holt, Rinehart and Winston, 1961), pp. 260, 261.

[4] Charles W. McCundy, "Justice Field and the Jurisprudence of Government-Business Relations," *Journal of American History* 61 (March 1975): 978, 979 and *Munn* v. *Illinois*, 94 U.S. 113 (1876).

pendent of courts and legislatures—was one of the major developments in this area under public service liberalism. As many industries and industrial problems generally became more complex, both courts and legislatures became less competent to deal with many issues, such as setting a railroad rate for corn syrup from Decatur to Baltimore. Summarizing the views of the influential reformer Charles Francis Adams in his plea for the creation of a railroad commission, economic historian Thomas K. McCraw writes: "In order to close the gap between public and private interests, analytical expertise must somehow be made a permanent part of the government."[5] At the same time that expertise was served, the independent regulatory commission device removed potentially inflammatory and divisive issues from both the political machinations of legislatures and the overly critical view of courts.[6]

No business shows these transformations and tensions more clearly than the telephone industry. From the expiration of the basic patents in 1893–1894 until the reinstitution of monopoly, the telephone industry attests to the creativity of public service liberalism in adjusting to changed conditions and utilizing increased information on what would best serve the public. Telecommunications policy, as we will see, moved from monopoly stemming from patent policy to competition and then back to monopoly due to regulatory policy. These swings began before the dawn of the Progressive Era and continued throughout the period.

The End of the Patent Monopoly

At 7:00 A.M. on September 17, 1945, the fuses in the offices of the Keystone Telephone Company in Philadelphia were pulled. The event marked the close of the second competitive era in telephony, for Philadelphia was the last major locale to have competitive local loop service. Although the Bell System did not control all of the local loop service in the nation, it occupied a monopoly position in most major markets, providing approximately 23.5 million telephones, while all other companies provided about 4.3 million telephones. AT&T was the network manager of the system, interconnecting every Bell System telephone and all except fourteen thousand independent phones. Yet in 1907 the Bell System faced competition in many cities, as well as rivals in the provision of long distance services. Most important, in 1907 the Bell System had approxi-

[5] Thomas K. McCraw, *Prophets of Regulation* (Cambridge, Mass.: Harvard University Press, 1984), p. 15.

[6] On the issue of regulatory commissions to remove but not solve conflicts, see Stephen Skowroneck, *Building a New American State* (Cambridge: Cambridge University Press, 1982), p. 149.

mately 3,013,000 telephones, whereas the independents, in that peak year of their strength, had 3,106,000.[7]

Although many independent operating companies survived—but not as Bell competitors—Keystone's rise and fall typified the telephone companies that challenged Bell. Keystone was incorporated in 1900 and began business with one operator. One year later, over the objections of the Bell System, it received a franchise to compete with Bell in Philadelphia. The city fathers granted the franchise to Keystone on the theory that competition would reduce telephone rates. By 1907 Keystone had garnered forty-one franchises in New Jersey and Philadelphia and was constructing an elaborate conduit network under the streets of Philadelphia. Later, in order to utilize more of the conduit network's capacity, it leased space in it to Philadelphia Electric. The company's ambitiousness was further seen in 1917, when it unsuccessfully sought to integrate a new nationwide telephone system, combining Keystone, Postal Telegraph (then Western Union's principal domestic telegraph rival), and several independent telephone companies. In addition, the promoters of this plan contemplated forming new operating companies that would challenge the Bell System in such strongholds as Boston and New York, where Bell had no rivals.[8]

Keystone clearly was an able competitor. Yet by 1941 it was badly beaten, and informed public sentiment then favored the monopolistic provision of service. The company had paid no common stock dividends and was hopelessly in arrears in its preferred stock payments. More than 93 percent of its thirteen thousand subscribers (only five hundred of whom were residential) also used Bell service. But the Bell Telephone Company, at the time it absorbed Keystone, had approximately 465,000 telephones in Philadelphia. More important, Keystone experienced virtually no growth during the last twenty years of its existence.[9] Yet even at the time of its passing Keystone offered business services that Bell did not provide, such as "call through" whereby an incoming call could be connected directly with an individual extension telephone of a private branch exchange (PBX), bypassing the PBX operator. Notwithstanding that Bell made no commitment to continue "call through," and other innovative Keystone services, and over the opposition of Sears, Roebuck and other Keystone customers, the FCC approved the acquisition. The agency declared: "This dual service involves duplication of line and line

[7] The data is taken from U.S. Department of Commerce, Bureau of the Census, *Historical Statistics of the United States* (Washington, D.C.: Government Printing Office, 1975), pt. 2, p. 783.

[8] "Keystone's Knell," *Business Week* (February 15, 1941): 60–62.

[9] Peter Schauble, "Dual Telephone Service Ends in Philadelphia," *Bell Telephone Magazine* 24 (Winter 1945–46): 311–16.

structures, central offices and duplication of costs for maintaining and operating separate facilities whose function could be served by a unified plant. . . . Elimination of unnecessary and uneconomic duplication of telephone plant and maintenance expense . . . should promote greater convenience of telephone use and efficiency of telephone service."[10]

The FCC's conclusion must be considered in the context of the fact that only a few years earlier it had issued a series of reports that were extremely critical of AT&T and its subsidiaries; indeed, AT&T and others considered some of the FCC's reports unduly biased and hostile toward the Bell System.[11] Nevertheless, the FCC accepted the then prevalent view that duplicate, competitive telephone service was undesirable for a number of reasons. The benefits of monopoly outweighed the disadvantages, according to the commission. In this respect the FCC was consistent with the general consensus of state regulatory bodies and the Interstate Commerce Commission (ICC), which previously had jurisdiction over interstate telephone affairs. Regardless of which political party was dominant, all state commissions had reached the 1918 conclusion of the California Commission that "there should be one universal service, as this will enable complete interchange of communication . . . this, in addition to the usual advantages of consolidation of utility properties resulting from the elimination of duplicate property and duplicate operating expenses."[12] Indeed, as the Indiana regulatory body urged, unification and consolidation of competing services should be encouraged even if it is resisted.[13]

One should note that the issue in these matters was not whether the telephone was subject to public service regulation. That was a given during the period of intense competition as well as after its decline. The issue was whether the public would be better served by competitive telephone firms or local monopolies. For this reason, as the Minnesota regulatory authority observed in 1918, while monopoly was preferable, there was no ironclad rule and there could be unusual situations in which competition would be in the best interests of the public.[14] But in general, as the ICC concluded in a 1921 matter, telephone competition results in economic losses from dual maintenance, dual plants, dual switchboards,

[10] *New Jersey Bell Telephone Co. et al.*, 9 FCC 261, 267 (1943).

[11] See "Wrong Number," *Business Week* (March 28, 1936): 11–12; "FCC's Trial Balloon on AT&T," *Business Week* (April 9, 1938): 17–18; and *Brief of Bell System Companies on Commissioner Walker's Proposed Report on the Telephone Investigation* (December 5, 1938).

[12] *In Re Pacific Teleph. & Teleg. Co.*, 15 Cal R.C.R. 993 (1918).

[13] *Central U. Teleph. Co.*, PUR 1920B, 813; and *Indiana Bell Teleph. Co.*, PUR 1922C, 348.

[14] *Northwestern Teleph. Exch Co.*, PUR 1918E, 481. To the same effect is the Pennsylvania decision in *Perry County Teleph. & Teleg. Co.*, PUR 1917A, 916.

dual operating forces, and dual office organizations, the expenses of which are ultimately borne by the public. If service or rate problems arise, the ICC argued, regulatory bodies exist to protect the public.[15]

A look at the rise of telephone competition after 1894 and the reasons for its falling into disfavor during the second decade of the twentieth century can provide some insight into the course of public service liberalism. As the Missouri Public Service Commission noted in 1919:

> Competition between public service corporations was in vogue for many years as the proper method of securing the best results for the public. . . . The consensus of modern opinion, however, is that competition has failed to bring the result desired. . . . Nearly all of the states in this country have adopted laws providing for the regulation of public service corporations. . . . It is the purpose of such laws to require public service corporations to give adequate service at reasonable rates, rather than to depend upon competition to bring such results.[16]

Yet when the second competitive era began, sentiment was overwhelmingly on the side of the independents, which grew rapidly both in new territories and at the expense of the Bell interests. Antimonopoly sentiment was rampant in the 1890s. Moreover, the powerful monopoly was under challenge from a host of humble inventors such as Daniel Drawbaugh. Its basic patent claims were upheld by the barest majority in the Supreme Court, and it was continuously involved in litigation. Most important, in those countries in which the basic patents had already expired, rates had fallen, sometimes quite dramatically. In Great Britain competition sprung up immediately after the expiration of the basic patents and led to rate reductions that were widely reported in the United States in 1891.[17] Closer to home, competition in Montreal led to the cheapest rates in North America in the same year. Significantly, in Montreal the Federal Telephone Company had signed up more subscribers than the Bell licensee, notwithstanding the latter's lengthy head start.[18]

Although the Bell interests had a multifaceted strategy designed to combat its competitors when the basic patents expired, its program was initially focused on the accumulation of new patents and enforcement through infringement suits. As we have seen, the Bell interests had been singularly successful in prosecuting this strategy since they began doing business. Managers were told in February 1894: "The American Bell Telephone Company owns or controls many others [patents] covering various forms of apparatus essential to the efficient transaction of the tele-

[15] *Consolidation of Ohio Bell and Ohio State Tel. Cos.*, 70 ICC 463, 465 (1921).

[16] *Johnson County Home Teleph. Co.*, 8 Mo. P.S.C.R. 637 (1919).

[17] "The Fate of the Telephone," *New York Times*, February 4, 1891, p. 2.

[18] "When the Patent Runs Out," *New York Times*, May 23, 1891, p. 4.

phone business."[19] Accordingly, managers were exhorted to promptly report all cases of potential infringement in their territories. The potential for infringement suits was substantial, since Bell in 1893–1894 held approximately nine hundred patents covering every aspect of telephony.[20]

Nevertheless, after 1894 the patent strategy no longer worked as it did previously to exclude competitors. The most important patent on which Bell relied to block competition was the one for the Berliner microphone improvement. Although not a commercially practical device without further advances (largely provided by the transmitter invented by Francis Blake of American Bell), Berliner filed a caveat for his invention in 1877. National Bell acquired his patent in the same year, but the Patent Office did not issue a patent on the Berliner device until 1891. Because a microphone device was essential for state-of-the-art telephonic transmission, the Bell interests sought to prevent their opponents from using such a device until 1908—the expiration date of the Berliner patent. A government suit brought to annul the patent, charging that American Bell fraudulently delayed the process, was dismissed by the Supreme Court in 1897.[21]

Nevertheless, the Bell victory was a narrow one. As the *New York Times* noted in reporting the decision, the Court explicitly limited its decision to the fraudulent delay issue, leaving open the issues of patent validity. The *Times* noted that American Bell's legal strategy indicated that the company had serious doubts about the patent's validity.[22] Those doubts were not misplaced, for when American Bell brought an infringement suit on the Berliner patent, the Court construed it so narrowly that it could not be used to block the independents from employing a practical transmitter. The U.S. Circuit Court for the District of Massachusetts limited the Berliner patent to a microphone with metallic contacts, which had since been supplanted by the far more efficient carbon contacts.[23]

While AT&T and its subsidiaries continued to prosecute patent infringement suits, these suits could no longer be used to prevent other firms from competing in local or long distance transmission or the manufacture and distribution of customer premises equipment or switching

[19] Southern Bell Telephone and Telegraph Company, Memorandum for Information of Managers, 15 February 1894, in AT&T Archives, Box 1268.

[20] Gerald A. Brock, *The Telecommunications Industry* (Cambridge, Mass.: Harvard University Press, 1981), p. 109.

[21] *United States* v. *American Bell Telephone Co*., 167 U.S. 224 (1897).

[22] "The Telephone Decision," *New York Times*, May 23, 1897, p. 15.

[23] FCC, Special Investigation Docket 1, Exhibit 1989, *Patent Structure of the Bell System, Its History and Policies and Practices Relative Thereto* (1936), p. 11.

and transmission gear. Although the quality of independent equipment or service was in some cases very inferior to that of the Bell System (whereas in others it was equal or superior to Bell's), it was good enough to compete viably. Thomas D. Lockwood, for many years the Bell interests' chief patent attorney, effectively admitted this in various internal reports and letters written during the competitive era. For a few years after 1894, he wrote, it was possible to prevent effective competition by large telephone exchanges because of Bell's multiple switchboard patents. But this was only a rear guard action. In 1907 Lockwood noted that a number of firms, such as Stromberg-Carlson, American Electric Telephone, the Sterling Company, Kellogg Switchboard and Supply, and the Dean Company "have all had skilled men constantly to follow up our advances, and to then contrive something which shall if possible do similar work, but dodge the patents."[24]

Because of the ability of competing manufacturers to produce adequate and sometimes superior equipment, Lockwood noted that competing companies, such as Keystone in Philadelphia, were then able to deploy large switchboards using the common battery system. Although Lockwood claimed that independent systems could not be as reliable and efficient as those operated by AT&T, they could still give "fair satisfaction." He therefore concluded that patents could no longer be a viable weapon to prevent or even "seriously hinder or hamper such manufacture and use."[25] This did not mean, of course, that AT&T would abandon the patent as a relic of the past. To the contrary, as we have seen, new inventions and innovations (many of which were patentable) played other roles, including keeping ahead of competitors and reducing the marginal cost of service so as to expand the system economically.[26]

American Bell and AT&T would have to devise new ways to compete against the independent movement. Although it had competed for a short time against Western Union earlier, that war was short-lived and settled largely on the legal battlefield. The experience of the first competitive era would be of little help in devising a strategy for the second competitive era. AT&T would now have to compete on economic and political terms, not just legal ones. It would have to show that a structure AT&T dominated would better serve the values of public service liberalism than a competitive one.

[24] The August 8, 1907, letter from Lockwood is reprinted in ibid., pp. 28–31. The quoted language is on p. 30.

[25] Ibid., p. 31.

[26] See Leonard S. Reich, *The Making of American Industrial Research* (Cambridge,: Cambridge University Press, 1985), pp. 5, 6, and Chaps. 6, 7.

The Rise of the Independents

The interlude between the Bell patent monopoly and the system embracing AT&T as network manager supervised by federal and state independent regulatory commissions may have been a necessary step in the evolution of the structure that evolved by the 1920s. Certainly the results are clear. Not only did the telephone growth rate accelerate, but even the Bell System's growth rate accelerated after 1894. Moreover, the independents moved into territories that the Bell System did not reach before 1894. Thus, the Bell System's average annual increase in telephone coverage was 6.26 percent in the 1885–1894 period. But in the period 1895–1906 the average annual increase was 21.54 percent. In the period 1907–1912, when the independent competitors of the Bell System were in decline (and Theodore N. Vail had returned to lead the company), the comparable figure was 9.5 percent.[27]

It is extremely difficult to disaggregate the various factors that led to the accelerated growth of the telephone industry after the basic patents expired, but we should note that the *absolute* increases in each five-year period from 1900 through 1930 were substantially greater than those before the turn of the century (see Table 5.1). In short, these data indicate that the telephone did finally attain a take-off status around the turn of the century. But the greater levels of sustained growth continued unimpeded after the end of the competitive period. Thus, even if telephone competition contributed to system growth, the growth continued at a rapid rate in the era of the regulated network manager system, as Table 5.1 demonstrates. After the inevitable decline during the Great Depression, growth in the post–World War II era—the heyday of AT&T dominance—was striking. Indeed, the more AT&T became a monopolist, the greater its accomplishments.

The 1900 AT&T Annual Report provides at least one reason for the greater increases after the turn of the century. The Annual Report observed that the number of exchange stations grew from 384,000 to more than 800,000 in the previous three years. It attributed the increase to dramatically superior methods of construction and operation. Metallic circuit party lines that led to lower rates were singled out as one of the most important reasons for the increase in both business and residence rentals. Moreover, measured service (in which rates are based on distance, duration, and/or time of day) was adopted in place of or in complement to flat rates. Measured service was designed to encourage persons and firms, unsure of how extensively they would use the telephone, to

[27] Federal Communications Commission, *Investigation of the Telephone Industry in the United States* (Washington, D.C.: Government Printing Office, 1939), p. 136.

TABLE 5.1
Bell and Independent Telephones in Selected Years (in thousands)

Year	*Total Telephones*	*Bell*	*Independents*	*Increase in 5-Year Period*
1895	340	310	30	112
1900	1,356	836	520	1,016
1905	4,127	2,285	1,842	2,771
1910	7,635	3,933	3,702	3,508
1915	10,524	5,968	4,556	2,889
1920	13,273	8,736	4,537	2,749
1925	16,875	12,622	4,253	3,602
1930	20,103	15,983	4,120	3,228
1945	27,867	23,547	4,320	5,939
1950	43,004	36,795	6,209	15,137
1955	56,243	48,028	8,215	13,239
1960	74,342	62,989	11,353	18,099
1965	93,656	78,632	15,024	19,314

Source: Historical Statistics of the United States, pt. 2, pp. 783, 784.

become subscribers.[28] Among the developments put into place during this period that greatly simplified customer premises equipment and its use were the common battery system and the first commercial four-party selective ringing system, developed in 1896, which generated unique ringing signals for each party on a four-party line.[29]

Nevertheless, the advent of competition clearly caused significant changes in the telephone industry. In order to appreciate these changes one must first realize that far from all of the new companies were engaged in head-on competition with an AT&T licensee. There were three main groups of independents. The first group consisted of commercial corporations. In some instances, these companies developed territories that had not been exploited by the Bell licensees, and in other cases they undertook to do business in areas that Bell licensees already occupied. The second group was composed of mutual systems, which were oper-

[28] AT&T, 1900 Annual Report, p. 9.

[29] M. D. Fagen, ed., *A History of Engineering and Science in the Bell System* (New York: Bell Telephone Laboratories, 1975), pp. 122, 123.

ated cooperatively by the persons enjoying the service. Thus, the subscribers were primarily interested in controlling the cost of the service rather than profiting from it. Mutual companies rarely competed with Bell licensees. Farmer lines made up the third group. These were built by farmers and consisted of circuits strung within a rural area connected to the farmhouses of subscribers. Equipment was primitive in this early period; iron wire and sometimes even barbed wire were used to transmit conversations. Farmer lines almost never competed with AT&T or other commercial corporations.[30]

Our focus, then, is on the commercial systems (which in 1902 operated 97.5 percent of the total wire mileage). A small number of these were established even before the expiration of the Bell patents, but 3,039 commercial companies and 979 mutuals were formed between 1894 and 1902. Few of the newly formed independents had telephone experience or significant knowledge of electrical engineering. The result, according to Harry B. MacMeal, a celebrant of independent telephony, was that "there was no standard practice. . . . Construction, particularly in the smaller towns and rural communities, was apt to be haphazard."[31] But in considering the service rendered by independents, one must also bear in mind that the quality of Bell licensee service and their rates varied considerably from area to area. Subscribers were sometimes disgusted at the Bell licensees' service. In Rochester, New York, for example, subscribers of the local Bell licensee engaged in a strike in 1886 and 1887, refusing to use their instruments until the Bell licensee rescinded its proposed rate increase and rate structure change. The licensee yielded after an eighteen-month strike.[32] Variation in the service provided by licensees was, as we have seen, an important reason that AT&T undertook to fully control them.

The new competition also had an undeniable effect on lowering rates (and therefore widening use). A 1909 letter from the AT&T comptroller to President Vail, summarizing the impact of competition, conceded: "It seems then that with competition development is somewhat greater than without. Of course, part of this greater development is to be ascribed to the lower rate prevailing under competition."[33] Total operating revenue per Bell telephone station was eighty-eight dollars in 1895. By 1900 it had dropped to sixty-three dollars, and by 1907 to forty-three dollars.

[30] J. Warren Stehman, *The Financial History of the American Telephone and Telegraph Co.* (Boston: Houghton Mifflin, 1925), pp. 52, 53.

[31] Harry MacMeal, *The Story of Independent Telephony* (Chicago: Independent Pioneer Telephone Association, 1934), p. 79.

[32] Ibid., pp. 110, 111; and F. L. Howe, *This Great Contrivance* (Rochester, N.Y.: Rochester Telephone Corp., 1979), p. 12.

[33] Charles G. DuBois to Theodore Vail, 20 August 1909, AT&T Archives, Box 1375.

Unit profits similarly plunged from approximately thiry-four dollars in 1893 to about twenty dollars in 1900. Corporate profits in the years between the end of the patent monopoly and the turn of the twentieth century were below those prevailing in the last years of the patent monopoly.[34] A leading business journal remarked in 1899: "Many friends of the Bell System deplore the policy of high rates . . . and the gain of 50 percent within a comparatively few months in the number of Bell telephones in use . . . shows that the public will respond to more reasonable treatment and lower rates."[35]

Nevertheless, these trends are not as unequivocal as the above data would, on first impression, indicate. Clearly obfuscating a simplistic version of the benefits of competition is that Bell rates had been reduced in about the same proportions under noncompetitive conditions as under competitive conditions. Second, in 1909 independent rates had been raised in sixteen of the twenty-seven cities and reduced in only one studied in a confidential internal AT&T report. According to the AT&T analysis, the original independent rates in many places were unremunerative and were arbitrarily set to undercut Bell's. They were revised upward so that in the twenty-seven cities studied the average residential minimum rate was higher than Bell's in 1909, and the average minimum business rates were about equal to Bell's.[36] A follow-up study that excluded Chicago and New York, because their great size and telephone densities would disproportionately influence results, compared eleven cities without competition to nineteen with competition and came to even more striking results. Rates in both groups of cities had been reduced, but rates in the exclusively Bell areas were lower on January 1, 1909, than those in competitive areas (see Table 5.2).[37]

Although these data were compiled from AT&T sources, they are highly reliable. They were not intended to be public or self-serving. Rather, they were compiled by the AT&T statistical staff for the honest appraisal of top AT&T officials to whom the reports were exclusively circulated. They show, of course, that rates under competition were lower than those under conditions of *patent* monopoly. That is not a surprise. Prices for basic inventions generally decline after the expiration of the

[34] Robert Garnet, *The Telephone Enterprise* (Baltimore: Johns Hopkins University Press, 1985), pp. 108, 109; and FCC, *Investigation of the Telephone Industry*, pp. 132, 133.

[35] "American Bell Telephone and Proposed Competition," *Commercial and Financial Chronicle* 69 (1899): 1224.

[36] "Statement Showing Growth in 'Bell' and Independent Telephone Development with Changes in Exchange Rates in Various Cities of the United States Arranged by Five Year Periods from 1894 to 1909," 10 August 1909, AT&T Archives, Box 1375.

[37] "Effect of Competition on Development and Rates," 20 January 1910, AT&T Archives, Box 1375.

TABLE 5.2
Effect of Competition on Minimum Rates (in dollars)

	Jan. 1, 1894		*Jan. 1, 1899*		*Jan. 1, 1904*		*Jan. 1, 1910*	
	Business	*Residential*	*Business*	*Residential*	*Business*	*Residential*	*Business*	*Residential*
Bell average in noncompetitive cities	65	52	46	34	35	24	32	22
Bell average in competitive cities	69	52	40	29	36	23	40	25
Independents	—	—	43	29	39	24	41	25

Source: "Effect of Competition on Development and Rates," January 20, 1910, in AT&T Archives, Box 1375.

temporary patent monopoly. But this tells us nothing about the comparative advantages and disadvantages of competition and the regulated network manager system that evolved because of widespread dissatisfaction with competition. Certainly the January 1, 1910, figures in Table 5.2 show that Bell was not subsidizing its competitive efforts with outrageously high profits derived from areas in which the company had a monopoly (including the two largest American cities). Rather, these data tend to show that after the independents began to price telephone service realistically, there were efficiencies associated with the monopolistic provision of local loop service and with being part of a larger integrated network.

This, of course, was the view Vail advanced in the 1910 AT&T Annual Report. By that time the AT&T leadership had formulated the commission-regulated network manager system conception. That there was room in the conception for major independents such as Rochester Telephone helps in part to explain the accommodation that was reached between AT&T and the independents (at which we will look in the next chapter). One would have never guessed that such an accommodation would have taken place, however, given the warfare that occurred after the expiration of the basic patents. A flavor of the deep enmity is provided by one of the most influential tracts written on behalf of the independents, *A Fight with an Octopus*. In graphic terms it accused AT&T and its predecessors of virtually every known sin.[38]

The independent movement began gradually and then rapidly picked up steam, as Table 5.3 shows. In 1902, according to a leading study of the independent movement, there were 5,670 independent exchanges and approximately 100 manufacturers.[39] Although the sizes of the independents varied considerably, they were on average small. A 1905 AT&T letter prepared by Walter S. Allen, a key company official, concerned with strategy to combat the independents, estimated that the average capitalization of 1,916 independents examined that year was approximately $42,000.[40]

The major early opposition to the Bell interests was concentrated in the agricultural states of the Midwest (although Texas and New York also accounted for a large number). Documents prepared during the 1890s, the most important of which are reports made by roving American Bell officials, indicate the vigor of early competition in these states. In Iowa, for example, a rival to Iowa Union (the Bell licensee) claimed as early as June 1895 that they had already wired northern Iowa, southern Minne-

[38] Paul Latzke, *A Fight with an Octopus* (Chicago: Telephony Publishing Co., 1906).

[39] MacMeal, *Independent Telephony*, p. 132.

[40] Walter S. Allen to Frederick P. Fish, 1 March 1905, AT&T Archives, Box. 1006.

TABLE 5.3
New Independent Telephone Companies Established per Year, 1894–1902

Year	*Commercial Systems Established*	*Mutual Systems Established*
1894	80	7
1895	199	15
1896	217	21
1897	254	32
1898	334	75
1899	380	84
1900	508	181
1901	549	269
1902	528	295

Source: Bureau of the Census, *Special Report on Telephones and Telegraphs*, 1902, p. 9.

sota, and eastern Dakota. Western Electric Telephone Company, Bell's rival, announced its intention of linking thousands of miles and reaching such major points as Minneapolis. Typically, the Western Electric Telephone Company had grand plans for both the local loop and toll businesses.[41]

Again we must use Bell documents to explain its response to the Western Electric Telephone Company. Since these are internal documents, not propaganda tracts intended for public consumption, they may be considered reliable. More important, in some respects the admissions made in documents are contrary to AT&T's interests. In general, however, the Bell investigators found their competitors' service uniformly poor. Nevertheless, Bell officials conceded that their own rates were too high and could be sharply cut. Further, Bell investigators found that the ensuing rate competition had increased the total number of subscribers. The local evidence paralleled the aggregate trends that we have examined. For while Iowa opposition companies enticed subscribers from Bell licensees, the licensees were adding many more subscribers than were lost. Overall, in 1895 the Bell interests were confident that they would not be injured in Iowa. In part this stemmed from the substantial diversion of Bell funds previously used for dividends and now to be used for waging the competitive struggle.[42]

[41] J. F. Cass to Iowa Union Telephone Co., 22 June 1895, AT&T Archives, Box 1156.
[42] This summary is taken from materials in a folder titled *Competition in Iowa 1895–1896*,

But Bell's confidence in 1895 was soon undermined as more and more independent companies emerged and Bell's unit profits continued to decline. In understanding these developments one must focus not only on the economic front, but also on political considerations. A company could not simply enter the telephone business in the same way that one could enter retail trade, for example. Since telephone wires and cables had to extend great distances above and/or below streets, municipal permission and the exercise of eminent domain were required. Typically, municipalities (and sometimes states) granted franchises to companies undertaking activities that burdened public thoroughfares only if they obtained the right to regulate the companies in exchange for the franchise and the rights-of-way. For example, New York City elevated railway franchises were first granted in 1831. Gradually the city assumed regulatory authority over many of their activities, including safety and economic matters.[43]

During the earlier phases of urbanization, before the full-scale application of electricity, gas, telephone, and streetcars that began in the last quarter of the nineteenth century, the issue of how many franchises a city should grant was not usually a pressing one. As the applications for new authority to lay track, string wires, dig up streets, and so on multiplied, the issue of how many firms should be permitted to undertake each such activity came more to the fore. As Alfred Kahn remarked, "Why let several companies tear up the streets to lay competing gas or water mains or build their own telephone or electricity poles when one would suffice?"[44] The legal strategy devised under public service liberalism to deal with the question of how many franchisees should be allowed to operate particular services in a community was the certificate of public convenience and necessity, first instituted in Massachusetts in 1872.[45] Behind the policy issues were sharp political divisions (and sometimes corruption) in which potential entrants advocated the benefits of competition to better serve the public, whereas already franchised firms urged that monopoly would better serve the public.

The many legislatures and city councils facing these issues took a variety of positions, sometimes reversing themselves and sometimes taking different stands on various utility services. But it is clear that AT&T lost

---

AT&T Archives, Box 1156. The principal documents relied on are Arthur Fuller to C. Jay French, 11 September 1895; Iowa Union Telephone Co., *Notice of Reduction in Rates*, 15 July 1895; Arthur Fuller to C. Jay French, 24 October 1895; and *Interview With Mr. Cutter at Davenport, Iowa*, 18 September 1895.

[43] Max West, "Municipal Franchises in New York," in Edward Bemis, ed., *Municipal Monopolies*, (New York: Thomas Y. Crowell, 1899), pp. 371–73. See also Frank Parsons, "The Legal Aspects of Monopoly," in Bemis, *Municipal Monopolies*, pp. 466, 467.

[44] Alfred E. Kahn, *The Economics of Regulation*, 2 vols. (New York: John Wiley, 1971), 2:3.

[45] See William K. Jones, "Origins of the Certificate of Public Convenience and Necessity: Development in the States 1870–1920," *Columbia Law Review* 79 (April 1979): 426–518.

more battles than it won in this arena. Thus, the rapid proliferation of new competitive telephone companies in the years around the turn of the century, over the opposition of the Bell licensees, must be seen as evidence of the independents' political success during this period. In part, the success of the local competitors stemmed from close ties with local business and political leaders and relatively easy access to local capital.[46] Thus, *Telephony*, the independents' principal publication, could proudly report in March 1907 that Tennessee, after a political battle between the independents and the Bell interests, had enacted a law providing for telephone competition. Similarly, in September 1907 *Telephony* announced that the new Kansas City municipal government permitted an independent company, backed by local financiers, to establish a competing exchange.[47]

The evidence indicates that the independents during that period were permitted to operate in virtually every community in which they sought such rights. For example, in 1904 independents competed against Bell licensees in every significant Texas community except Dallas. In Illinois, independents were franchised in 996 communities in 1907, while independents competed against Bell in every significant place in western New York as early as 1903. Independents operated against the Bell licensees in virtually every California city and town, including such lucrative markets as Los Angeles.[48] In some of the very largest cities in the country, however, independents were unsuccessful in gaining entry. These cities' preferences for underground cables and the serious additional disruption in densely populated locations where the Bell licensees were already providing good service tilted public preference against a second franchisee. Similarly, a Bell applicant in a market already serviced by an independent was sometimes refused a franchise.[49]

In short, AT&T was not winning the fight against independents on political grounds, and whatever reasons may be ascribed to the development of the regulated network manager structure that began to take

[46] W. S. Gifford, Memorandum, "Independent Telephone Structure in the Territory of Bell of Pennsylvania," 28 October 1906, AT&T Archives, Box 1156 and FCC Special Investigation Docket 1, Exhibit 2096F, *Financial Control of the Telephone Industry* (1937), p. 56.

[47] H. S. Cranfield, "Bell Trust Beaten in Tennessee," *Telephony* (March 1907): 178–79; and C. D. Wright, "How Monopoly Lost in Kansas City," *Telephony* (September 1907): 145, 146.

[48] J. J. Nate, "Texas and Telephones," *Telephony* (1904): 332–34. E. J. Mock, "Story of the States—Illinois," *Telephony* (January 1907): 1–8; B. G. Hubbell, "Independent Telephony in the Empire State," *Telephony* 6 (1903): 210, 211; and A. B. Cass, "Independent Telephony in Southern California," *Telephony* (November 6, 1909): 459.

[49] See, for example, H. D. Fargo, "Story of the States—Indiana," *Telephony* (February 1907): 94.

shape after 1907, AT&T's undue political influence must be ruled out as an explanation. AT&T was losing most of the fights at the local level and, as we will see in the next chapter, was on the defensive at the national level.[50] Moreover, independent telephony had quickly come a long way from its modest beginnings, making Bell's task on economic grounds increasingly formidable. In 1894 independents were modest indeed. A typical independent was the Contoocook Valley Telephone Company in New Hampshire, a system begun by six people (two of whom were state legislators) and including twelve subscribers and a switchboard in a general store.[51] Montgomery Ward was even selling telephone system construction kits for rural areas.[52]

Scale increased dramatically in a short time. In 1897 a trade association of independent exchanges was formed, one of the purposes of which was to interconnect independent companies and gradually establish a long distance network that would be larger than AT&T's.[53] By 1899 a plan was under way to merge many companies (accounting for 90 percent of the independent telephone manufacturers sales) into a "trust." The attempt ultimately failed because Stromberg-Carlson would not enter the combination, but the event portended an important transition in the customer premises equipment and transmission and switching gear markets; Western Electric faced formidable competitors.[54] More ominous, the *Commercial and Financial Chronicle* reported in 1899 that "several large companies, however, have now been organized for the avowed object of operating long distance as well as local lines in opposition to the Bell System."[55] At least four major groups, each centered in a large city, had been formed for that purpose with substantial financial backing. Moreover, three of the four groups obviously knew how to engage in the political infighting that resulted in franchise awards, for their promoters were associated with the management of street railways and electric lighting in New York, Philadelphia, Baltimore, Cleveland, and other places.

The most imposing of these groups—Telephone, Telegraph and Cable Company—collapsed because of the withdrawal of some financial backing, but it did not appear likely in 1900 that AT&T would again achieve

[50] An important case study showing the Bell System's political weakness against the independents in the South is Kenneth Lipartito, *The Bell System and Regional Business* (Baltimore: Johns Hopkins University Press, 1989), pp. 84, 93, 96, 180, 181.

[51] Eleanor Haskin, ed., *Independent Telephony in New England* (Waitsfield, Vt.: New England chapter, Independent Telephone Pioneer Association, 1976), p. 191.

[52] Jerry and Mary Nagel, *Talking Wires* (Aberdeen, S.D.: North Plains Press, n.d.), p. 6.

[53] "American Bell Telephone," *Commercial and Financial Chronicle* 64 (1897): 1040.

[54] "For A New Telephone Trust," *New York Times*, March 23, 1899, p. 1.

[55] "American Bell Telephone and Proposed Competition," *Commercial and Financial Chronicle* 69 (1899): 1223.

overwhelming dominance in telephony. Yet within a few years that dominance was once again achieved, competition was reduced to a minor role, and public sentiment had shifted to almost universal support of the regulated network manager regime that governed telephony (with modifications) until the AT&T breakup. AT&T's strategy, the problems with competition, and the emergence of something better—the regulated network manager system—combined to demonstrate the flexibility of public service liberalism.

AT&T's Political Strategy

AT&T, in order to retain its dominant position, had to adopt both a technological-economic strategy and a political one that would justify its dominance under the goals of public service liberalism. On the economic front, the Bell interests moved their headquarters from Boston to New York because of New York State's more liberal corporate capitalization laws, and AT&T became the parent company. Further, AT&T garnered more and more control over the operating companies, prodding them to increase their capital investments and to employ the most modern equipment. An 1899 report in the *Commercial and Financial Chronicle*, a respected journal whose principal function was to provide cold-blooded advice to investors, concluded that the Bell licensees "without exception . . . have spent liberally from earnings not only for maintenance but for improvements. Largely by this means they replaced the small iron wires and short light poles used at the outset with heavier iron wires and large strong poles, and later substituted heavy copper wires for the iron and in the cities for the most part underground conduits for the overhead system."[56]

Nevertheless, American Bell and AT&T had to devise new, innovative strategies to meet and beat the opposition. They obviously had to redouble their efforts in standard activities, which they did. No evidence can be supplied that the Bell interests were lackadaisical toward the new competitors that arose after 1894.[57] But more than traditional responses were required.

The Bell economic response was multifaceted and can conveniently be divided into two periods: one from the rise of competition until 1907 and the other from 1907, when Theodore N. Vail reaffiliated with AT&T.

[56] "The American Bell Telephone and Its Recapitalization," *Commercial and Financial Chronicle* 69 (1899): 828.

[57] As examples of Bell's attempts to enlarge its capital base, see the following short pieces in the *Commercial and Financial Chronicle*: 57 (1894): 222; 68 (1898): 72; and 69 (1899):25.

When Vail retired as president in June 1919, the structure of the Bell System was largely in place.

AT&T's political-legal strategy in both phases was a complex one. It initially floundered badly. First, because of the close ties that competitors had to local elites, AT&T most frequently lost its battles to prevent competitive franchises at the local level. Second, as we have seen, Bell interests effectively lost the ability to thwart competitors through patent infringement suits. Third, the Bell interests had to worry not only about private competitors, but also about the possibility of nationalization. By the turn of the twentieth century, public ownership or government corporations competing with private telephone firms had appeared in most Western European countries. Agitation in England would lead to a virtually complete nationalization of the telephone industry by the end of 1911. In nearby Canada, after a long effort, the Manitoba provincial government acquired the extensive Bell System at the end of 1907. And in the United States many respectable voices called for nationalization of the telephone system, claiming that a nationalized industry would better serve public service goals. These proponents of nationalization were usually not socialists. Indeed, many conservatives embraced nationalization as the best means of assuring the attainment of public service goals in traditional public utility industries.[58]

Squeezed between the threats of competition and nationalization, AT&T needed a political position consistent with public service liberalism that would at once head off these threats, advance an acceptable public policy program that would garner substantial public support, and effectively be linked to its economic and technological goals. The political position would have to meliorate the inevitable complaints about its Goliath-like competitive response to the opposition (such as denying them interconnection into the Bell network). AT&T would, therefore, have to be keenly alert to the problem of public relations in an era when such terms as *monopoly* and *trust* excited public passions and put a company accused of such sins on the defensive.

The AT&T leadership did not, at once, settle on the best ways to solve these linked problems. Indeed, floundering, backtracking, and inconsistent policies occurred to some extent before the remarkable clarity that took over when Vail assumed control in 1907. Nevertheless, a key underpinning of the political component of that strategy—the theory of natural

[58] On Europe, see A. N. Holcombe, *Public Ownership of Telephones on the Continent of Europe* (Boston: Houghton Mifflin, 1911). On England, see J. H. Robertson, *The Story of the Telephone* (London: Sir Isaac Pitman & Sons, 1947); and on Manitoba, see James Mavor, *Government Telephones* (Toronto: Maclean Publishing Co., 1917). The American arguments in favor of public ownership are well presented in Edward W. Bemis, "Regulation or Ownership," in Bemis, *Municipal Monopolies*, pp. 631–80.

monopoly—can be traced back to before the turn of the century. The theory of natural monopoly was critical to the argument that competition was bad and should be replaced with a new regime—in this case the regulated network manager system.

It is impossible to determine which set of factors—political-legal, economic, or technological—was most important in the unfolding AT&T strategy. But because telecommunications was a public service industry, the political-legal aspect was very important. Clearly a large role had to be reserved for government intervention to assure the achievement of public service goals. But AT&T and other regulated interests did not view every regulatory proposal with unmitigated joy. Public service companies were not feigning when they bitterly complained about specific rules or laws or some activities of the new commissions. As we will see in later chapters, such commissions are not readily controlled by regulated interests. Thus, when President Taft proposed in 1910 to expand the ICC's regulatory jurisdiction, the influential *Commercial and Financial Chronicle* complained that the ICC has "always acted as the champions of the shipper, and in determining cases coming before them have had a bias against the carrier."[59] Clearly, Bell officials would have to chart a crafty course as they designed a strategy on regulatory policy and a theory to support the strategy.

The divisions, and sometimes shifts, that generally prevailed in the business community on progressive proposals occurred in the telephone industry as well. The independents placed their faith in the doctrine of competition, since competition had allowed them to erode the Bell licensees' market shares. The independents largely rejected the concept of natural monopoly because they correctly feared that AT&T and its allies would become the natural monopolists. And they rejected the views of the National Civic Federation (NCF), an influential elite club consisting of business leaders and other important people, which sought to sharply limit competition. The NCF approved of large-scale enterprises because of their purported greater efficiency and ability to serve the public compared to smaller firms. According to the NCF, the potential of large firms to abuse the public would be controlled by strict regulation and other techniques such as federal incorporation.[60]

AT&T's burden, then, was to demonstrate convincingly and gain public support for the view that it should be treated as a responsible monop-

[59] "Mr. Taft's Recommendations regarding Railroad Regulations," *Commercial and Financial Chronicle* 90 (1910): 74.

[60] On the independents' views, see "The National Civic Federation and Trust Legislation," *Telephony* (November 6, 1909): 458. On the National Civic Federation, see James Weinstein, *The Corporate Ideal in the Liberal State* (Boston: Beacon Press, 1968), passim; and Gabriel Kolko, *The Triumph*, pp. 131–34.

olist that could be regulated effectively by independent commissions or other governmental instrumentalities. It would need a theory that would at once criticize competition and point to a new regime that would better serve the public.

AT&T and the Natural Monopoly Concept

As early as 1898 Central Union Telephone Company, one of the most important Bell licensees, was distributing a pamphlet containing material decrying telephone competition as wasteful and unnecessarily duplicative. The pamphlet, consisting of letters, an editorial, and articles disapproving of telephone competition, argued in essence that new entrants in areas already covered by an existing telephone company must undercut the existing firm's rates. This inevitably leads to a rate war. In turn, service degenerates because the increasingly lower rates prevent the outlay of money for repairs and improvements. Additionally, the low rates prevent companies from setting aside the fixed amounts required to purchase the next generation of capital equipment. According to the Central Union material, the inevitable consequences of this scenario are: (1) one company buys the other, (2) one goes under, or (3) they divide the territory between them. Thus, competition, although briefly beneficial to telephone customers, is deleterious in the long run since it inexorably leads to deteriorated service and the wasteful expenditure of resources on the competitive struggle, rather than on improving and extending service.[61]

The Central Union arguments could be written off as self-serving. The stroke of genius that the Bell interests employed to capture considerable support for their position was to tie their position to a theory that was playing an increasingly important role in progressive public policy. The theory of natural monopoly was different in important respects (although similar in others) to the current theory bearing the same name. Today an industry or activity is a natural monopoly if production is done most efficiently by a single firm. This most often occurs when an industry's average costs continuously decline with increasing rates of output. This condition is usually associated with high fixed costs relative to total costs. Thus, the largest producer is necessarily the lowest-cost one and is always capable of undercutting smaller rivals. When these cost characteristics are combined with such others as (1) lack of close substitute products, (2) nonstorability, (3) nontransferability of the service, and (4) the

[61] Central Union Telephone Co., *Some Facts as Others See It: Local and Long Distance Service* (Indianapolis: Central Union Telephone Co., 1898); and materials included in AT&T Archives, Box 1277.

need for more or less permanent fixed physical connections, an industry is denoted as a natural monopoly. When an industry meets these criteria, entry is barred to would-be competitors, and the extant monopolistic firms are heavily regulated.[62]

The understanding of "natural monopoly" at the turn of the century, although containing elements of the modern conception, was somewhat different and best understood in historical context. Prior to the development of the natural monopoly idea, it was not uncommon for governmental authorities in many localities to franchise two or more public utility firms for each service. (Again we should note that public service, franchise, and natural monopoly are very different notions.) Thus, New York had franchised six lighting companies in 1887, Chicago had franchised forty-five electric light companies by 1907, and Pittsburgh had granted more than two hundred streetcar franchises by 1902. Most cities had, prior to the turn of the twentieth century, granted multiple franchises for each of the public utilities, including telephone. A city as small as Raleigh, North Carolina, for example, had three telephone companies in the period from 1900 to 1908. In electricity, gas, and street railways the results of competition were almost always the same: competition proved a failure. In the words of one of the leading scholars on public utilities: "The unavoidable and well-nigh universal verdict has been unfavorable to duplication. Cutthroat competition favored the public for a time with low rates, but invariably at the expense of deteriorated service. Financial exhaustion of one or more of the companies eventually brought about a complete consolidation, or an agreement as to rates or territory."[63]

The consistent manner in which this experience occurred in these industries all over the United States, and in other nations, simultaneously led to a demand by some members of the business elite and other influential people for some form of regulated monopoly in public service industries. At the same time scholars were fashioning, on the basis of such events, the natural monopoly conception. Although the conception can ultimately be traced to John Stuart Mill, its theoretical foundations at the turn of the twentieth century are most clearly associated with the distin-

[62] See especially James C. Bonbright, *Principles of Public Utility Rates* (New York: Columbia University Press, 1961), pp. 4, 8, 11–13; Richard Schmalensee, *The Control of Natural Monopolies* (Lexington, Mass.: Lexington Books, 1979), p. 3; and Kahn *Economics of Regulation*, 2: 119–23.

[63] Burton N. Behling, *Competition and Monopoly in Public Utility Industries* (Urbana: University of Illinois Press, 1938), p. 20. See, generally, Behling, *Competition and Monopoly*, Chap. 2; and Eliot Jones and Thomas C. Bigham, *Principles of Public Utilities* (New York: Macmillan, 1931), pp. 68–88. On the Raleigh, North Carolina, telephone situation, see Edwin A. Clement, *The North Carolina Telephone Story* (Raleigh,: NCITA, 1978), p. 17.

guished University of Wisconsin political economist, Richard T. Ely, whose theories and policy ideas influenced many Progressive Era politicians. Ely's conception of natural monopoly is *not* based on comparing production and distribution costs under monopolistic and competitive conditions and then choosing the lower cost alternative. Rather, his key point is that monopolies will arise whenever the combination of competing interests results in a greater gain for all the parties to a transaction than the gains that accrue through the sum of the parties' independent efforts. Thus, a net gain will accrue to the combined parties when average unit costs for a combined company are lower than the average unit costs of the separate firms. This is similar, of course, to the modern conception of natural monopoly.

But as Ely points out, the declining cost situation is only one of several that meets his conception of natural monopoly. Ely illustrates with the case of the telephone:

> The importance of unity must sooner or later overcome all obstacles standing in the way of combination of the various telephone interests. . . . Two telephone companies cannot perform the same service which one can perform, inasmuch as complete unity is lacking. The object of the telephone is to bring people together, and the more completely it does this, the better it performs its functions. Two or more competing telephone plants, however, separate people, and thus operate antagonistically to the purpose for which the telephone was established.[64]

Thus, even if the unit costs of providing telephone service per subscriber rise with each added subscriber (because the number of connections required to interact with all existing subscribers rises more rapidly than the number of added subscribers), telecommunications, according to Ely, is still a natural monopoly as long as each company operates a separate, unconnected network. This argument concludes that as the network is enlarged, the company with the largest network will tend to have the lowest unit service costs. And because a combined network has greater gains than the individual firms that compose it, telephony is a natural monopoly.

Monopoly and the Regulatory Commission

The Bell companies, for obvious reasons, adopted the natural monopoly position and began distributing to the public material prepared by scholars and trade associations as well as newspaper articles advancing that

[64] Richard T. Ely, *Monopolies and Trusts* (New York: Macmillan, 1900), p. 64.

viewpoint. Nevertheless, until Theodore Vail's return to AT&T in 1907, company officials were divided on accepting an important consequence of the natural monopoly argument—independent commission regulation, which would protect the public against the potential abuses of monopoly. A 1906 New York Telephone Company internal memorandum prepared before Vail's return to AT&T illustrates the split:

> It seems to be the consensus of opinion that with the present state of public sentiment on the question of corporations and monopolies, the city officials cannot be expected to take a position against competition unless some form of regulation or control can be established for the protection of the public against excessive charges.
>
> A prominent ex-city official who has had large experience in municipal matters now retained by the Telephone Company . . . maintains that sooner or later we are bound to be placed under the control of some public authority, and that by coming forward now and agreeing to it, we may be able to obtain terms as to the method of this regulation which will be much less onerous than are likely to be established if we oppose the movement until it is forced upon us.
>
> My own judgment is that the establishment of a commission, even in the most favorable form likely to be obtained, would be a great embarrassment to the company . . . and the public interest would not in the end be benefitted.[65]

There were many reasons that Vail assumed AT&T's presidency in April 1907, but one was the need to clarify the company's political strategy. Earlier annual reports, such as that for 1902, expressed doubt about the value of competition and avoided the issue of regulatory commissions. The 1902 report employed such evasive language as: "The public has also in some instances during the past year . . . seriously considered whether . . . there should be two telephone exchanges serving one community, and have come to the conclusion that a duplication of telephone service was undesirable."[66] But the 1907 Annual Report, the first written since Vail's return, was unequivocal in its support for the commission form of regulation: "It is not believed that there is any objection to [public control] provided it is independent, intelligent, considerate, thorough and just, recognizing, as does the Interstate Commerce Commission . . . that capital is entitled to fair return, and good management or enterprise to its reward."[67]

The 1910 Annual Report was even more enthusiastic about independent regulatory commission regulation.

[65] Memorandum, *Telephone Competition in New York City*, 14 February 1906, AT&T Archives, Box 1082.

[66] AT&T, 1902 Annual Report, p. 9.

[67] AT&T, 1907 Annual Report, p. 18.

Public control or regulation of public service commissions has come and come to stay. Control or regulation exercised through such a body has many advantages over that exercised through regular legislative bodies or committees. . . . It would in time establish a course of practice and precedent for the guidance of all concerned. . . . Experience also has demonstrated that this "supervision" should stop at "control" and "regulation" and not "manage," "operate" nor dictate what the management or operation should be beyond the requirements of greatest efficiency and economy.

State control or regulation should be of such character as to encourage the highest possible standard in plant, utmost extension of facilities, highest efficiency in service . . . rigid economy in operation.[68]

But just as the public service company owes obligations to the public that the public utility commission is expected to guarantee, so also the commission must protect the telephone company in order to assure that the company is able to effectively discharge its obligation. The 1910 Annual Report continues:

If there is to be State control and regulation, there should also be State protection—protection to a corporation striving to serve the whole community (some part of whose service must necessarily be unprofitable) from aggressive competition which covers only that part which is profitable. . . . That competition should be suppressed which arises out of the promotion of unnecessary duplication, which gives no additional facilities or service. . . . State control and regulation, to be effective at all, should be of such a character that the results from the operation of any one enterprise would not warrant the expenditure or investment necessary for mere duplication and straight competition. . . . Two local telephone exchanges in the same community are regarded as competing exchanges, and the public tolerates this dual service only in the fast disappearing idea that through competition in the telephone service, some benefit may be obtained. . . . Two exchange systems in the same place offering identically the same list of subscribers . . . are as useless as a duplicate system of highways or streets not connecting with each other.[69]

The 1910 Annual Report established the quid pro quo that would replace competition. AT&T and its affiliates would willingly submit to commission regulation that would be committed to guaranteeing an efficient and progressive telephone service. The commissions would be expected, further, to prod the telephone system to attain high standards and expand the system. In exchange, AT&T committed itself to be a network manager of the Bell operating companies and noncompeting independents. AT&T rejected, however, interconnection with independent companies

[68] AT&T, 1910 Annual Report, pp. 32, 33.
[69] Ibid., p. 33.

that competed with either a Bell licensee or an affiliated independent on the ground that such interconnection was redundant. AT&T promised unremittingly to improve equipment and operating procedures and to continue expansion into uneconomic, sparsely settled, and difficult to reach territories. Most important, AT&T committed itself ultimately to attain universal service so that virtually everyone who desired a telephone could have one and could communicate with everyone else. Obviously, the public utility commissions would be expected to regulate telephone pricing so that the subsidy flows would allow these goals to be achieved.

One year later AT&T was able to report that its strategy was working well. Not only was the competitive regime failing, but AT&T's ideas were becoming accepted. Twenty-eight states had public utility commissions with jurisdiction over telephone systems. Its 1911 Annual Report asserted that the commissions "are of a very satisfactory character," in part because they do not favor competing and duplicate systems. The Nebraska commission was singled out for praise because it "has approved our plan for cooperating with the independent interests in giving universal service."[70] In short, AT&T advocated not that it become sole provider of telephone service, but rather the regulated network manager of a system that included independents as well as Bell licensees and sublicensees.

Economically the difference was small, but politically the distinction between sole provider and network manager was substantial, since it assured the support of noncompeting independents, which often had important ties to local elites. Moreover, the far more palatable formula undermined the argument for competition. In New York City, for example, New York Telephone was able to prevent the licensing of a rival telephone company. In that battle it enlisted the critical support of local business associations, which argued that the telephone is a natural monopoly. Equally important, the Bell interests supported the move in New York State to create a public service commission, notwithstanding the bitter antagonism and early success of the independents in opposing the creation of one.[71]

[70] AT&T, 1911 Annual Report, p. 22. Bell support for commission regulation was contingent upon monopoly provision of service. See argument of J. W. Gleed to the Kansas Judiciary Committee on the Public Utilities Bill, 27 February 1909, in AT&T Archives, Box 1187.

[71] "Will Seek to Amend the Utilities Law," *New York Times* September 24, 1907, p. 5; "Urges State Control," *New York Times*, May 20, 1908, p. 3; "Assembly Defeats Two Hughes Bills," *New York Times*, April 9, 1909, p. 2; "Charter Set Back by Political Plot," *New York Times*, April 17, 1909, p. 3; "Control Wire Companies," *New York Times*, May 25, 1910, p. 6; and "Vail for State Control," *New York Times*, July 10, 1911, p. 12.

Acceptance of the regulated network manager system was, of course, contingent upon AT&T and its licensees demonstrating not only that competition failed, but that its system was a feasible one and that AT&T was the appropriate network manager. Continual technological progressiveness was essential. The two major themes that AT&T developed to show that the system would succeed under its leadership were universal service and the development of long distance. As Frederick Fish, Vail's predecessor as AT&T president, testified: "Whatever might be the cut-throat rates of the opposition companies, a very substantial proportion of the subscribers had to have the Bell instruments because of the long distance connections. That was the mainstay of the business."[72] The long distance program in turn required that AT&T make important marketing and technological breakthroughs or acquire important patents of those who did. It realized this even before the new century. As early as 1893, it had leased private line long distance service to a major customer (the Pennsylvania Railroad) and had begun its efforts (albeit unsuccessfully) on radio telephony.[73]

It was, however, the almost simultaneous invention of the loading coil by George A. Campbell, a Bell in-house scientist, and Michael I. Pupin (whose patent AT&T acquired in 1900) that gave AT&T its decisive advantage in long distance transmission.[74] The loading coil patents greatly increased the previous twelve-hundred-mile practical limit of long distance telephony and placed AT&T well ahead of existing or potential competitors in toll service. Progress continued at a heady pace thereafter in the development of amplifiers, repeaters, and other inventions that were intended to further improve long distance transmission quality. In 1912, for example, a major breakthrough occurred when the vacuum tube, invented only a few years earlier, was first employed as an amplifier. During the same period considerable progress was made in placing cable underground; in late 1906 the AT&T New York–Philadelphia underground cable was opened. Materials used in loading coils, too, were improved by Bell scientists, and long distance transmission became more efficient.

A milestone of AT&T's efforts in long distance occurred in January

[72] Testimony of Frederick Fish, Abstract of Record, vol. 2, p. 985, *William A. Read* v. *Central Union Telephone Co.*, Term no. 34, General no. 2364, Appellate Court, Ill., 1st Dist. 1917. This abstract of record was printed, but the case never reached a hearing or decision by the appellate court.

[73] M. D. Fagen, ed., *A History of Engineering and Science in the Bell System* (New York: Bell Telephone Laboratories, 1975), p. 362; and AT&T, *Events in Telecommunications History*, (New York: AT&T, 1979), p. 14.

[74] James E. Brittain, "The Introduction of the Loading Coil: George A. Campbell and Michael I. Pupin," *Technology and Culture* 11 (January 1970):6–57; Neil H. Wasserman, *From Invention to Innovation* (Baltimore: Johns Hopkins University Press, 1985), chap. 4.

1915, when the first transcontinental telephone line between New York and San Francisco was opened. At the opening ceremony Alexander Graham Bell, who was in New York, said once again to Thomas A. Watson, "Mr. Watson, come here. I want you."[75] By that time competition was slain and the regulated network manager system was ascendant. But the requisite commitments to science and technology had important effects on AT&T's organization.

The company moved from reliance on the individual inventor to large-scale organized efforts. It made major efforts to assure communication between its scientists and engineers on the one hand and managers and professionals on the other. Under the leadership of John J. Carty, appointed chief engineer by Vail, close working relationships between AT&T's scientists, engineers, administrators, and patent attorneys were established. Carty explicitly adopted a strategy of steady improvement so that the company would be in the forefront of future communication trends as well as then current ones.[76] This attitude toward closely integrating research and development into other corporate activities, unusual in the first two decades of the twentieth century, had payoffs in areas not directly related to long distance and, thus, provided further evidence that AT&T was the appropriate network manager under the principles of public service liberalism. For example, in 1888 only fifty pairs of no. 18 gauge wire could be placed in a full-sized cable sheath. In 1902 the size limit was one hundred pairs of no. 19 gauge wire. But by 1918 the limit had been dramatically increased to 455 pairs of no. 19 gauge wire.[77] In short order each part of the network adopted the superior wire. Standardization and the efficient flow of information in the form of advice and assistance to licensees on engineering and other matters were intended to assure Bell System technical attainment of its universal service goals at every point in the network.

Vail's return to the leadership of AT&T in 1907 did not mark a change of direction so much as a clarification of the direction toward which the company had been pointed. Early on, technological leadership, high quality, a vast network, and other factors were used to argue that AT&T should dominate telephone communication. The 1902 AT&T Annual Report proclaimed: "The users of telephones recognize the enormous advantage of a complete service such as is afforded by the Bell Compa-

[75] Fagen, *Engineering and Science*, chaps. 8–10, and *Events in Telecommunications History*, pp. 17–21.

[76] Leonard S. Reich, *The Making of American Industrial Research* (Cambridge, Cambridge University Press, 1985), pp. 160–80; and Wasserman, *Invention to Innovation*, chap. 6.

[77] Cited in FCC Special Investigation Docket. 1, Exhibit 2096G, *Effect of Control upon Telephone Rates and Service* (1937), p. 48.

nies."[78] That is, AT&T's dominance best served the public service goals of the industry. It then announced a commitment to toll service that would cover the entire nation. This, in turn, the Annual Report continued, would require a large increase in investment. The 1905 Annual Report further elaborated the commitment to technological and scientific development and linked this effort to the network manager system. It announced that it was adapting service to large and small users, businesses and residential subscribers, large and small cities, and rural areas. In pursuance of this policy the report announced that combinations of individual and party lines had been introduced and special equipment had been devised for hotels, department stores, apartment houses, private branch exchanges, charge service, coin boxes, and special farm and ranch service.[79] Finally, AT&T was learning to blend central control with greater local company input into equipment design and other matters.[80]

Thus, even before Vail's succinct summarization of the Bell System's program as universal, intercommunicating, and interdependent telephone service, company officials had made a comparable commitment; it was necessary to thwart competition. In turn, this led to a firm reliance on technological development (and, as we have seen, scientific development as well) as a key element in corporate strategy. Given the number of components of a complete telephone service and the diverse types of service, each requiring different types of equipment, the technological commitment required extraordinary breadth as well as intensity.[81]

The Competitive Threat

While AT&T might adopt its political, economic, technological, and organizational strategies to demonstrate that it was capable of becoming *a* system manager, this was still a long way from demonstrating that it should be *the* system manager. Not only did it have to react against individual companies in manufacture and transmission, but against alternative systems as well. In this respect AT&T policymakers had to carefully consider the times.

The period from the end of the patent monopoly until 1907 overlaps

[78] AT&T, 1902 Annual Report, p. 5.

[79] AT&T, 1905 Annual Report, p. 5.

[80] Lipartito, *Regional Business*, pp. 162, 163.

[81] Lillian Hoddeson, "The Emergence of Basic Research in the Bell Telephone System, 1875–1915," *Technology and Culture* 22 (July 1981): 526–31. As an example of the continuity of AT&T research and development consider the coin box. Important Bell system developments were made in 1899, 1901, 1902, 1903, 1909, 1911, and 1920. Fagen, *Engineering and Science*, pp. 160–62.

with a time when one of the most dramatic changes in American economic history was taking place. From 1899 to 1919 "the number of wage earners almost doubled, wages increased almost fivefold, and the value of products almost sixfold."[82] Aided by urbanization and industrialization, telephones per thousand increased from 7.1 in 1897 to 123.4 in 1920. The total number of telephones rose from 515,000 to 13,273,000 in that period.[83] Large investment banking firms, of which J. P. Morgan & Company was the most famous, supplied much of the capital for this expansion. Under their leadership the stock market reached maturity.[84] Investment bankers, who usually retained a major stock position in the expanding corporations, triggered the rise of professional managers (such as Theodore Vail) with little proprietary stake in the enterprises they governed.

The implications of these changes for the American Bell and AT&T leadership were profound. First, they had to consider the possibility of creating competing consolidated, fully vertically integrated companies, put together by investment bankers, that could approach the size of AT&T. Indeed, with large injections of funds, some of them could overtake AT&T. As previously noted, a "trust" of the major equipment independents was in the offing in 1899. Further, in that same year, as we saw, the *Commercial and Financial Chronicle* reported the establishment of four different developing systems in local loop service and long distance, each of which could vertically integrate. The largest of the rival groups was the Telephone, Telegraph and Cable Company (TTCC), which had close connections with important investment bankers.[85] Clearly, Bell management had cause for concern that TTCC was a major threat. One should consider that in 1901, under J. P. Morgan's direction, the U.S. Steel Corporation was formed by consolidating 11 firms that were once 170 independent companies. Could not the same pattern occur in the telephone industry?

The acquisition policy that AT&T undertook must be understood in this context. Acquisitions and attempted acquisitions of key competitors in equipment manufacture or local telephone service should be viewed as a strategy not to monopolize those segments of the industry, but rather to thwart or slow down the opportunities of the opposition companies to build a rival system. Moreover, major acquisition opportunities usually depended on fortuitous events not likely to be repeated. AT&T at-

[82] Harold V. Faulkner, *The Decline of Laissez Faire, 1897–1917* (New York: Holt, Rinehart and Winston, 1951), p. 115.

[83] *Historical Statistics*, pt. 2, pp. 783, 784.

[84] W. J. Schultz and M. R. Caine, *Financial Development of the United States* (Englewood Cliffs, N.J.: Prentice-Hall, 1939), p. 441.

[85] FCC, *Financial Control*, pp. 58–60.

tempted to block investment banker financing of large rivals. It knew that local loop operations and equipment manufacturers had readily available sources of *local* capital; it was a fully vertically integrated system that constituted the primary threat, not competition at each level that might be loosely tied together.[86]

Only in the case of long distance did the Bell interests see a reasonable opportunity for it to achieve a monopoly position, at least for very long distances, although not for shorter toll service. The loading coil patents virtually assured that AT&T would have a higher-quality long distance service than any opponents. In addition, these patents and AT&T's head start in long distance helped to assure a more extensive network. For these reasons focusing resources on long distance became a top priority. As early as 1901 George Leverett, a top AT&T policymaker and attorney, wrote to President Fish that long distance "is destined . . . to be very much more important in the future than it has been in the past. Such lines may be regarded as the nerves of our whole system. We need not fear the opposition in a single place, provided we control the means of communication with other places."[87] Based on the strong commitment to long distance, which continued in the Vail administration, the AT&T strategists derived another important piece of the company's overall economic strategy—sublicensing. Under this concept AT&T offered noncompeting independents connections into the Bell network provided they met the standards set by Bell engineers. The network thus expanded even though AT&T was not required to provide much capital for expansion into new geographical areas that were frequently rural and difficult to serve.[88]

While the sublicensing arrangements provided a way to expand the network without major capital commitments, other forces were at work to create a major capital crunch—one that could only be resolved by resorting to investment bankers. Because of the unrealistically low rates offered by independents (which we will examine later), AT&T was often forced to aid its licensees financially. In the long run this helped to achieve AT&T's objective to control the policies of the licensees, since stock was taken in return for funds. But in the short run these demands

[86] The documents reprinted in ibid., pp. 175–88, make considerably more sense in this interpretation than the arguments advanced by the staff authors who ignored the ready availability that telephone companies had to capital sources, especially at the local level.

[87] George Leverett to Frederick P. Fish, 17 October 1901, p. 6, in AT&T Archives, Box 1375.

[88] Ibid., pp. 1–8 and the following material in Box 1375: Joseph P. Davis to Frederick P. Fish, 23 October 1901; E. J. Hall to Frederick P. Fish, 22 October 1901; and Thomas Sherwin to Frederick P. Fish, 22 October 1901. See also Garnet, *Telephone Enterprise*, pp. 125, 126.

drained AT&T's resources and it had to turn to investment bankers, notably J. P. Morgan, for an injection of funds. But this in turn led to the return of Vail in 1907, the company's recovery, and the institution of widespread public stock ownership.

In the long run, then, investment bankers played a crucial but temporary role in AT&T's response to competition. Investment bankers were crucial in the transition from family control to professional management in many large corporations. As Daniel Bell perceptively observed, "The investment bankers, in effect, tore up the social roots of the capitalist order. By installing professional managers—with no proprietary stakes themselves in the enterprise, unable therefore to pass along their power automatically to their sons, and accountable to outside controllers—the bankers effected a radical separation of property and family."[89]

AT&T's Acquisition Maneuvers

Both the Justice Department in the 1974 case and the FCC staff in the 1930s investigation pointed to AT&T's acquisitions and attempts to acquire other companies as strong evidence of the company's monopolistic intent during the period from 1893–1894 to the full emergence of AT&T's dominance. The inference was that competition collapsed not because of its own defects, but because of AT&T's devious conduct. A close look at the major acquisitions is called for, because the argument suggests that public service liberalism failed to create a sound public policy regime.

AT&T's acquisitions were not a major part of its strategy and only occurred opportunistically, not as part of an overall plan to acquire or drive out every competitor at every level. Four episodes are of major consequence here: those concerning TTCC, Kellogg Switchboard and Supply, Stromberg-Carlson, and Western Union (which will be considered in the next chapter). The TTCC episode is the most complex.

As noted earlier, in 1901 TTCC was one of the principal rivals of AT&T. It had excellent financial connections and was a major threat to establishing an alternative vertically integrated network. The threat to AT&T was even greater than that, however. Five of the most important AT&T licensees (Cleveland Telephone, Northwestern Telephone, Southwestern Telegraph and Telephone, Michigan Telephone, and Wisconsin Telephone), making up more than 15 percent of the Bell System lines, were controlled by Erie Telegraph and Telephone, with which AT&T had no financial connection. Given the size and importance of these properties, AT&T could not afford to have these companies move from it to a

[89] Daniel Bell, *The End of Ideology*, rev. ed. (New York: Free Press, 1962), p. 43.

rival that was already committed to the "combining [of] the independent telephone companies of the country into a nationwide organization, including a nationwide long distance service."[90] But in 1900 the worst, from the AT&T perspective, occurred: TTCC announced that it had acquired control of Erie through stock purchase. At the same time TTCC continued its policy of acquiring operating companies.

Ultimately, however, the Erie acquisition proved to be a sand trap for TTCC. Less than one year after the acquisition, Erie, unable to float a new stock issue, was forced to incur a substantial debt to finance improvements and expansion. Erie stock declined from over $110 in March 1900 to $40 in September 1901, reflecting the company's serious difficulties. It was soon clear that Erie could not repay its debt and that a plan of reorganization would be required. As the uncertainty over Erie's fate continued, its stock dropped lower and lower, and AT&T, seeing its opportunity, bought more and more of it—some as low as $15 per share. The plan of reorganization, prepared by the investment banking house of Kidder, Peabody and Company and AT&T, effectively allowed the latter, through a new holding company, to gain control of the five operating companies, which was consistent with AT&T's general plan to control its operating companies. TTCC, notwithstanding its strong initial backing (including from such luminaries as John Jacob Astor), found that some of its original financial supporters withdrew from the enterprise. Despite the ready availability of funds for *sound* ventures in a rapidly expanding capital market, TTCC was unable to attract more investment. Consequently, TTCC gradually sold controlling interest to an intermediary acting on behalf of Bell interests, who then resold the stock to a consortium of Bell companies. (Acquisitions through such intermediaries were very common at the time.) Eventually the Bell consortium purchased TTCC's minority shares and dissolved the company.[91]

Just as the TTCC-connected acquisitions were intended to head off alternative vertically integrated networks, so also were the attempted ac-

[90] FCC, *Investigation of the Telephone Industry*, p. 131.

[91] Ibid., p. 132; FCC, *Financial Control*, chap. 4; and Stehman, *The Financial History*, pp. 97–104.

The FCC investigation, with its unremitting hostility to AT&T, assumes that AT&T's later financial backers were instrumental in the decline of TTCC. But it supplies no evidence that J. P. Morgan, then deeply involved in much bigger deals, such as the construction of the U.S. Steel Corporation, forced other potential backers to shy away from TTCC. Most important, the vast expansion of finance during the period in question, coupled with the formation of many new companies challenging older ones in their fields, belies the FCC staff's paranoid vision (common in the 1930s) that J. P. Morgan single-handedly ruled the world of finance and that TTCC could not have obtained new investors. See especially Vincent P. Carosso, *Investment Banking in America* (Cambridge, Mass.: Harvard University Press, 1970), chaps. 3, 4.

quisitions of Kellogg and Stromberg-Carlson, which, in any event, failed. The alternative explanation, common at the time, was that AT&T attempted to make those acquisitions in order to foreclose the independents from purchasing the equipment needed to carry on their businesses. As we will see, the argument makes no economic sense. We should first appreciate that telephone equipment manufacturing was part of a larger category of manufacturers of electrical equipment and supplies. Although Westinghouse and General Electric were the largest such firms, the industry contained many others. Most important, there was simply no way that AT&T could have prevented such firms (or new ones, for that matter) from entering a lucrative market such as telephone equipment manufacturing. Second, an attempt to form a telephone equipment trust that failed in 1899 listed fifteen companies, excluding Stromberg-Carlson. Another independent source lists thirty manufacturers. Nor do these estimates include every domestic manufacturer, any foreign manufacturer, or any potential manufacturer. There were, indeed, numerous other manufacturers.[92] Third, anyone who examines early issues of *Telephony*, the voice of the independent operating companies, or other early trade magazines, cannot fail to be impressed by the large number of companies advertising their telephone equipment. Thus, AT&T was far from being in a position to monopolize the equipment business and prevent independent operators from obtaining equipment, even if it wanted to.

Why, then, did AT&T attempt to acquire Kellogg and Stromberg-Carlson, the two most substantial independent manufacturers? First, it is not clear who approached whom in both cases. Indeed, AT&T President Fish testified that AT&T did not want to purchase Stromberg-Carlson. But the badly foundering U.S. Independent Telephone Company (USITC), which controlled several operating companies and Stromberg-Carlson, insisted that it would only sell the operating companies to Bell if it also acquired Stromberg-Carlson. The reason was simply that Stromberg-Carlson was USITC's most expensive asset. There is nothing that discredits Fish's statements. The New York State attorney general brought a successful action in state court to temporarily enjoin the acquisition under the state's antitrust law, which AT&T did not appeal.[93] If AT&T strongly desired to keep Stromberg-Carlson, it presumably would have appealed.

The attempted Kellogg acquisition is a strange story. In 1901 Milo G. Kellogg, who founded the company bearing his name, was in ill health and granted Wallace L. De Wolf the power of attorney to dispose of Kel-

[92] See "National Civic Federation and Trust Legislation," p. 458; Weinstien, *Corporate Ideal*; and Kolko, *The Triumph*, pp. 131–340.

[93] Howe, *Great Contrivance*, pp. 19, 20 and FCC Special Investigation Docket 1, Exhibit 2096D, *Control of Independent Telephone Companies* (1937), pp. 23–26.

logg's controlling shares of the company. De Wolf sold the shares to a front man allegedly acting on behalf of AT&T. Kellogg, who recovered after a California stay, brought suit in the Illinois courts against De Wolf, AT&T, and various agents to set aside the stock sale. The underlying issue raised in the case was not whether AT&T had violated the federal antitrust laws, but rather whether AT&T had acted *ultra vires*—in excess of the powers granted under its corporate charter in making the purchase. In its first decision in the case, the Supreme Court of Illinois considered no evidence and simply concluded that the plaintiffs seeking to annul the sale had stated a cause of action and that the matter should go to trial.[94]

The case again reached the Supreme Court of Illinois in 1909 after a trial in which the lower court found that the plaintiffs' allegations were true and that the sale of Kellogg stock was void. One issue at this point was whether AT&T had sought to monopolize telephone equipment through the acquisition. But since this was not the underlying question, the Illinois courts did not treat it rigorously. Thus, no evidence was adduced on market shares. Nor did the Supreme Court of Illinois show that the acquisition would foreclose independent telephone operating companies from reasonable alternative sources of supply. Rather, using fallacious reasoning the Supreme Court of Illinois asserted: "*If* the independent manufacturers should go out of business or pass under the control of the American Company, the independent exchanges would be reduced to the alternative of going out of business or becoming subsidiary to the American Company" (emphasis supplied).[95] But the reasoning hinges on an "if" that did not occur and is a linear extrapolation of a trend that is simply assumed and not proved.

In fact, if the Supreme Court of Illinois had carefully considered the matter, a Bell attempt to monopolize telephone equipment would have been futile. MacMeal's standard history of independent telephony lists more than thirty telephone manufacturers that came into existence before the dawn of the twentieth century, many with substantial financial backing.[96] In addition, there were numerous manufacturers of telegraph and electric equipment who could easily move into telephone equipment, an industry no longer burdened by *basic* patents. Clearly, then, entry into the industry was easy, and the transaction costs of negotiating a monopoly were beyond the resources that Bell could then muster.

That AT&T was willing to take advantage of a fortuitous and unique chain of events stemming from Mr. Kellogg's illness and (perhaps) wrong

[94] *Dunbar* v. *American Telephone & Telegraph Co.*, 79 N.E. 423 (1906).

[95] *Dunbar* v. *American Telephone & Telegraph Co.*, 87 N.E. 521 (1909).

[96] MacMeal, *Independent Telephony*, chap. 5.

choice of a person to manage his affairs cannot be used as the basis for concluding that AT&T was bent on monopolizing local telephone service by foreclosing the supply of equipment to independents. A quote from AT&T President Fish to the effect that it would have used the acquired company to its own advantage by not selling to opposition companies was strained under this view to signify a grand design to drive all competitors from business. But after all, in a competitive environment firms are expected to compete vigorously, not assist one another. If this entails not selling a product or service to a competitor, so be it.

The attempted acquisitions, then, played no role in AT&T's coming dominance of telephony. Coincidentally, in the few years from 1906 to 1909, between the first and second Kellogg cases, major strides were made in replacement of the competitive regime by the regulated network manager system. Public service liberalism's task was to create the appropriate supervisory institutions.

The Independent Regulatory Commission

Independent regulatory commissions were first established in Rhode Island and New Hampshire in 1844. Their concern was railroads. These commissions were largely advisory, with the power to conduct investigations and make recommendations. Not until 1871, when the Illinois Railroad and Warehouse Commission was granted enforcement powers over the maximum rates of railroads and warehouses, did the first state commission with moderately strong powers emerge. Within the next fifty years, all of the other states established railroad regulatory commissions, while the federal government established the Interstate Commerce Commission in 1887—the first national regulatory commission. The first state commission with jurisdiction over local public utilities was the Massachusetts Gas and Electric Commission, established in 1885. Gradually commissions began to enlarge their responsibilities as well as to assert jurisdiction over additional public services.[97]

What is true about railroad regulation may not be true about telephone regulation. But there is one point of similarity between railroads and telephones that should be highlighted. Once the commitment to strongly regulate the economic conditions of an increasingly complex business is made, the tendency toward employing the commission form is inexorable. Neither legislators nor courts can exercise the *continuing* supervi-

[97] See, generally, Charles M. Kneier, *State Regulation of Public Utilities in Illinois* (Urbana: University of Illinois Press, 1927), chaps. 1, 2; Irston R. Barnes, *Public Utility Control in Massachusetts* (New Haven: Yale University Press, 1930), chap. 1; and Jones, Origins of the Certificate, pp. 431, 432.

sion that is required. Nor do they have the expertise to regulate the many interrelated facets of complex and changing businesses. Rates, routes, extension of lines, abandonments, quality, safety, finance, future planning, and so on are all related to one another, and regulations in one facet will affect the others. Further, regulations are inoperative, unless someone continuously enforces them. *Individual* litigants (such as many railroad shippers and passengers) are often unwilling to incur the costs needed to bring an action to enforce a regulatory statute, because the amounts that they would individually recover are too small. Obviously agency enforcement solves the problem. Finally, legislatures and courts, charged with more general duties, cannot adjust as quickly to changing industry conditions as more specialized agencies can.[98]

Although states regulated telephones early, their statutes before the turn of the century usually were limited to rates. An example is an 1884 Indiana law limiting charges per telephone to three dollars a month.[99] Fluctuating general price levels (usually downward) in the last third of the nineteenth century and a diversity of service offerings rendered such simplistic formulas patently unsupportable and paved the way for the more flexible kind of regulation that independent commissions can provide. Nevertheless, by 1900, although twenty-three states had laws regarding telephone service, only three states (Louisiana, Mississippi, and North Carolina) had established state commissions with jurisdiction over telephone rates and practices. By 1911 twenty-seven states had such commissions, and by 1920 there were only three states without a state commission asserting jurisdiction over telephone rates and practices. There were also numerous municipal commissions, and the ICC was given regulatory power over telephone, telegraph, and cable companies in interstate commerce under the 1910 Mann-Elkins Act, which required interstate telephone rates to be "just and reasonable."[100]

The earliest state commissions with telephone jurisdiction were in the South. In North Carolina, the first state with commission telephone regulation, support initially came from farmers, and the Board of Railroad Commissioners was granted jurisdiction over telephone companies in 1891. At this early stage, American Bell adopted an equivocal, mildly hostile attitude toward commission regulation.[101] But as the new century

[98] See the excellent discussion in William K. Jones, *Cases and Materials on Regulated Industries* (Brooklyn: Foundation Press, 1967), pp. 33–36.

[99] Stipulation/Contention Package, episode 3, par. 15, *United States* v. *AT&T*.

[100] Ibid., episode 3, par. 18; Arthur Stedman Hills, *The Origin, Growth and Work of Public Utilities Commissions* (New York: AT&T, 1911), p. 15; and I. L. Sharfman, *The Interstate Commerce Commission* (New York: Commonwealth Fund, 1931), pt. 1, p. 53.

[101] See Stipulation/Contention Package, episode 3, par. 16; and Clement, *North Carolina Telephone Story*, pp. 20–23.

progressed AT&T became an enthusiastic supporter of commission regulation, especially after the reemergence of Vail as president in 1907. He proclaimed in 1913: "We believe in and were the first to advocate state or government control and regulation of public utilities."[102] Not quite true. Nevertheless, AT&T certainly shared in a broad consensus of opinion that favored commission regulation of utilities. AT&T shaped the sentiment to its own advantage, arguing that it should become a stringently, but fairly, regulated network manager. In this view it was joined by commercial interests, investment bankers, industrialists, and important social reformers.

The principal opponents of the commission-regulated network manager system were the independent telephone companies, which saw that they would be the major losers. The influential National Civic Federation (NCF) urged: "Regulation is inconsistent with competition. . . . The . . . right of a commission to let in a competing company . . . ought not to be surrendered on the part of the public but it ought to be used only when all other means of safeguarding the public interest have failed."[103] Because AT&T President Fish was on one of the NCF's committees, *Telephony*, in answer, could only decry the natural monopoly theory and cast aspersions at the NCF.[104]

Competition Rejected

One of the more influential reports rejecting telephone competition was prepared in July 1905 under the auspices of the powerful Merchants Association of New York in opposition to the 1905 independent Atlantic Telephone Company application for a second New York City franchise. The New York City franchise battle was a critical one. Obviously a network that wanted to compete with Bell would be severely handicapped without a New York City franchise. The arguments advanced in the Merchants Association report (reprinted and widely circulated by Bell licensee New York Telephone Company) cogently indicate why the weight of elite opinion moved definitively in the direction of local monopoly's provision of telephone service:

[102] Statement by Theodore Vail, "Mutual Relations and Interests of the Bell System and the Public," August 1913, in AT&T Archives, Box 1187.

[103] Quoted in Jones, "Origins of the Certificate," pp. 452, 453. See also Weinstein, *Corporate Ideal*, pp. 24–26.

[104] "The National Civic Federation and Trust Legislation," *Telephony* (November 6, 1909), p. 458. See also "Charter Set Back by Political Plot," *New York Times*, April 17, 1909, p. 3; and Lipartito, *Regional Business*, p. 185.

> During this committee's inquiry, statements have been received from merchants in various cities, where competitive telephone service exists, to the effect that competition has resulted in an increased instead of a decreased burden of cost, and in a divided service which has materially obstructed intercommunication. Business and professional men find it impracticable to dispense with the established telephone system, no matter what inducements of apparently low price may be offered by a new company. General intercommunication is the essential requirement, especially for businessmen. This cannot be assured by either competing company in case of a divided service. The use of both systems and an increased outlay is therefore compulsory.[105]

The Merchants Association report went on to say that no advantage is gained by such duplication, since service is generally not displaced but rather duplicated. Thus, intercommunication possibilities are not enlarged, but the costs of duplicate customer premises equipment, lines, and switching and transmission plant are doubled. The additional costs are ultimately passed on to customers while, at the same time, economies of scale are foregone. Further, because the smaller exchange is only able to compete on the basis of lower price (and with a lower degree of efficiency), it has a strong incentive not to properly amortize plant and equipment. To the smaller, struggling company the burdens of today outweigh the benefits of tomorrow, and it focuses exclusively on the short run. Consequently, "Equipment will be permitted to deteriorate, operating expenses will be reduced below the proper limit and the efficiency of service will be lowered."[106] Finally, the committee argued that a new telephone company would necessarily burden the public by disrupting the streets in its construction. Since no compensating benefit was found, the committee recommended against granting a franchise to a competitor of New York Telephone.

The Merchants Association of New York followed its initial report with a supplemental one completed in September 1905. The influential *Supplemental Telephone Report* included findings of the most comprehensive nationwide survey conducted to that time on the effects of telephone competition on business firms. It included the testimony and opinion of business people and academics (such as economist Richard Ely), as well as a compilation of data covering large and small cities. Once again, the association concluded that telephone competition was injurious to the business subscriber. But this time the theoretical views advanced in the earlier report were strengthened by the detailed data developed in

[105] *Telephone Competition from the Standpoint of the Public* (New York: New York Telephone Co., 1906), p. 6, in AT&T Archives, Box 1082.

[106] Ibid., p. 107.

the investigation. The association's principal additional finding was that where rival systems exist, business firms must subscribe to both in order to be in touch with all suppliers, customers, and potential customers. Thus, while initially the new entrants charge lower rates, a business firm's *total* expenditures for telephone service increases, even when the older company reduces its rates. Further, the association found, without specifying exact numbers:

> Experience has shown that the rates of the independent companies are at first made without regard to the cost of rendering service, that, in consequence a great number of independent companies have become bankrupt, that several of the principal independent companies doing business in the large cities have been reorganized because of failure to earn expenses, and finally that where the test has been sufficiently long continued, several leading companies have been compelled to advance their rates, in some instances approximately to those of the Bell Company which they had previously denounced as exorbitant.[107]

The association, therefore, concluded that since the independents' rates are made without regard to sound accounting practices on repair costs, which tend to increase at the later stages of equipment life, and depreciation, their rates must inevitably increase—as indeed they had been. While the Bell System had an obvious self-interest in publicizing this charge, Lincoln Telephone and Telegraph (LT&T)—one of the leading surviving independents—admitted that it engaged in such unsound practices during the pertinent years. Conceding that the company in 1906 had not set aside a reserve for depreciation, it was forced to either increase its rates or go bankrupt.[108] Aside from Lincoln's experience, the argument tends to be confirmed when we note that the number of independent companies not connecting with the Bell System increased until 1907 and then declined, even though telephone density was increasing and the telephone was rapidly reaching new areas and subscribers.[109]

Accounting data of long-disappeared companies is, of course, impossible to obtain, but the testimony of LT&T, whose president was the leading figure in the independent telephone association, tends to confirm the Merchants Association finding: "This general lack of experience and knowledge in the science of telephony was a serious handicap to the officers of nearly every independent company. Some of them, in the first flush of apparent large returns, paid big dividends to their shareholders

[107] Special Telephone Committee, Merchants Association of New York, *Supplemental Telephone Report, Further Inquiry into Effect of Competition* (1905), p. 15.

[108] Lincoln Telephone & Telegraph Co., *The History of L.T.&T* (Lincoln, Neb.: Lincoln Telephone and Telegraph Co., 1955), chap. 3. (The book is not paginated.)

[109] *Historical Statistics*, 2:783.

and were soon to find themselves in serious financial trouble as the problems of increased maintenance and depreciation made themselves felt."[110] LT&T was reorganized and changed its ways, but other independents were not and suffered the consequences.

Even though the Merchants Association of New York concluded that Bell's service was more valuable (and therefore could command higher rates than the independents) because it allowed intercommunication among many more subscribers, the Supplemental Telephone Report did not conclude that Bell service was uniformly good, that the independents were uniformly bad, or that Bell should be granted a nationwide monopoly in telecommunications. Rather, the performance of the Bell local companies was bad in some places and competition had prodded better Bell service. In other areas Bell service had been high prior to the advent of competition. In still others the Bell licensee had improved its service, despite the absence of competition or even its prospect. Overall, however, the Merchants Association concluded that competition was not the answer. Moreover, a survey of business opinion throughout the country showed no support for dual telephone systems. Indeed, business opinion concluded that the use of resources for competitive purposes depleted the capital of the existing company so that improvements and expanded service were slowed down.[111]

These arguments had been made by the Bell interests even before the turn of the century.[112] But it was not until the experience of the first decade of the twentieth century that the initial sentiment in favor of competition was reversed and virtually all informed opinion turned against it. A 1906 Boston investigation of competition in several American cities conducted by George W. Anderson, an eminent utility attorney (but commissioned by the Bell licensee) was critical of some Bell licensees and generally complimentary of the impetus *initially* supplied by competition. Anderson nevertheless concluded: "The competitive movement in telephony . . . is now entirely a speculative, and not a sound business, movement. . . . Whatever may be said as to the effect of competition six or eight years ago in educating the public into an increased use of the telephone and showing the Bell companies in a clearer light, the capabilities of their industry . . . the competitive movement as it is today . . . is

[110] LT&T, *History of L.T.&T.*, chap. 4. See also New England Telephone & Telegraph Co., *Competition in Telephony* (Boston: New England Telephone & Telegraph Co., n.d.), pp. 18–23, in AT&T Archives, Box 1082; and W. S. Allen, untitled memorandum on low rates of independent telephone service, 11 December 1902, in AT&T Archives, Box 1272.

[111] Merchants Association of New York, *Supplemental Telephone Report*, pp. 17–31.

[112] See materials collected in Central Union Telephone Co., *Some Facts as Others See It on Local and Long Distance Telephone* (Indianapolis,: Central Union Telephone Co., 1898), in AT&T Archives, Box 1277.

utterly unsound."[113] Instead, Anderson recommended commission regulation of local telephone monopolies as the best alternative. But such regulation would have to make sure that New England Telephone & Telegraph, the system manager in New England, would extend its service as far and as fast and render the service in as efficient and progressive a manner as was reasonably possible.

Competition eventually failed even in those places in which public service regulation initially permitted it. In Wisconsin, for example, the Progressive-sponsored 1907 Public Utility Act brought telephone companies under the jurisdiction of the Wisconsin Railroad Commission but did not include them within the antiduplication-of-facilities clause. The independents had, thus, won an important battle in the legislature over the objections of the Wisconsin Telephone Company, the Bell licensee. Nevertheless, the political victory was not followed by an economic one. The lack of proper accounting procedures, the most important being the failure to provide for sufficient reserve funds for depreciation, undid many of the independents after 1907. They were frequently unable to rebuild plants when this became necessary and sometimes they were even unable to make requisite repairs. Thus, after 1907 many independents were eliminated as the Wisconsin Telephone Company expanded.[114] By 1915 the Kansas Supreme Court could write that competing telephone companies "place a useless burden on the community, cause sorrow of heart and are altogether undesirable."[115]

Telephone competition was widely viewed as a failure. But that did not necessarily lead to the adoption of the state commission-regulated network manager system, for forces hostile to public service liberalism sought public ownership or municipal regulation or to transform commissions into institutions that would plan every facet of a public service company's activities by government professionals. AT&T's behavior, as we will see in the next chapter, combined with other factors to preserve the system of regulation by commission that had been developed in telephony. Public service liberalism had flexibly adjusted from monopoly to competition to monopoly again.

[113] G. W. Anderson, *Telephone Competition in the Middle West* (Boston: New England Telephone & Telegraph Co., 1906), pp. 38, 39, in AT&T Archives, Box 1082. See also Stehman, *Financial History*, chap. 4.

[114] John Lester Miller, "A History of the Telephone Industry as a Regulated Industry in Wisconsin" (Ph.D. diss., University of Wisconsin, 1940), pp. 88, 89; and Fred L. Holmes, *Regulation of Railroads and Public Utilities in Wisconsin* (New York: D. Appleton, 1915), pp. 202, 203.

[115] *Janicke* v. *Washington Mut. Tel. Co.*, 150 P. 633, 634 (1915).

6

Public Service Liberalism and the New Political Economy

The New Political Economy

THE DECLINE of competition in the telephone industry did not necessarily mean that AT&T would emerge triumphant as the dominant—indeed, overwhelming—factor in that industry. The decline of competition could have led policymakers in many directions, including efforts through the antitrust laws and legislation to revitalize competition. In fact, as we will see, this alternative was attempted, and it failed. Again, important voices were heard favoring such other alternatives as nationalization or municipal ownership. Telephone and telegraph nationalization had occurred in other major countries and, in the context of the time, was not necessarily considered socialistic. Municipal ownership of other utilities had been taking place. Thus, even as competition declined, other alternatives to the system in which AT&T would become a highly regulated network manager loomed as serious possibilities. AT&T's triumph, and that of the regulated network manager system, must be seen as a triumph of the adaptability of public service liberalism in the face of dramatically changing political-economic conditions.

One overarching fact dominated the Progressive Era. Whether vilified as trusts or lauded as the most modern form of business organization, the large-scale integrated corporation was in rapid ascendancy. At one extreme, demagogues played to irrational fears of a public at the mercy of a few trusts. But more serious observers, such as Herbert Croly and Louis Brandeis, were carefully considering how the new economic enterprises would fit into the American political framework. The organizational and technological capabilities of the new corporations were incorporated into the views of influential modern liberals. AT&T, was an important focus of the clash between public service liberalism and modern liberalism.

The period from 1902, the end of the great merger wave, until 1920 witnessed marked changes in the size, organization, and activities of American corporations. By the end of the 1920s, as Alfred Chandler shows, the modern business enterprise dominated the American industrial landscape. Many enterprises had become vertically integrated so that they covered activities from the extraction of raw materials through the marketing of finished products. Sometimes vertical integration was accomplished through internal growth, but often it was achieved through merger or acquisition.[1] At the same time many corporations enlarged their product offerings into closely related ones. For example, Du Pont, a gunpowder maker, moved into the production of paints, varnishes, and other chemical products. Product extension occurred through both internal expansion and acquisition.[2] As we will see later in this chapter, AT&T typified this movement as it sought to acquire Western Union, which dominated the telegraph business.

At the same time because of transportation improvements large firms extended their markets to cover ever increasing portions of the United States. In 1903 the National Good Roads Association received the support of leaders of the major political parties, and around 1910 motor trucks began to compete with horse-drawn vehicles. The Ford Model T would spread the use of automobiles dramatically. At a quieter level, the electric self-starter, which would supplant hand cranking of vehicles and further expand their deployment, was introduced in 1911. Above the ground, the U.S. government ordered a plane from the Wright Brothers in 1908. Even trains exhibited major breakthroughs during this period with electrification and the construction of more powerful locomotives that could more rapidly move freight from coast to coast.[3] In the same period, equally dramatic changes occurred in communication.

Improvements in transportation and communication reduced the delivered costs of goods and services, thereby contributing importantly to the creation of a national market in product after product. At the same time improvements in communications technology and the expansion of the network allowed firms in every business to coordinate a diversity of activities, such as ordering, from a central location. In this way, too, communication and transportation improvements greatly contributed to the creation of the national, multiproduct, vertically integrated corporation. The modern corporation, as Alfred Chandler observed, became a center not only of the production and distribution of goods, but also of scientific research, technological development, sophisticated capital funding, credit provision, installation, repair, and after-sale service.[4] AT&T not

[1] Alfred D. Chandler, Jr., *The Visible Hand* (Cambridge, Mass.: Harvard University Press, 1977), chaps. 10, 11.

[2] See Alfred D. Chandler, Jr., and Stephen Salsbury, *Pierre S. Du Pont and the Making of the Modern Corporation* (New York: Harper & Row, 1971), pp. 381–86.

[3] John W. Oliver, *History of American Technology* (New York: Ronald Press, 1956), pp. 478–95.

[4] Chandler, *Visible Hand*, pp. 347, 364.

only contributed to the development of the modern corporate enterprise, but was the quintessential reflection of it as well.

As we might expect, these remarkable changes in the shape and structure of business enterprise generated responses that portended the development of modern liberalism. To some of the Progressive Era influentials, government planning or restructuring should reshape the economy according to some grand plan without regard to the specific goals that public service liberalism would have required in each industry. Thus, some political leaders, such as Louis Brandeis, fearful that small and medium-sized enterprise would be wiped out, called for a return to small-scale enterprise through antitrust action and the enactment of general statutes that would protect weaker companies confronted by the market power of stronger ones. Other sweeping theories placed faith in nationalization. Many big business leaders seeking market stability in the face of the inroads made by smaller competitors also called for the enactment of statutes that would restrict competitive methods.

Notwithstanding their differences, those persons who advocated nationalization, those who sought to restructure industry according to preconceived notions favoring small-scale enterprise, and big business advocates of stability had one thing in common: they deviated sharply from public service liberalism. Instead of initially asking what the appropriate goals of a particular industry should be, whether free competition could best attain them, and then devising interventionist policies based on experience, these precursors of modern liberalism generated sweeping theories of government control. But public service liberals, too, had to react to the increasing scale of firms and the enlargement of markets from local to regional and national size. Clearly, national regulation had to supplant or complement local regulation.[5]

The acid test for public service liberalism would be whether it could make the necessary policy adjustments in lieu of the changing shape of industries and firms. Railroads, still the lifeblood of intercity transportation, were a central focus of the battle between public service liberalism and modern liberalism seeking a strong role for government in public affairs. A major battle that generated widespread attention erupted in 1911 in two critical ICC general rate-making proceedings covering large territories. In the eastern and western rate advance proceedings, railroads operating in the covered territories sought general rate increases reflecting a rise in their costs. They were opposed by a variety of shipping interests, some of whom engaged Louis Brandeis as their attorney. Brandeis advanced the then novel argument that the rate increases should be

[5] See, for example, Alan Stone, *Economic Regulation and the Public Interest* (Ithaca, N.Y.: Cornell University Press, 1977), chap. 2 and materials cited therein.

denied because the railroads were woefully inefficient. If they employed Frederick Taylor's scientific management principles, Brandeis argued, the railroads could simultaneously raise wages and lower costs by becoming more efficient.[6]

The ICC rejected the Brandeis argument in part because scientific management was in an experimental stage and the precise application of such techniques to railroads was not explicitly delineated. But equally important, the ICC reaffirmed the public service conception by urging that railroads, as public service companies, were constrained to serve values in addition to efficiency, some of which may undermine efficiency. In the words of Harvard professor William J. Cunningham, then one of the most influential theorists of railroad policy: "A railroad is a public service corporation. The public rightfully demands that adequacy of service shall outrank the payment of dividends. . . . It has two functions, public service and profit making; it may not neglect service to favor profits. Necessarily, therefore, methods are employed in the interest of public service even tho they involve economic loss."[7]

But while scientific management lost the battle, Brandeis's argument did catch the public's imagination. Coupled with general antirailroad sentiment and the sensitivity of the professional and managerial middle class to rising costs and prices, Brandeis's argument set off, in historian Samuel Haber's phrase, an "efficiency craze."[8]

While Brandeis's version of scientific management would deeply influence the Woodrow Wilson administration, Herbert Croly's conception of scientific planning in place of competition had a similar effect on many Republicans. In his view large corporations strongly directed by government agencies would serve the national economic interest. But, he cautioned, the private sector needed "rigid and comprehensive official supervision."[9] Whether in the Brandeis, Croly, or still other versions, an intellectual revolution was at hand. Scientific management was the key to planning through bureaucracies controlling those corporations. This was seen as the solution to policy problems. Although industries differ,

[6] See *Advances in Rates—Eastern Case*, 20 ICC 243 (1911); *Advances in Rates—Western Case*, 20 ICC 307 (1911); I. L. Sharfman, *The Interstate Commerce Commission* (New York: Commonwealth Fund, 1936), 3B:46; and Oscar Kraines, "Brandeis' Philosophy of Scientific Management," *Western Political Quarterly* 13 (1960): 191–201.

[7] William J. Cunningham, "Scientific Management in the Operation of Railroads," *Quarterly Journal of Economics* 25 (1911): 551. See *Advance in Rates—Eastern Case*, 20 ICC at 279–80.

[8] Samuel Haber, *Efficiency and Uplift* (Chicago: University of Chicago Press, 1964), p. 52 and chap. 4.

[9] Herbert Croly, *The Promise of American Life* (1909; rpt., Cambridge, Mass.: Harvard University Press, 1965), p. 362.

of course, the communications industries offer a microcosm of the American political economy in a period of turmoil.

As the telephone became a national institution, opportunities loomed just as pitfalls lurked. The task of the company's leaders, especially Theodore Vail, was to navigate in these dangerous political waters by showing that AT&T's private interests conformed to the new public service demands of the state. Not only would it have to cooperate in new ways with the national government, but AT&T would have to show that the industry structure and regulatory institutions it desired would better serve than any other alternative a variety of goals, including growth, technological advance, productivity advance, and service improvement. AT&T sought to show that its private interest accorded with public service liberalism. In the face of the threat, advocates of public service liberalism in the telephone industry had to devise an institutional structure that would embrace business-government cooperation and the ability to resolve tensions and conflicts efficiently.

Revolution at the Top

Theodore Vail's return to AT&T was connected with one of the most serious economic crises in American history—the Panic of 1907. GNP fell from $30.4 billion of 1958 constant dollars to $27.7 billion from 1907 to 1908. In the course of the Panic several overextended New York trust companies were unable to meet their commitments, leading to a depositor run on numerous banks. Interest rates increased and stock trading declined precipitously.[10] From the perspective of the Bell System, the decline could not have occurred at a worse time. In July 1906 AT&T, already in the throes of a cash flow crisis, curtailed providing funds to licensees for capital expenditures.[11]

In many ways AT&T's success was the cause of its crisis. Bell System expansion, which was rapid when Fish became president in 1901 until he departed in 1907, required increasing resort to new funds. Inevitably this meant increased investment banker participation in AT&T financing. Not unnaturally, investment bankers wished to assure that their funds were used wisely. Accordingly, even though they did not participate in day-to-day management, the investors sought to assure that those who did were highly qualified professionals. For this reason they demanded the return of Vail, the preeminent telephone company manager. Investment banker

[10] See, generally, W. C. Schluter, *The Pre-War Business Cycle* (New York: Columbia University Press, 1923).

[11] Robert Garnet, *The Telephone Enterprise* (Baltimore: Johns Hopkins University Press, 1985), p. 127.

activism during this period was a prelude to AT&T becoming the quintessential example of people's capitalism in which, at the time of the AT&T breakup, there were more than 3 million shareholders, and shareholding was so widely dispersed that large stockholder domination of management was virtually impossible.[12]

When Vail assumed office in 1907 there were many who would not have predicted a bright future for the company. While a $20 million bond issue was sold without difficulty to a Boston syndicate in 1904, the company's later needs required the participation of a major syndicate headed by very large investment bankers, most of whom were located in New York. On February 8, 1906, AT&T agreed to sell $100 million of convertible bonds to a syndicate headed by J. P. Morgan & Company. Under the complex agreement the bonds would be convertible to stock beginning in 1909. But in late 1906 the issue was already in trouble. On July 23, 1906 the *Wall Street Journal* reported that the syndicate did not then intend to offer the bonds and would not do so "until the general tone of the bond market has shown a marked improvement."[13] In October AT&T raised its dividend rate to help the bond sale. The stock price still remained below the conversion price. On October 25 the *Wall Street Journal* reported that the syndicate, which had held the bonds for more than six months, was becoming anxious about the situation. On January 8, 1907, the investment bankers obtained an important concession from AT&T on the conversion terms in anticipation of the sale, which was planned soon. The syndicate began its sale on January 31, 1907, but it was a fiasco. During 1907 and 1908 only 10 percent of the $100 million of bonds had been sold. The failure was apparent almost immediately since desirable bonds are usually subscribed at once. In any event the continuing decline of AT&T stock made the convertibility feature even less attractive.[14]

The failure of the convertible bond sale compelled the syndicate members to take a closer look at the company. Indeed as early as January 1907 one of the bankers suggested that Fish appoint a committee to probe the various problems of the company. The committee investigation and the company's financial problems resulted in resignations from AT&T's ex-

[12] For example, a 1974 Senate study showed that the seven largest New York banks *together* held only 2.9 percent of the company's shares in 1972. Chase Manhattan, the bank with the largest percentage of such shares, held only 1.1 percent of the more than 41 million AT&T shares. See U.S. Senate, Committee on Government Operations, *Disclosure of Corporate Ownership* (Washington, D.C.: Government Printing Office, 1974), p. 84.

[13] Quoted in Federal Communications Commission, Special Investigation Docket 1, Exhibit 2096A, *Control of American Telephone & Telegraph Co.* (1937), p. 59.

[14] Ibid., pp. 54–63; and N. R. Danielan, *AT&T: The Story of Industrial Conquest* (New York: Vanguard Press, 1939), pp. 60–65.

ecutive committee in early April 1907. This was followed by Fish's resignation of both his presidency and a directorship on April 23. Vail, who had refused the presidency six years earlier, accepted it now, determined to rectify the company's problems and clarify its positions on major policies. The immediate problem was, of course, to ascertain what had gone wrong financially and to restrict expenditures until a new method of obtaining funds—obviously not through the sale of convertible bonds—could be found.[15]

What Vail found was a general decline in *unit* revenues, even as the telephone industry was booming. A January 26, 1907, *New York Times* article titled "Great Telephone Growth" noted that the Bell companies had experienced "a growth in six years of 171 percent in the number of employees, of 289 percent in the number of stations, and of 349 percent in the total number of miles of wire."[16] Yet at the same time revenue per station had declined precipitously. In 1895 operating revenue per station was 88 dollars. By 1900 the comparable figure was 63 dollars and by 1907 it had declined to 43 dollars. (It should be added that the decline was not reversed until 1916; indeed, in 1914 and 1915 operating revenue per station was 41 dollars.) The decline in operating revenues per station exerted substantial downward pressure on net earnings per station, which also began its decline in 1894 and 1895 and was not *generally* reversed as a trend until 1920.[17]

These trends, of course, exerted considerable pressure on AT&T to both expand the system in order to increase *total* profits in the face of *unit* profit declines and at the same time to innovate rapidly, thereby decreasing unit costs. In turn, expanding the system (and lowering costs) implied a commitment, as we have seen, to long distance, discovering new uses for the telephone and improving the quality of transmission. This ambitious program meant that AT&T would have to raise considerable funds above what could be raised internally. When the convertible bond sale failed, the strategy of the Fish management failed with it.

Vail's first short-term problem was to raise funds in the face of the bond failure. Since the bond market could no longer be used for this purpose, the stock market was the inevitable answer. But how would it be possible to market a new issue of stock? In May 1907 Vail made one of his first major moves by announcing to the legion of AT&T shareholders a stock issue of approximately 220,000 shares, each current stockholder being

[15] FCC, *Control of AT&T*, pp. 63–68; Danielan, *AT&T*, pp. 65, 66; Garnet, *Telephone Enterprise*, p. 127; and Albert Bigelow Paine, *In One Man's Life* (New York: Harper & Bros., 1921), chap. 37.

[16] "Great Telephone Growth," *New York Times*, January 26, 1907, p. 12.

[17] Federal Communications Commission, *Investigation of the Telephone Industry in the United States* (Washington, D.C.: Government Printing Office, 1939), pp. 132–35.

entitled to buy one of the new ones. The sale proved an immediate success. Most of the new stock was purchased almost immediately, rights were sold at high levels, and the company raised more than $20 million in short order. The short-term cash flow problem had been solved, even though AT&T stock was trading at a comparatively low price during this period.[18]

The lesson that the company learned, and applied thereafter, was to finance its activities in a variety of ways—from profits, stock issues, convertible bonds, notes, and other debt instruments. This strategy became especially important as Vail unveiled his grand plans. In addition, as AT&T acquired greater control of its operating companies, the latter became financially more secure and began offering their own bonds. Finally, as part of the general attentiveness to public relations that Vail impressed upon the company, efforts were made to distribute AT&T's stock widely. Thus, at the end of 1916, 43,000 employees and 70,555 others held stock in the company with an average holding of fifty-four shares.[19]

A long-term problem that Vail would have to face was competition. As we saw in the last chapter, one facet of his response was to enthusiastically endorse independent commission regulation as the best way to protect the public interest when AT&T became *the* network manager. AT&T would have to put its own house in order for the new strategy to succeed. The comprehensive plan that Vail sought to follow when he became president was outlined in a confidential memorandum he prepared in 1901, about the time that Fish became AT&T's president. Vail followed his 1901 views with remarkable consistency when he assumed the presidency. The centerpiece of Vail's strategy was what he termed "consolidation," in which AT&T would more closely control the activities of its component companies and the relations between them. He noted the high transaction costs resulting from attempting to settle the disputes between AT&T, licensees, sublicensees, and others, including rights, traffic, and revenue divisions. Vail proposed that "consolidation would render the actual operation of the business much more simple, as with proper organization, there would be one central authority with its direct lines of representation and responsibility, from the centre to the very furthest limits of the business."[20]

[18] Paine, *One Man's Life*, pp. 230, 231.

[19] J. Warren Stehman, *The Financial History of the American Telephone and Telegraph Company* (Boston: Houghton Mifflin, 1925), pp. 163, 164.

[20] See T. N. Vail, "Copy of his views on the general policy which should govern the Company, etc., etc.—as written to Gov. Crane about the time Mr. Fish became President," p. 3 in folder titled "Vail, T.N.—Policy and Plans for Expansion of Business—Circa 1901,"

In brief, then, Vail sought consolidation not for purposes of "power" or "control," but rather for the purposes of reducing transaction costs—the costs of negotiating, drafting, and safeguarding agreements *ex ante* and policing agreements and settling disputes *post ante*.[21] That is, Vail urged that AT&T employ administrative integration rather than market arrangements because he believed that the transaction costs of the firm would be lower if AT&T organized its transactions hierarchically within the firm, rather than through arm's-length dealings with autonomous firms.[22] Given the great need for technological coordination in an industry experiencing rapid scientific and engineering progress, a much larger number of transactions is necessary than in an industry with slow change and a lower need for technological coordination. Vail essentially argued that under AT&T's circumstances, administrative hierarchy was preferable to market transactions.

For this underlying reason, when Vail assumed the presidency a set of policies were devised to reorganize the company internally in a more closely integrated manner.[23] Under his leadership, AT&T centralized the system of collecting data, analyzing and approving local operating company budgets. Long-range planning was developed. The company was reorganized along functional lines in order to increase the level of technical expertise. Standardization of equipment and practices was pushed vigorously in order to enhance efficiency and cut costs. Testing and inspection procedures, previously undertaken to a large extent by licensee companies, were now centralized. Most important, the program already under way to gain full control of the licensee companies was intensified. As E. J. Hall, one of Vail's principal lieutenants, wrote to him in 1909: "When we acquire the ownership of all the stock of any company we are in a position for the first time to say just how [it] should be handled. Up to this time, we have not been in that position with relation to any company."[24]

But Vail realized that there can be dangers in overcentralization. It can result in the central authority making innumerable decisions based on limited and sometimes wrong information because of ignorance of local conditions. Vail therefore opted for a federal structure. Under that struc-

---

AT&T Archives, Box 1080. The memorandum will hereafter be cited as the 1901 Vail Memorandum.

[21] Oliver E. Williamson, *The Economic Institutions of Capitalism* (New York: Free Press, 1985), pp. 20, 21.

[22] See, generally, R. H. Coase, "The Nature of the Firm," *Economica*, n.s. 4 (1937): 386–405.

[23] This paragraph summarizes the material on AT&T internal reforms described in Garnet, *Telephone Enterprise*, chap. 9, and the extensive documentation cited therein.

[24] E. J. Hall to Vail, 27 September 1909, in AT&T Archives, Box 1010.

ture, which prevailed until the breakup, AT&T would determine basic policy and organizational principles. AT&T would be primarily responsible for developing scientific, technological, and other specialized information and disseminating that information to the operating companies. They, in turn, would be organized by state, where state regulatory authorities required that, and by regional lines otherwise. The operating companies would make the pertinent decisions based on the application of the central policies and organizational principles to the problems within their respective jurisdictions. The operating companies would, in turn, critically evaluate the results obtained by applying these central policies and principles, transmitting their findings to the central authority. In turn, AT&T would modify or reverse the central policies and principles when appropriate and continue the policies that were successful. Such, in any event, was the underlying theory that AT&T and the operating companies employed.[25] In time other large successful companies adopted such a federal structure.

Internal corporate improvement of the type outlined would, according to Vail in the 1901 memorandum, be the best way to overcome the opposition. "The worst of the opposition has come from the lack of facilities afforded by our companies—that is, either no service or poor service."[26] Vail attributed these failings simply to a cash shortage. Money, he then argued, should also be used to head off the formation of an alternative network. At the same time Vail, however, was prepared to be conciliatory toward those independents willing to cooperate with the Bell System. Restrictions on independent interconnection were significantly relaxed in accordance with Vail's conception of AT&T as network manager, not network monopoly. Thus, in 1906, the last full year prior to Vail's assumption of the presidency, only 297,218 independent telephones interconnected into the Bell System. By 1908 the comparable figure was 1,188,235, and in 1912 it was 2,496,257. Nonconnecting independent telephones, almost 2.28 million in 1907, declined to approximately 1.43 million in 1912 and continued to decline thereafter. Again, whereas there were almost twice as many non-Bell connecting independent telephones than Bell connecting independent telephones in 1907, by 1910 there were more Bell connecting than nonconnecting independent phones.[27]

Additionally, in 1908 the manufacturing contract with Western Electric was amended to allow the latter to sell some equipment to telephone

[25] That the federal system worked well was the conclusion in Robert Sheehan, "AT&T—A Study in 'Federalism,' " *Fortune* (February 1965): 143 et seq.

[26] 1901 Vail Memorandum, p. 5.

[27] See "Telephone Development in the United States," in AT&T Archives, Box 2028.

companies that were not Bell licensees.[28] Prices to the nonlicensees were, it should be noted, higher than those to licensees. Nevertheless, this new policy marked a radical departure from the past and a further conciliatory move toward the independents. The new equipment policy was shrewd in several respects. In the short run, it improved Western Electric's profit as the Bell manufacturing arm gained new customers.[29] In a more fundamental sense, the gesture reinforced the image of the Bell System as paternalistic network manager that Vail was trying to project in conjunction with his regulatory program. The program aided the standardization that was required as Bell companies interconnected more and more with independents, and it fit neatly into Vail's federal structure program, which required the major divisions of AT&T and the associated companies to be independently efficient. For the same reason Western Electric was then abandoning unprofitable product lines, taking on new supply functions, and internally reorganizing.[30]

Operating Companies Acquisitions and the Bell System Structure

Perhaps the most controversial part of Vail's new program was the acquisition of independent operating companies and Western Union. These acquisitions were consistent with Vail's overall program and were not undertaken in a hostile attempt to destroy competitors, except insofar as independents threatened to create a rival network. In order to understand the acquisitions it is first necessary to focus more closely on the composition of the independent movement in the early years after Vail's return to the AT&T presidency. Indeed, one of Vail's first requirements was accurate intelligence on the size and strength of independent companies. A January 1909 AT&T report shows that though there were then 12,000 independent companies not connecting with the Bell System, most were quite small. Even the largest of them were no match for AT&T and its associated companies. Thus, at the time of the report only 40 of these companies had issued stocks and bonds worth more than $1 million. These 40 companies had approximately 765,000 stations. The re-

[28] AT&T, 1907 Annual Report, p. 11; and AT&T, 1910 Annual Report (condensed), reprinted in *Commercial and Financial Chronicle* (April 25, 1912): 811.

[29] AT&T, 1909 Annual Report, reprinted in *Commercial and Financial Chronicle* (March 19, 1910): 782.

[30] See the sections on Western Electric in the 1909 and 1910 AT&T Annual Reports; and Stehman, *Financial History*, pp. 132–34.

mainder of these 12,000 companies, whose issued stocks and bonds were less than $100,000, had approximately 1,147,000 stations.[31]

In short, the independents, though controlling a large number of stations, did not constitute as formidable a threat as one would believe by examining only aggregate data. As Table 6.1 shows (AT&T's internal figures vary to some degree from U.S. Bureau of the Census data), independents controlled more stations than AT&T. But when one adds to AT&T Bell sublicensees and connecting companies, AT&T was well on its way to becoming network manager.

Except in the extraordinarily unlikely event that the large number of small independents and small number of large independents could overcome the negotiating obstacles to creating a rival system, AT&T no longer had to fear that prospect, as it did earlier in the century. Rather, its more imposing problem was political—first to convincingly demonstrate the advantages of the regulated network manager system and second to show that it was the appropriate manager. Extending its lines and expanding its sublicensing and interconnecting arrangements were two ways of moving toward Vail's goals of one interconnecting system with universal service. For this reason Vail established a policy of becoming

TABLE 6.1
Bell and Independent Companies and Stations on January 1, 1909

Company Type	*Number of Stations*
1. AT&T and licensees	3.2 million
2. AT&T sublicensees and connecting companies (approx. 6,000 companies)	1.1 million
3. Independent companies not connecting with Bell System (approx. 12,000 companies)	2.3 million
a. 40 companies with more than $1 million in stocks and bonds: 765,000 stations	
b. 36 companies with $500,000 to $1 million in stocks and bonds: 114,000 stations	
c. 165 companies with $100,000 to $500,000 in stocks and bonds: 274,000 stations	
d. 11,759 companies with less than $100,000 in stocks and bonds: 1,147,000 stations	
TOTAL	6.6 million

Source: C. G. DuBois to Theodore N. Vail, 25 January 1909, in AT&T Archives, Box 1006.

[31] C. G. DuBois to Theodore Vail, 25 January 1909, in AT&T Archives, Box 1006.

especially liberal toward integrating farmer lines in isolated areas into the Bell System. This was not an easy task. As the Pennsylvania Public Utilities Commission (PUC) found, this required sending Bell's skilled technicians to rural areas to construct new lines, install telephones, upgrade equipment, and inspect existing plant and equipment so as to assure full compatibility with the Bell System. These services were, according to the Pennsylvania PUC, provided free to the farmer lines.[32]

Within this context we can now see that Vail's vigorous acquisition policy was conciliatory, not hostile. The numerous smaller telephone companies that were not connected into the Bell System, especially competitors of Bell companies, could only begin to feel more and more isolated as the Pupin coil and other improvements made regional and national interconnection an increasingly valuable part of telephone service. Many local companies could not reach numerous subscribers in even their own communities, never mind adjacent or distant ones. And even regional toll companies were at a significant disadvantage for business customers who sought to use the telephone nationally.

Thus, one must surmise that the controlling shareholders of independent companies faced an increasingly difficult situation. As Keystone Telephone's decline illustrates, nonconnecting independents were at a serious disadvantage that could only get worse. Since any rational investor investigating the situation would realize this, an offering bid for such a company's stock would reflect the probable decline. Thus, a nonconnecting independent's best recourse was to reach an accommodation with the Bell System. This could be done either through an interconnecting or sublicensing agreement or through the sale of the company to AT&T or a Bell licensee, which would guarantee interconnection into the Bell System. Seen in this light, Vail's vigorous acquisition policy was as conciliatory as the other components of his program, especially the increasingly liberal interconnection policy.

The acquisition policy was carried out within the framework of reorganizing the operating companies into state and regional units. Although details vary from operating company to operating company and additional changes were frequently made until shortly before divestiture, Southwestern Bell's (SWB) development typifies the process.[33] SWB can trace its lineage to twenty pre-twentieth century ancestors, the four most important of which were doing business in, respectively, (1) St. Louis, (2) Kansas City, Missouri, (3) Arkansas and Texas, and (4) Oklahoma and part of Kansas. By 1912 these four large companies had absorbed the smaller

[32] Cited in Stehman, *Financial History*, pp. 129–32.

[33] Details on Southwestern Bell's history are taken from David G. Park, Jr., *Good Connections* (St. Louis: Southwestern Bell Telephone Co., 1984), passim.

ones and were supervised by a general staff. The integrated operation, known as the Southwestern Bell Telephone System, was formally restructured as the Southwestern Bell Telephone Company in April 1920. SWB and its predecessor companies not only acquired smaller firms, but in a few instances also sold components of its system to independents.

The process of acquiring small companies continued after 1911, subject to the constraints imposed by the Kingsbury Commitment, which will be discussed later. Thus, one of the largest acquisitions was that of the Southern Telephone Company of Fordyce, Arkansas, which had nine-thousand subscribers in Arkansas and Louisiana. Acquisitions were made not only of small companies with which the Bell System did not compete, but of competitors as well. In 1912 one of SWB's predecessors acquired the Little Rock Telephone Company, the only remaining competitor in that city. SWB's acquisitions in the 1920s accelerated, and many of the acquisitions were of sizable companies, such as the 1925 purchases of the Home Telephone Company of Joplin, Missouri, and the Dallas Telephone Company, which until then ran a competing company in that city. During the Great Depression of the 1930s, SWB again acquired many small companies in Kansas and Texas that were about to go under because of shrunken revenues.

While this pattern of many small acquisitions typifies the acquisition policy of the Bell System, there were instances in which Bell sold operations to independents. For example, a Bell company in Rochester and Rochester Telephone, one of the largest surviving independents, competed so vigorously in the period around 1907 that neither company made a profit for a time. Lengthy negotiations between Bell and independents in three major upstate New York cities (Buffalo, Jamestown, and Rochester) were finally concluded in 1917 when the various companies agreed that the Bell licensee would acquire the independent company in Buffalo and sell its properties to independents in the other two cities.[34] Rochester Telephone, it should be noted, undertook a series of acquisitions of smaller companies, just as the Bell System did. In any event, this trade typified the way many of the remaining independents survived and the improving relations with the Bell System that began during the Vail years.

The Independents Search for a Strategy

Some of the Bell acquisitions, notably those in Ohio and Indiana, roused suspicion and anger because they were made with the financial assistance

[34] F. L. Howe, *This Great Contrivance* (Rochester, N.Y.: Rochester Telephone Corp., 1979), pp. 22, 24.

and intervention of J. P. Morgan & Company and other investment bankers.[35] Of course, while some of the independents accepted Bell overtures of sublicensing, interconnection, or acquisition, others initially resisted, seeking recourse by exerting pressure on state attorneys general and the Justice Department to bring antitrust suits. At times independents brought private antitrust suits. Thus, even though Vail was holding out the olive branch to independents, the level of anti-AT&T activity was mittee to fight Bell acquisitions under the antitrust laws. The independents' spokesman asserted that they would bring, or pressure state and ents' spokesman asserted that they would bring, or pressure state and federal judicial officers to bring, "a series of suits paralleling in national importance the recent Government actions against the Standard Oil Company. . . . For instance in the State of Indiana there are eighteen cities in each of which the Bell organization is charged with violation of the antitrust laws. . . . In Ohio there are twenty-two. . . . The National and State Governments have grounds for bringing similar suits against the Bell Telephone Company in almost every state in the Union."[36]

Notwithstanding the independents' exuberance to use the antitrust laws to attack the Bell System, the results were sufficiently disappointing by 1911 to suggest to many of the anti-Bell independents that another strategy was required. Theodore Gary, one of the most influential independent leaders, acknowledged that the campaign was not only largely unsuccessful, but actually hurting many independents. He acknowledged that many independents sought to sell their properties to the Bell System companies. The uncertainty and turmoil created by pending litigation (often brought against the wishes of the acquired independents) left independents up in the air. Acknowledging that Bell companies were bound sooner or later to acquire many independents, Gary urged that the independent movement should adopt a new strategy. He argued that they should attempt to assure that independents obtain the best terms possible from AT&T and its licensees. Noting that the independents that had become Bell sublicensees were content with their status, Gary further diminished anti-AT&T sentiment within the independent movement and paved the way to the accommodation that would take place in a few years.[37]

Although many independent leaders, including Frank H. Woods, president of the independents' national association, adopted Gary's views, arguing that compulsory interconnection laws and federal and state regu-

[35] Federal Communications Commission, Special Investigation Docket 1, Exhibit 2096D, *Control of Independent Telephone Companies* (1937), pp. 29, 30.

[36] "To Bring Telephone Suits," *New York Times*, January 22, 1909: p. 16.

[37] Harry B. MacMeal, *The Story of Independent Telephony* (Chicago: Independent Pioneer Telephone Association, 1934), pp. 187, 188.

lation of telephone companies were better protection against possible Bell abuse than antitrust laws, some independents were intent on fighting on.[38] To these independents sensing the changing ideology of many government administrators to modern liberalism, the local skirmishes were now subsidiary to persuading the attorney general to bring a federal antitrust case that would break up AT&T and reconstruct the industry pursuant to a government plan. As early as August 1912 the *New York Times* reported that the Department of Justice was about to bring suit for the dissolution of AT&T. The report was premature, but in late November 1912 a number of fiercely anti-Bell independents sought to persuade Attorney-General Wickersham to bring a Sherman Act action against AT&T based on the Department of Justice's lengthy investigation of the company.[39]

A major federal antitrust suit was filed in August 1913 charging Bell with attempting to create a long distance monopoly in the Pacific Northwest region, but that case was not the critical event that brought about a truce and a new regime in the telephone industry.[40] Rather, the critical event was an overall settlement of many issues in a document known as the Kingsbury Commitment, which we will look at later. The event that virtually forced the Department of Justice to act against AT&T was the company's acquisition of a substantial portion of Western Union shares. To some of the new government professionals, it justified breaking up or nationalizing AT&T. But as we will see, the acquisition had a sound technological basis.

The Western Union Acquisition

Why did AT&T acquire control of Western Union? Again, an examination of the episode shows that a convergence of incentives and fortuitous circumstances led to the event. The starting point in understanding the stock acquisition is a technology called the "phantom circuit," the commercial possibilities of which were realized in the 1902–1905 period. The phantom circuit had its origins in the problems of multiplexing—using the same wire (or other pathway) for several communication channels. As we saw, the search for a multiplex telegraph was one of the triggers that led to the invention of the telephone. But the interference problems of telegraphy were minor compared to those of telephony. Yet the obvious

[38] Ibid., p. 183.

[39] "To Sue Telephone Merger," *New York Times*, August 23, 1912; p. 1; and "Fighting Telephone Co.," *New York Times*, November 30, 1912, p. 15.

[40] See Original Petition, *United States* v. *AT&T*, Equity no. 6082 (D. Oreg., 1913) and 1 Decrees and Judgments in Civil Antitrust Cases 483 (1914).

economies of telephone multiplexing led AT&T to expend considerable amounts of effort on solving the various engineering problems. One of the first proposed solutions (initially devised in the 1880s by Frank Jacob) was to superimpose an additional circuit (C) on two circuits already in use (A and B). The third circuit was called the phantom circuit. In the simplest arrangement, the current from the third telephone was divided by a resistance bridge into two equal parts, each of which was transmitted over one wire of the pair used for one conversation. Other apparatus combined the components at both ends. The return path is akin to the transmission path (see Figure 6.1).[41]

Bell engineers worked intensively on the problem of making the phantom circuit economically and technically feasible. Not until 1902, however, was the first phantom circuit put into commercial operation between Lewiston, Maine, and Berlin, New Hampshire. The development of a new type of coil in 1903 greatly improved performance, but much remained to be done. Then, based on an 1895 report, AT&T unveiled a commercially viable system, with excellent acoustic performance. The new system was based on "quads"—twisted pairs of wires, twisted together in a cable. So successful was AT&T's research efforts that after 1911 all AT&T cable used in toll calls was of the quadded type. Progress on the phantom circuit in the early years of the twentieth century was

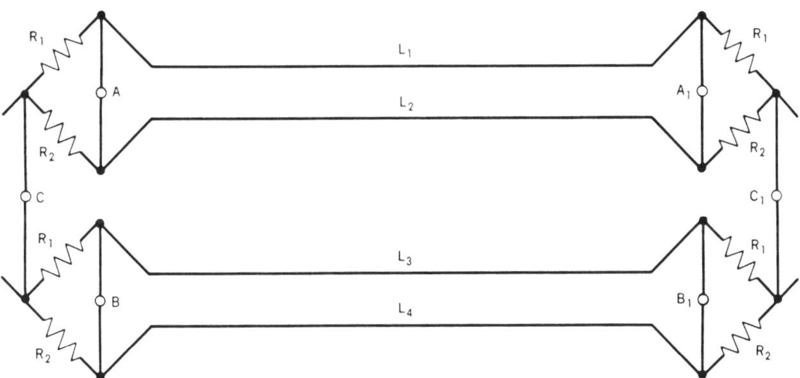

Figure 6.1. The Phantom Circuit. (Source: M. D. Fagen, ed., *A History of Science and Engineering in the Bell System: The Early Years (1875–1925)* [New York: Bell Telephone Laboratories, 1975], p. 237.)

[41] Details on the origin and early development of the phantom circuit are contained in M. D. Fagen, ed., *A History of Science and Engineering in the Bell System: The Early Years (1875–1925)* (New York: Bell Telephone Laboratories, 1975), pp. 236–40; and Frederick Leland Rhodes, *Beginnings of Telephony* (New York: Harper & Bros., 1929), chap. 13.

sufficiently rapid to encourage company officials to believe that commercial viability was only a matter of time—and a short time at that.

As work progressed, another important principle with commercial repercussions was recalled. As early as the 1880s, it had become clear that the great differences in frequencies between telegraph and telephone signals would allow the two types of signals to be transmitted at the same time if the two frequency bands could be kept separate. Such a "composite" circuit was developed in the 1880s.

Combining the composite circuit principles with the newly emerging phantom circuit principles had an important implication. The same sets of wires could be used to convey both telephonic and telegraphic information. To Vail and other farsighted executives of AT&T these technological developments suggested a way to more fully utilize the capacity of AT&T wires. The same toll wires could be used to convey telephone and telegraph messages. For this reason, AT&T's acquisition of a telegraph company made sound economic sense. A 1909 internal AT&T memorandum on acquiring Western Union stated: "The composite or phantom spare wire facilities of these telephone wires would provide for the needs of the Telegraph Company for increased traffic for probably many years."[42] The memorandum then proceeded to spell out other advantages to AT&T of acquiring Western Union. These included: (1) local telephone companies would be able to use Western Union wires; (2) they and AT&T could use Western Union poles for stringing additional wire; (3) Western Union held superior rights-of-way adjacent to railroad lines; (4) the Bell System could add many telephone pay stations at railroad stations since Western Union had an elaborate wire system next to many railroads; (5) the Bell System could add telegraph stations where only telephone stations were then located; and (6) the Bell System could use telephone pay stations to place and pay for telegraph orders "thus adding many points to the Telegraph system at which it is now impracticable or unprofitable to establish telegraph offices."[43]

The 1909 memorandum further pointed out that a telephone-telegraph merger could be a prelude to a large number of additional communications uses, not narrowly confined to telegraphy or voice-grade telephony. The prophetic memorandum contemplated using a single communications instrument and its wires for burglar and fire alarms, safe protection, time clock service, financial market news, and a general news ticker service "making duplication of wires underground and overhead unneces-

[42] Undated 1909 Memorandum, "Possible Advantages of a Telephone and Telegraph (WU) Combination," in AT&T Archives, Box 65, p. 1. See also FCC, *Control of AT&T*, appendix 7, sheet 31.

[43] "Possible Advantages," p. 1.

sary."[44] The great capacity offered by phantom circuit technology, especially as it further progressed, could be used to effect substantial cost savings in construction, purchases (especially wires), testing, repair, and maintenance. In short, the acquisition of Western Union, if it had been allowed to stand, might have effected substantial cost savings and prevented that company's dismal subsequent performance. Government would have recognized the coming convergence of communications technologies and its economic justifications. Instead, AT&T divested its Western Union stock.

Western Union, like other communications companies, had owned stock before 1909 in several companies that were AT&T licensees. But this did not imply a close relationship between the companies. More important, the Mackay Companies, which were the largest shareholders of AT&T stock, controlled the Postal Telegraph Company, Western Union's principal competitor. Nevertheless, Postal Telegraph did not control AT&T, as we have seen, and as Clarence Mackay, to his regret, admitted in a November 27, 1909, letter to Vail.[45] While Vail had earlier considered acquiring Postal Telegraph for the reasons discussed above, a unique opportunity was afforded to acquire a controlling interest in Western Union. AT&T had decided to finance New York Telephone's consolidation plan in 1909 in part by selling additional stock to existing shareholders at 140. Western Union, which owned approximately one-third of New York Telephone stock, had fallen on hard times and could not raise sufficient cash. Indeed, it desperately sought cash. Under an ensuing agreement, Western Union disposed of its New York Telephone shares to AT&T. During this same period AT&T purchased approximately 30 percent of Western Union's shares from its president, George Gould, and others. Western Union, for its part, received not only a badly needed flow of working capital, but also capable management and the advantages of close integration with AT&T.[46]

After the Western Union acquisition the Mackay Companies divested themselves of AT&T stock, and Vail assumed charge of Western Union's affairs, instituting such innovations as lower-cost day letters and night letters to utilize the capacity of the telegraph system at off-peak hours, when it would otherwise be underutilized. Nevertheless, the long-run benefits of the relationship were not to take place. The size of the marriage provided considerable ammunition to independents hostile to the Bell System; indeed, it was their golden opportunity to launch a broadside against the "telephone trust." Moreover, in the process of acquiring

[44] Ibid., p.2.

[45] Reprinted in FCC, *Control of AT&T*, p. 47.

[46] Stehman, *Financial History*, pp. 147–54.

Western Union stock, AT&T had made a new enemy in the powerful Mackay interests. Fearful of the adverse impact that the AT&T–Western Union relationship would have on their Postal Telegraph system, they had a great interest in dissolving the alliance. Indeed, Postal Telegraph brought an antitrust action against a New Mexico Bell licensee in 1910.[47] Thus, the Western Union acquisition was the incident that forced drastic public action.

While state involvement in telecommunications regulation was increasing during this period, so also was federal involvement. This was occurring at several different levels—in the federal courts through private antitrust action, in Department of Justice investigations and proceedings, and in the enactment of the first federal regulatory statute covering telecommunications. All of this activity would lead to the Kingsbury Commitment. Although it was later drastically modified, the commitment would usher in a period of peace and stability in telephony. But first it is necessary to examine the important role of the ICC in shaping telephone policy.

The Expansion of Federal Intervention: The ICC

In 1887 Congress enacted the Interstate Commerce Act, the first *national* economic regulatory statute. That statute created the Interstate Commerce Commission, whose job it was to regulate many of the railroad industry's practices. As we have seen, under public service liberalism the characterization of an industry as a public service did not necessarily mean that the *structure* of such an industry should be monopolistic or competitive. Thus, the ICC in a report on consolidations stated that it was unlawful for one road to acquire control of parallel and competing lines; competition between them is the policy of the nation.[48] Similarly, the Department of Justice successfully prosecuted several railroad combinations as restraints of trade under the Sherman Act.[49]

When telephones were first subject to federal regulatory commission jurisdiction (by the ICC) under the 1910 Mann-Elkins Act, the intention was not to declare the telephone industry a "natural monopoly" exempt from antitrust considerations. Although state regulation was moving more rapidly in that direction, federal policymakers lagged. Some held

[47] "Telephone Trust to Dissolve, Giving up Western Union Control," *New York Times*, December 20, 1913; p. 1.

[48] *In the Matter of Consolidations and Combinations of Carriers*, 12 ICC 277, 305 (1907).

[49] Among the important examples are *Northern Securities Co.* v. *United States*, 193 U.S. 197 (1904), and *United States* v. *Reading Co.*, 226 U.S. 324 (1912).

to the natural monopoly theory, others subscribed to competition. Forces in the Post Office and elsewhere argued for public ownership and in the process bolstered the natural monopoly argument. All of this, however, took place in a context in which federal jurisdiction over telephones under the commerce clause—article 1, section 8, of the Constitution—was sharply limited because it applied only to "the transportation of electric fluids" across state lines and very few such phone calls were made.[50] Whereas the states therefore had jurisdiction over most telephone matters, federal jurisdiction could be asserted over AT&T's interstate (but not intrastate) long distance service.

ICC regulation over telephones came about almost as an afterthought. The primary focus of the Mann-Elkins Act was on railroads. The bill was drafted by the attorney general to fulfill President Taft's campaign pledge to further regulate the railroads. Taft primarily sought to exempt traffic agreements from the antitrust laws, but Congress rejected this proposal and, under the influence of many Democrats and insurgent Republicans, prohibited carriers (unless the ICC otherwise authorized) from charging more for the aggregate of short hauls than for the equivalent long haul. The new law also gave the ICC authority to suspend rate changes during an investigation of their reasonableness.[51] These hotly debated issues were primarily on the minds of policymakers.

Federal telephone regulation had been considered prior to the Mann-Elkins floor debates. For example, House and Senate Committees had considered in 1908 bills to prevent discrimination in telephone and telegraph rates.[52] The telephone amendments to the Mann-Elkins bill were the first time that the issue reached the House and Senate floors. None of the speakers doubted that the telephone industry could be so regulated, since it was a public service industry. Although Georgia Representative Bartlett, who offered the House amendment, described telephone and telegraph companies as "monopolies," the language was used loosely and, perhaps, in an inflammatory manner. Rather, the key point was that these companies should be required to make "reasonable" charges, and that the ICC should be allowed to judge reasonableness.[53] Since federal regulation (unlike state regulation) would be directed almost entirely against AT&T because it was one of the very few telephone companies in interstate commerce, the National Independent Telephone Association was on record favoring the amendment, claiming (without any evidence submitted) that AT&T discriminated in rates.[54]

[50] See 45 *Congressional Record* 5537, 7265 (1910).

[51] I. L. Sharfman, *The Interstate Commerce Commission* (New York: Commonwealth Fund, 1931), pt. 1, pp. 52, 53.

[52] Stipulation/Contention Package, episode 3, par. 92, *United States* v. *AT&T*.

[53] 45 *Congressional Record* 5533 (1910).

[54] Ibid., 6973, 6974.

The principal point that divided the proponents and opponents of the amendment was whether a floor amendment was the appropriate way to bring under regulation a method of communication uniformly conceded by legislators to be vital to the commerce of the nation. Opponents argued that there were such substantial differences between transportation and communications that full hearings and a bill specifically adopted to communications should be carefully prepared. Others argued that the ICC was already burdened sufficiently with its transportation regulation duties. Therefore, they said, placing a second major area of responsibility in the lap of the ICC would almost be an invitation for the agency to treat communications matters in a cursory manner. Still other legislators objected to the legislative draftsmanship, claiming that the telephone language was ambiguous. Nevertheless, the most striking aspect of the floor debate was what was not said. No one suggested that telephone companies should not be subject to federal regulation.[55] Interstate telephony was uniformly viewed as too important a public service to further evade public regulation.

Different versions of the telephone amendment (and much else in Mann-Elkins) were passed in the House and Senate. While the Conference Committee was considering its task of reconciling the two versions, Attorney General Wickersham suggested that the committee ought to include a clause giving the ICC authority to compel interconnection between companies—one of the key independent demands. Rejecting the suggestion, the Conference Committee produced a version that was acceptable to both the House and the Senate, which became law. The Mann-Elkins Act required telephone and telegraph rates to be "just and reasonable" and nondiscriminatory. It required the ICC to declare transgressing rates unlawful and to prescribe new ones. The act was silent on the subject of telephone industry structure. Consequently, the Department of Justice could bring antitrust actions against telephone companies, just as it could (and did) against railroads. Thus, at this stage, from AT&T's perspective, much remained to be settled.

The Department of Justice Acts

The strident complaints of some of the independents to the Department of Justice, as well as AT&T's Western Union stock acquisition, virtually compelled the attorney general to do something—even if it was the time-honored bureaucratic response of conducting an investigation or a study. Other than tirades against AT&T's comparatively greater size, the inde-

[55] See ibid., 5533–37; 6972–77, 7264–7666.

pendents' complaints centered on the advantage that the Bell System had in its larger, more elaborate network. Some letters complained about particular acquisitions, accusing AT&T of seeking to gobble up every independent telephone company. In fact, internal Bell documents indicate that the operating companies were quite circumspect—indeed wary—in their acquisition policies, fearful of acquiring what is currently termed a "turkey." Purchases were made of weaker companies, for example, only on a showing that they offered a good alternative switching route for toll service.[56] Although, prior to 1913, the independent complaints were used by some Justice Department officials to urge state attorneys general to file antitrust actions against AT&T, they undertook no formal action.[57]

Nevertheless, it is clear that in late 1912 Attorney General Wickersham was still considering how to treat the situation. Thus, on August 29 he sent a letter to AT&T Vice President Kingsbury concluding that the Michigan Railroad Commission's approval of several Bell acquisitions "seems to provide very carefully for the public interest." Nevertheless, Wickersham continued somewhat elusively, the Department of Justice was engaged in a "wider investigation of questions concerning the effect of the Sherman Law on the telephone situation." Thus, the department's response to the Michigan acquisitions was not to be taken as "binding."[58] In other words, as Wickersham confided in an August 29 memorandum to the antitrust assistant attorney general, he had not yet made up his mind. Notably, however, based on earlier conferences with Kingsbury, he was "inclined to think that AT&T is prepared to meet the Government frankly and to abandon any contracts or situation which we conclude to be in violation of the Sherman Law."[59] That view was based not only on the conversation with Kingsbury, but also on the fact that Vail had even earlier (August 6) sent a letter to the Bell associated companies suggesting:

> On Account of present conditions it seems wise that for the present all acquisition of independent telephone properties should be suspended.
>
> We advise and request, therefore, that all negotiations now pending should be delayed and that no new negotiations should be commenced. Any completed purchases, mergers or sales should be suspended as far as contractual conditions will permit.[60]

[56] See, particularly, the report on the acquisition of the Delaware Telephone Company in C. E. Yost to N. C. Kingsbury, 2 June 1913, in AT&T Archives, Box 21.

[57] Stipulation/Contention Package, episode 3, par. 112, *United States* v. *AT&T*.

[58] Quoted in ibid., par. 113.

[59] Ibid., par. 114.

[60] Quoted in FCC, *Control of Independent Telephone Companies*, p. 31. See also N. C. Kingsbury to C. E. Yost, 9 June 1913, in AT&T Archives, Box 21.

At the beginning of 1913 Attorney General Wickersham was prepared to move, concluding that certain *proposed* midwestern acquisitions by AT&T would probably violate the Sherman Act. These proposed acquisitions constituted two networks that crossed state lines. While those states contained the largest number of competitors of Central Union, the principal AT&T licensee, it is pertinent to evaluate just how formidable this competition was—especially in view of AT&T's conciliatory attitude toward the Justice Department. In 1913, AT&T virtually conceded all that the Justice Department wanted. The most plausible explanation of this conciliatory behavior is that AT&T and its licensees had little to fear from the independents in 1912 and 1913. As noted earlier, the independents often failed to maintain sound cost accounting procedures (especially with respect to depreciation and other allowances), and that failure ultimately caught up with them in the forms of bankruptcy, very low profit rates, or requests to public authorities for substantial rate increases. In 1907, when Vail resumed office, there were 2,280,000 independent telephones *not* connected into the Bell System. By 1912 there were 1,430,000.[61] The possibility of independents launching a competing toll network was greatly diminished.

Although data on long defunct companies is impossible to obtain, one detailed 1911 article on midwestern competition observed that independents in that region had been steadily losing ground to Central Union and that independent interconnection into the Bell System increased their profitability and valuation. Further, many independents had gone bankrupt or were in dire financial straits. Focusing on one of the midwestern groups that Bell interests were asked not to acquire, the article reported that it was in J. P. Morgan & Company's hands. The article reported that Interstate Telephone & Telegraph—Central Union's most formidable Illinois competitor—had gone into receivership with liabilities far in excess of assets. At the same time that the independents were experiencing great difficulties, Central Union was growing stronger.[62]

AT&T, in short, decisively held the upper hand. Since the Bell network was expanding and faced no substantial toll competition, the act of taking over independent companies was an act of conciliation more than an act of monopolization. Thus, when AT&T agreed not to acquire further independent companies, it was not giving up very much. Moreover, even as the attorney general accused AT&T of probable Sherman Act violations, he doubted the wisdom of competition in the telephone industry to achieve public service goals. John H. Wright, manager of an

[61] Bureau of the Census, *Historical Statistics of the United States* (Washington, D.C.: Government Printing Office, 1975), pt. 2, p. 783.

[62] J. N. Kins, "Some Consequences of Telephone Competition," *National Magazine* (February 1911): 563–70.

independent telephone company, confided that Wickersham told him that competition was an inappropriate policy solution in the telephone industry. A regulated monopoly or public ownership, according to Wright's recollection of the conversation, was a much better solution.[63]

In that respect the attorney general's views were consistent with the ascendant views of state regulatory authorities toward competition. Although competition was acceptable in some situations (such as inferior service by an existing franchisee), every state that considered the issue opted for monopoly as the natural, appropriate state of affairs in public utilities. Although decided a few years after the events discussed here, the California Railroad Commission's views typifies those prevailing in 1913 at the state level: "The subscribers almost unanimously demand a consolidation into one system. . . . The telephone being a natural monopoly, there should be one universal service as this will enable complete interchange of communications between all telephone users in the community; this in addition to the usual advantages of consolidation of utility properties and duplicate operating expenses."[64] Or as the Missouri Public Service Commission succinctly put it, "Competition between public service corporations was in vogue for many years as the proper method of securing the best results. . . . The consensus of modern opinion, however, is that competition has failed to bring the result desired."[65] According to the Nebraska commission, telephone service is less efficient when competitive.[66]

Notwithstanding the innovative regulatory developments occurring at the state level, Wickersham, in accordance with his duty to enforce the Sherman law, decided to act. On January 7, 1913, he transmitted a letter to the ICC asking it to initiate an investigation of the telephone industry's rates and practices. Wickersham suggested that many of the issues raised would be more more effectively dealt with by regulation than by antitrust action. Wickersham stated that there were four major issues: (1) Bell acquisitions, (2) the alleged refusal by Bell companies to interconnect with independents, (3) rate increases by Bell companies after acquiring independents, and (4) the diversion of telegraph messages by telephone companies associated with AT&T from Postal Telegraph to Western Union contrary to senders' wishes.[67] During this time, the assistant attorney general for antitrust, clamoring for action, urged Wickersham that the ICC was incapable of correcting "the evils which are *likely* to be devel-

[63] Stipulation/Contention Package, episode 3, par. 116, *United States* v. *AT&T*.

[64] *Pacific Tel. & Tel. Co.*, 15 Cal. R.C.R. 993, 994 (1918).

[65] *Johnson County Home Teleph. Co.*, 8 Mo. P.S.C.R. 637,643–44 (1919).

[66] *Washington & W. Teleph. Co.*, 15 Ann. Rep. Neb. 5, R.C. 72 (1922).

[67] Stipulation/Contention Package, episode 3, pars. 117, 118, 119, *United States* v. *AT&T*; and FCC, *Control of Independent Telephone Companies*, pp. 32, 33.

oped" (emphasis supplied).[68] Assistant Attorney General Fowler had apparently already made up his mind that the claims of the complaining independents and the Mackay group were correct.

Woodrow Wilson's election to the presidency in 1912 led to the appointment of James C. McReynolds (later a Supreme Court justice) as the new attorney general on March 5, 1913. Fowler, whose division had been investigating telephone conditions in the Northwest, discovered that Pacific Telephone & Telegraph (PT&T), AT&T's West Coast licensee, had acquired the Northwestern Long Distance Company, which had competed with AT&T in long distance service from Corvallis, Oregon, through Portland and Seattle to Port Angeles, Washington. Northwestern's toll service was modest in comparison to AT&T's, even in the Pacific Northwest. Fowler, urging action on the newly appointed attorney general, conceded: "It, of course, is not an absolute certainty that the Northwestern Telephone Company would have been a success, but it certainly was not given by the Bell Company half a chance to demonstrate whether it could succeed or not, and its early downfall . . . is directly due to the intermeddling of the Bell interests."[69] At the time that Fowler suggested that Northwestern could be a "success," it was insolvent. Fowler recommended an action seeking structural relief that would set up a reorganized Northwestern Company independent of the Bell interests.

Fearing that McReynolds might share his predecessor's preference for regulation in the telephone industry, Fowler argued that it was the department's job to enforce the antitrust laws. "I do not think it is for this Department to pass upon the economic question whether one or more telephone systems are preferable in a city or throughout the country" if an antitrust violation is found.[70] In Fowler's view the consolidation of two competing lines under the circumstances of the case was a clear violation of the Sherman Act.

On July 24, 1913, the Department of Justice filed a suit based on the Northwestern situation, charging that AT&T and PT&T had conspired to monopolize trade in telephonic transmission in four western states. The case, which was ultimately settled because of the Kingsbury Commitment, was terminated by consent decree in 1914. Under it the Bell System agreed to divest itself of Northwestern, an independent company in Spokane, and other properties. The Bell System also agreed to provide

[68] Quoted in Stipulation/Contention Package, episode 3, par. 120, *United States* v. *AT&T*.

[69] Quoted in ibid., par. 121. See also Geoffrey M. Peters, "Is the Third Time the Charm?" *Seton Hall Law Review* 15 (1985): 253–55.

[70] Quoted in Stipulation/Contention Package, episode 3, par. 122, *United States* v. *AT&T*.

long distance access to Northwestern and other independents and to provide local customers with equal access to whatever long distance carrier (AT&T, Northwestern, or any other) that the customer chose.[71]

The agreement had little practical impact. The citizens of Spokane voted in a 1914 referendum to allow the local phone company to be consolidated into the Bell System. More important, in 1922 Northwestern joined PT&T in applying to the ICC for a certificate allowing PT&T to acquire the independent toll company. Noting that Northwestern was unable to attract sufficient business, that it was mired in debt, and that PT&T had sufficient toll lines and other equipment to handle all business for the foreseeable future, the ICC approved the acquisition. The decision stated, in part: "It is apparent that no useful purpose will be served by the continuance of the present duplicated service, which can only result in placing an added burden on the telephone using public in an amount equivalent to the expense required to maintain the Northwestern's separate organization and plant."[72]

The Kingsbury Commitment

Uncertainties piled up because of the policy dissonance between states (favoring the regulated network manager system) and the federal government. The federal government was divided into different segments, some favoring public service liberalism, while others were favoring such modern liberal proposals as planning through intensive regulation, restructuring into smaller units, and even nationalization. AT&T desperately sought clarification and a single policy. Nathan Kingsbury, the AT&T official most concerned with these issues, contacted Attorney General McReynolds almost immediately after McReynolds was installed in office. The attorney general at first told Kingsbury that he was disinclined to offer advice on whether to make acquisitions and whether these would be considered unlawful. The Justice Department had neither the time, ability, nor inclination to "run" AT&T's business. Kingsbury responded: "The Government had already, in a sense, entered into partnership with business, and that over 30 of the states had done likewise [and that the federal government] had established commissions, especially the Interstate Commerce Commission, for the very purpose of advising, regulat-

[71] Ibid., par. 132, and Peters, "The Charm," p. 255.

[72] *Acquisition of Control of N.W. Long Distance Tel. Co.*, 71 ICC 530, 531–32 (1922); and Peters, "The Charm," p. 255.

ing, and supervising the business of certain public service corporations."[73]

Kingsbury went on to underscore AT&T's commitment to universal service so that every person in the United States would be able to converse with everyone else. Kingsbury concluded by laying the groundwork for the subsequent agreement. It would never be necessary to bring an action against AT&T, he asserted, because if the department concluded that there was evidence of wrongdoing, AT&T would correct the practice complained of. McReynolds agreed to inform AT&T of alleged transgressions of the Sherman Act and allow the company to correct them.

McReynolds contacted independent companies and others to ascertain their precise complaints and held a series of meetings with Kingsbury in the autumn of 1913 to discuss the complaints. In the course of the discussions with Kingsbury, McReynolds made a number of suggestions that were intended to restore stability and peace to the telephone and telegraph industries. By letter dated December 16, 1913, Kingsbury advised McReynolds that AT&T would comply with the Department of Justice's recommendations. The Kingsbury Commitment was embodied in an AT&T letter on December 19 to the attorney general setting out the agreement finally reached. Although the commitment was modified in important respects by government wartime control that began on August 1, 1918, and was further modified in the early 1920s, it remains important for the illustration of AT&T's acceptance of public service liberalism and its triumph in this industry. It led to termination of the ICC investigation and settlement of the Oregon suit. The commitment was enthusiastically endorsed by President Wilson as an example of business-government cooperation.[74]

Under the commitment AT&T agreed to divest all of the Western Union stock it held so that Western Union control and management would be entirely independent of AT&T's. Second, AT&T and the Bell companies agreed not to acquire in the future any competing telephone operating companies or any companies that *may* operate in competition with any part of the Bell System. The commitment was silent on the acquisition of noncompeting companies so that unless barred by state law or a regulatory ruling, Bell System companies could acquire such companies. Third, in the case of companies that AT&T and the Bell System

[73] Quoted from Kingsbury's memorandum on an April 8, 1913 conversation in FCC, *Control of Independent Telephone Companies*, p. 35.

[74] AT&T's letter to the attorney general, his reply, and president Wilson's letter to the attorney general are bound together in a paginated government document referred to as *Letter to the Attorney General* (Washington, D.C.: Government Printing Office, 1914). President Wilson's letter is on p. 8.

had already acquired but had not yet integrated into the Bell System, "The question as to the course to be pursued in such cases will be submitted to your Department and to the Interstate Commerce Commission for such advice and directions."[75] Thus, AT&T would not be challenged on acquisitions of companies that had become integrated into the Bell System. Finally, one should add that there were no restrictions placed on any independent's acquisitions of other telephone companies.

The acquisition sections of the Kingsbury Commitment should not be construed as an endorsement of free and open competition. The telephone business was still overwhelmingly local and thus beyond the federal government's jurisdictional reach. McReynolds explicitly refused to endorse competition in local markets. In his 1914 report to Congress he stated that the government's theory in the AT&T matter "does not mean that where there are two telephone systems in a city or town there can never be a consolidation into a single system. . . . This interpretation leaves local communities generally free to have one telephone system."[76] Although local monopoly and long distance competition can be reconciled, the fact was, in the succinct phrase of AT&T's 1913 Annual Report, that "the general tendency in this country is to the 'one system idea of public utilities under regulation.' "[77] Clearly the states were moving in that direction, and in any event the anticombination provisions of the Kingsbury Commitment proved unworkable and were abandoned in the 1921 Willis-Graham Act.

The most lasting segment of the Kingsbury Commitment concerned interconnection. As noted earlier, Vail's strategy was to furnish interconnection to noncompeting independents. In that respect it was an important element of the regulated network manager system. Nevertheless, one of the principal independent complaints that led to the antitrust action was the claim that AT&T was not furnishing interconnection to competing independents. As the *New York Times* reported, "Independent telephone interests, finding it arduous to compete with local Bell companies because the trunk lines of the combination were closed to them, complained bitterly."[78] As we know, they did more than complain to the Department of Justice, often bringing suits themselves or exerting pressure in state legislatures to enact compulsory interconnection laws.

[75] Ibid., p. 4.

[76] Quoted in Defendant's Third Statement of Contentions and Proof, vol. 1, p. 166, *United States* v. *AT&T*. See also Stipulation/Contention Package, episode 3, par. 135, *United States* v. *AT&T*.

[77] AT&T, Annual Report, reprinted in *Commercial & Financial Chronicle* 98 (March 21, 1914): 30.

[78] "Telephone Trust to Dissolve Giving up Western Union Control," *New York Times*, December 20, 1913, p. 2.

Courts tended not to compel interconnection unless a statute explicitly required it. The courts generally upheld the right of the Bell companies not to interconnect on either freedom of contract grounds or on the theory that otherwise a local telephone company would artificially enhance its value and enlarge its network without incurring investment by demanding interconnection with every telephone system in the country.[79] The public utility rule requiring that a public service company serve all customers without discrimination did not apply to serving competitors.[80]

Rebuffed in most courts, the independents went to state legislatures and there met with considerable success because they were able to persuade legislators that public service goals would be advanced through mandatory interconnection.[81] By 1913 twenty-seven states had enacted laws governing interconnection between companies. Most of the statutes required the state public utility commission to find that the public convenience and necessity would be served before ordering interconnection. Although a company could avoid interconnection by showing that it would degrade service, cause irreparable damage to one of the companies, or was technically impracticable, the trend was clear. As Wisconsin illustrates, in the period when states were enacting interconnection laws that obviously pleased the independents, they were also enacting antiduplication laws for new systems where an existing one could reasonably serve a community. Obviously, this aspect of the law would please AT&T.[82] The underlying rationale states used in both situations was that telephone service would be made more valuable to the subscriber by these rules. Under the interconnection laws subscribers could talk to more people, and under the antiduplication laws telephone service would be more efficient by avoiding waste.

AT&T, seeing the handwriting on the wall, agreed in the Kingsbury Commitment to further liberalize its already liberal interconnection policies. At the same time the interconnection sections of the commitment were another step in the genesis of AT&T as network manager. AT&T promised to promptly make arrangements "under which all other telephone companies may secure for their subscribers toll service over the

[79] See, for example, *Home Telephone Co.* v. *Sarcoxie Light & Telephone Co.*, 139 S.W. 108 (Mo., 1911); and *Memphis Tel. Co.* v. *Cumberland Tel. & Telegraph Co.*, 231 F. 835 (6th Cir., 1916).

[80] *Pacific Telephone & Telegraph Co.* v. *Anderson*, 196 F. 699 (E.D. Wash., N.D., 1912); and *State* v. *Cadwallader*, 87 N.E. 644 (S.Ct. Ind., 1909).

[81] An exception to the general trend of judicial decisions on the subject is described in "Boon to Rivals in Telephone Field," *New York Times*, July 19, 1907, p. 6. The article makes eminently clear that the issue was a battle between AT&T and the independents.

[82] Stipulation/Contention Package, episode 3, par. 130, *United States* v. *AT&T; and Fred H. Holmes, Regulation of Railroads and Public Utilities in Wisconsin* (New York: D. Appleton, 1915), pp. 199–205.

lines of the companies in the Bell System" subject to several conditions that were reasonable.[83] The employees of the Bell System were to make the toll line connections and operate and maintain the connections. Bell System lines were to be used for the entire connection between points unless the Bell System had no such lines, in which case other arrangements could be made. Thus, by this concession AT&T integrated local independents into the network, while at the same time largely foreclosing the development of alternative long distance systems (except in the Pacific Northwest, for reasons already discussed). The interconnection provisions of the Kingsbury Commitment, as Gerald Brock has observed, linked the independents' interests with AT&T's because the extensiveness and quality of the Bell lines enhanced the communications and use of independent subscribers.[84]

Peace had been declared and at very little price to AT&T because the company was already on record—in its 1912 Annual Report, for example—favoring liberal interconnection. More important than the company's pronouncements, the trends on interconnection were not dramatically different in the few years before. The trend toward more interconnection simply continued after the Kingsbury Commitment.[85] The commitment also proclaimed AT&T's civic spiritedness and, thus, helped to head off nationalization, an approach favored by some of the new professionals moving toward modern liberalism.

The Specter of Public Ownership

Before the establishment of Soviet power in 1917, the combination of economic inefficiency and totalitarianism that public ownership seems to bring in every regime where it prevails was not known. Accordingly, one found some respectable advocates of modern liberalism calling for public ownership of public utilities. Segments of the business community joined in the call. Based on the tide of nationalizing telephone companies in other countries, from AT&T's perspective at the time the threat of public ownership was a serious one.[86] And the threat in the United States occurred at two levels—the municipal and the national. Municipal owner-

[83] *Letter to the Attorney General*, p. 4.

[84] Gerald W. Brock, *The Telecommunications Industry* (Cambridge, Mass.: Harvard University Press, 1981), p. 155.

[85] See AT&T 1911 Annual Report (condensed), reprinted in *Commercial & Financial Chronicle* 94, (March 23, 1912): 851.

[86] On the trend toward public ownership of telephone systems, see A. N. Holcombe, *Public Ownership of Telephones on the Continent of Europe* (Boston: Houghton Mifflin, 1911), passim.

ship, unlike nationalization, was a threat within the traditional context of public service liberalism.

Municipal ownership of gas, electric, and water utilities had become a reality in many cities. The basic arguments for public ownership in these cases were that adequate provision of such services to all who needed them was inherently unprofitable and that municipal ownership provided a closer form of supervision over utilities than public regulation of private companies.[87] For this reason, during a period in which municipal regulation was often a prelude to municipal ownership, AT&T strongly preferred state regulation at the intrastate level to municipal regulation.[88] For example, in Los Angeles (which regulated telephones) a municipal ownership league sprang up in 1916 to present an alternative to the merger of the two competing private companies.[89] In contrast, state regulation, as exemplified in Wisconsin, the archetypal progressive state, had moved in the direction of privately owned local monopolies. In 1913 the Wisconsin Railroad Commission announced that the antiduplication law would be strictly enforced with exceptions permitted only in unusual circumstances.[90] States regulated through public utility commissions that were removed from partisan politics.[91]

Cities, apprehensive of their loss of control over franchising, joined some telephone independents, fearful of the antiduplication trend in state regulation, in opposing state public utility commission regulation.[92] The problem that municipal control could not solve was the regulation of rates and service requirements for calls between cities and towns in the same state. In a state such as California, with numerous cities and towns clustered around major centers like Los Angeles and San Francisco, ef-

[87] See Clyde Lyndon King, *The Regulation of Municipal Utilities* (New York: D. Appleton, 1914), chap. 2.

[88] See, for example, D. C. Carroll to E. C. Bradley (vice-president of PT&T), 24 June 1911, in AT&T Archives, State Regulatory Files (California). See also Kenneth Lipartito, *The Bell System and Regional Business* (Baltimore: Johns Hopkins University Press, 1989), p. 185.

[89] *Los Angeles Record*, October 19, 1916, P.1, found in AT&T Archives, State Regulatory Files (California).

[90] *Proposed Extension Clinton Tel. Co.'s Line*, 13 Wis. R.C.R. (1913). See, more generally, *La Crosse Gas & Elec. Co.*, 2 Wis. R.C.R. 3, 5 (1907). "Active and continuous competition between public utility corporations furnishing the same service to the same locality seems to be out of the question. This has been shown by experience."

[91] See Ann Merkovitz, "Early Bell History in Illinois 1877–1920 Period of Municipal Regulation" (August 1978), pp. 32–42, in AT&T Archives, State Regulatory Files (Illinois).

[92] On independent opposition see "Assembly Defeats Two Hughes Bills," *New York Times*, April 9, 1909, p. 2; and "Charter Set Back By Political Plot," *The New York Times*, April 17, 1909. On the opposition of cities in California, especially Los Angeles, to state regulation, see various materials in AT&T Archives, State Regulatory Files on the 1911 public utility statute.

fective municipal regulation would have been extremely cumbersome. And municipal ownership of telephone companies regulated by state authority would have solved nothing. Ultimately, all states abandoned municipal telephone control and adopted state regulation as the best structure to supervise the telephone business and attain public service goals.

At the national level public ownership was one variant of modern liberalism's program. The major advocates of nationalization were located in the Post Office Department. They were aided by several legislators, the most important of whom was Maryland Representative David J. Lewis. In October 1913 the threat loomed as more menacing. The *New York Times* reported: "Notwithstanding efforts at profound secrecy, it has become known here that the Wilson Administration is engaged in preparing the groundwork for nationalization of the nation's telegraph and telephone lines."[93] Advocates of the idea analogized the parcel post system for carrying packages to the telephone and telegraph. The argument was that the parcel post was effectively competing with private express companies.

President Wilson was, in fact, far from persuaded that the nationalization proposed by Lewis and the postmaster general was the correct course. Not the least problem that nationalization advocates faced was how to pay for the system, which they estimated at more than $900 million, a sum that many viewed as extremely conservative. Clarence Mackay, president of the Mackay Company, asserted that the cost would be about $2 billion. Moreover, the exuberant Mackay voiced a theme that has been heard many times since: "And as to service—Government service would be a joke as compared with present service. If you don't believe it just try the Government service—telegraph and telephone—in Europe."[94] Apparently, Representative Lewis's claims that the Post Office Department was a far more efficient organization than AT&T was met with some skepticism by President Wilson, for the president pointedly congratulated AT&T and the attorney general for demonstrating the benefits of government-business cooperation in the Kingsbury Commitment.[95]

Whereas Lewis, angered by the Kingsbury Commitment, promised to continue leading the fight to nationalize (or postalize) the telephone industry, Postmaster General Burleson appreciated that the fight was, for

[93] "Federal Wires New Wilson Plan," *New York Times*, October 2, 1913: p. 1. See also Stipulation/Contention Package, episode 3, par. 146, *United States* v. *AT&T*.

[94] Quoted in "C. H. Mackay Derides Federal Ownership," *New York Times*, December 19, 1913, p. 17. See also "Wilson Gets Facts on Wire Control," *New York Times*, October 3, 1913, p. 2.

[95] "Lewis Opens Fight for U.S. Telephones," *New York Times*, December 23, 1913, p. 5; and "Telephone Trust to Dissolve," *New York Times*, December 20, 1913, pp. 1, 2.

a while, lost because Wilson was pleased with the Kingsbury settlement.[96] Vail turned the whole debate to his advantage, for when the postmaster general's report on government ownership was released in early 1914, it repeated an argument (although in the context of public ownership) that was dear to AT&T—that telephony was a natural monopoly.[97] Vail conceded the natural monopoly part of the argument, but he obviously won on the issue of government ownership "because we know that no government owned telephone system in the world is giving as cheap and efficient service as the American public is getting from all its telephone companies."[98]

World War I, as historian Ellis W. Hawley observed, thrust many new professionals into positions of power. Older mechanisms of resource mobilization were insufficiently slow to meet the sudden wartime exigencies. Planning under the coordinating control of these professionals was the order of the day. In 1918 a superagency, the War Industries Board, was placed in charge of this effort for the remainder of the emergency, with such powers as price setting and rationing. Modern liberalism received its first—albeit temporary—tryout during World War I, not because of the efforts of a cabal, but rather because of the wartime exigencies.[99]

Ironically, wartime government control of the telephone and telegraph systems from August 1918 to August 1919 would prove a great boon to AT&T. Government control, imposed because of American entry into World War I and the perceived need to integrate communications into the war effort, placed telephone properties under the control of Postmaster General Burleson. Although an operating board aided Burleson, actual operation of the telephone system was still conducted by private management. Nevertheless, the period had some short- and long-run results. In the short run, the postmaster general increased both long distance and local rates. At the state level, applications for increases were filed by company attorneys under the guise of acting on behalf of the Post Office.[100] In the long run, the postmaster general eroded the Kingsbury Commitment by announcing on August 17, 1918, that government op-

[96] "Burleson Won't Press It," *New York Times*, December 23, 1913, p. 5; and Horace M. Coon, *American Tel & Tel* (New York: Longman, Greens, 1939), pp. 136–44.

[97] Postmaster General, *Government Ownership of Electrical Means of Communication* (Washington, D.C.: Government Printing Office, 1914), pp. 10, 11.

[98] AT&T, 1913 Annual Report, p. 29.

[99] Robert D. Cuff, *The War Industries Board* (Baltimore: Johns Hopkins University Press, 1973), pp. 9, 150; Ellis W. Hawley, *The Great War and the Search for a Modern Order* (New York: St. Martin's, 1979), pp. 22–26; Robert H. Wiebe, *The Search for Order* (New York: Hill and Wang, 1979), pp. 23–26.

[100] FCC, *Control of AT&T*, pp. 82–102.

eration "will undoubtedly cause the coordination and consolidation of competing systems whenever possible."[101]

Despite the fact that the Bell System sold more telephone stations than they acquired in 1918 and 1919, the number of transactions was very high. The agreements reached frequently involved competing systems, not the noncompeting systems that were not barred by the Kingsbury Commitment. Thus, in 1918 telephone competition ended in western New York as Bell acquired the independent system in Buffalo and gave up its operations in Rochester and Jamestown. In that deal the Bell System acquired approximately thirty-two thousand independent telephones and gave up about thirty-five thousand, but there was a more important consideration than the number of telephones swapped. AT&T was de facto becoming the uncontested network manager of a unitary telephone system. And as the number of waivers of the terms of the Kingsbury Commitment increased, it became eminently clear that the agreement was not working in important respects.[102]

AT&T's Triumph

In late 1919 the Wilson administration was drawing to a close, soon to be replaced by twelve years of Republican rule under three presidents. Although many issues during the 1920s were divisive and controversial, there was surprising unanimity on the telephone. Republicans and Democrats, the Bell System and independents, reached a consensus that public service liberalism was best served by the regulated network manager system based on partnerships between Bell and independents, as well as between regulatory commissions and telephone companies. It was considered the regime that would best achieve efficient and universal service. In short, virtually every major actor had come to agree with Vail's 1908 formulation "One policy, one system, universal service." Accordingly, when Representative John A. Moon asked Burleson for legislative recommendations after the termination of government control, the postmaster general called for a new law that would recognize the ill effects of telephone competition and allow any telephone merger on consolidation to take place, subject to ICC approval.[103]

Even before the first statute that would lead to this result—the Trans-

[101] Quoted in FCC, *Control of Independent Telephone Companies*, p. 43; and Stipulation/Contention Package, episode 3, par. 151–54, *United States* v. *AT&T*.

[102] Howe, *Great Contrivance*, pp. 24–27; FCC, *Control of Independent Telephone Companies*, pp. 42–44; Stipulation/Contention Package, episode 3, par. 137–39 and Defendants' Third Statement of Contentions and Proof, vol. 1, pp. 166, 167, *United States* v. *AT&T*.

[103] Stipulation/Contention Package, episode 3, par. 159, *United States* v. *AT&T*.

portation Act of 1920—was enacted, technology had been conspiring to produce one unified system in each community and throughout the nation because interconnection between rival local systems had proven to be a technological failure. The Los Angeles experience illustrates this. As we noted earlier, many states and communities had enacted mandatory interconnection statutes. Los Angeles in 1915 had two major companies (Home and PT&T), each with an approximately equal number of stations, and three minor companies, none of which connected with the others. On June 1, 1915, Los Angeles voters expressed their demand for mandatory interchange by a vote of approximately sixty-three thousand to fifteen thousand.[104] But the technological obstacles proved greater than advocates apparently supposed. Interconnection costs would be so high, according to an independent telephone engineer appointed by the Municipal League of Los Angeles, as to double rates. Moreover, the switchboards and trunks used by the two systems were so incompatible that one of the existing systems would have had to have been scrapped or both scrapped and a brand new system installed. Even the voltage used by the two systems differed. Though not an insurmountable difficulty, it would have required the installation of additional switchboards and costly apparatus at existing switchboards.[105]

The Board of Public Utilities, Los Angeles, although critical of the service provided by AT&T before Home's competition, was inexorably led to conclude that interconnection of two competing companies in the same locale was impractical and that consolidation was the better solution.[106] Even though the Los Angeles board favored an independent local monopoly rather than one controlled by AT&T, an agreement reached in 1917 between the two companies allowed the Bell company to acquire the independent. State regulation of local monopolies through a commission with extensive power to order changes in rates, service, extensions, and franchises was now considered more effective than competition.[107]

Thus, in 1920, when Congress again considered the telephone question, experience had decisively shifted public sentiment to favor local monopolies and a regulated network manager. Telephone competition was now conceived as a nuisance; the Kansas Supreme Court went so far as to say that it caused "sorrow of heart."[108] Although the Transportation Act of 1920 largely concerned railroads, it did grant the ICC power over

[104] Cloyd Heck Marvin, "The Telephone Situation in Los Angeles" (Master's thesis, University of Southern California, 1916), p. 10.

[105] Ibid., chap. 3.

[106] Board of Public Utilities, Los Angeles, Fifth Annual Report (July 1, 1913–June 30, 1914), pp. 59, 60; and Eighth Annual Report (July 1, 1916–June 30, 1917), pp. 77, 104.

[107] *Application of Santa Barbara Tel. Co.*, 4 Cal. R.C.R. 470, 506 (1916).

[108] *Janicke* v. *Washington Mut. Tel. Co.*, 150 P. 633, 644 (Kan, 1915).

telephone rate discrimination and rebating. It also permitted the ICC to supervise the accounting practices of telephone companies more closely, especially with respect to depreciation charges.[109]

One year later Congress enacted the Willis-Graham Act, extending ICC jurisdiction over telephone company mergers and consolidations. The statute gave the ICC authority to approve such mergers and acquisitions if the commission found that they "will be of advantage to the persons to whom service is to be rendered and in the public interest." The statute further exempted telephone mergers and acquisitions from the coverage of the federal antitrust laws, which by implication abrogated the portion of the Kingsbury Commitment dealing with AT&T acquisitions. Willis-Graham, it should be noted, did not relieve state public utility commissions of jurisdiction to approve or forbid mergers and acquisitions so that an ICC certification of a merger did not bar a state commission's later refusal.[110]

AT&T and independent companies joined in supporting the statute. Although most of the legislators supported Willis-Graham and both Senate and House committee reports made the antiduplication and natural monopoly arguments that had now become common, there were nevertheless some who still strongly held antimonopoly sentiments, notably Mississippi Congressman Johnson. But even legislators who opposed the statute on narrow, specific grounds agreed on the undesirability of duplicate systems (then existing in about one thousand of the United States' twenty-one thousand exchange points). The Willis-Graham Act passed by wide margins in both chambers.[111]

In the thirteen years between Willis-Graham's enactment and the creation of the FCC in 1934, the ICC considered at least 274 merger and acquisition cases, 232 of which involved a member of the Bell System. Two hundred seventy-one of these mergers and acquisitions were certified. Of that number ninety-two involved consolidations of local competitors. Most of the consolidations involved small companies with relatively few subscribers, and the percentage of small consolidations relative to large ones tended to increase as the decade went on. One study that examined all of the consolidations that took place between 1921 and 1929 discovered that 148 of the 182 (where size was known) that had occurred until then involved companies with fewer than 2,000 subscribers. During the same period sixteen medium-size consolidations (2,001–4,500 subscribers) accounted for only 45,966 subscribers. But the 182 large consolidations accounted for the lion's share of subscribers—564,009. (One

[109] Stipulation/Contention Package, episode 3, par. 163, *United States* v. *AT&T*.

[110] 61 *Congressional Record* 1983–1985.

[111] Ibid., 1982–94 (1921); and Stipulation/Contention Package, episode 3, par. 165–85, *United States* v. *AT&T*.

should consider that in 1929 there were almost 20 million telephones in the United States.)[112] Those consolidations, which were also considered by the thirty state commissions that could regulate such acquisitions and mergers, were approved at that level, too.[113]

The consolidation activities of the ICC clearly served to rationalize the telephone industry. Duplicate systems were eliminated. In most of the cases involving nonduplicate acquisitions a large company acquired a smaller one, integrating it into its operations. Most often, according to Daggett's study of ICC regulation, the acquisition occurred "because the smaller system lacked capital, credit and revenues to extend or even to maintain its plant, or because the owners of the small enterprise desire to withdraw from the telephone business and to devote themselves to other activities. . . . A plant may be entirely operated by a man and his wife. Or it may be that a hardware merchant and a stock raiser may undertake telephone responsibilities. . . . Sometimes the health of the telephone magnate may break down or he may wish to move away, or simply to retire."[114]

The ICC did not have separate bureaus for telephone and telegraph matters, but there is no evidence that it was unable to handle the consolidation matters that came before it in a fair, effective manner. Virtually all interests that sought it were represented by counsel in the ICC proceedings. Thus, state attorneys general, Bell companies, independent telephone companies, trade associations, subscribers, business organizations, chambers of commerce, and public utility commissions, as well as the parties in interest, appeared at ICC proceedings. Even popular referenda by affected subscribers were considered. Public hearings, on proper notice, were held, and the ICC prepared reports that integrated facts with the reasons to grant the consolidation in the particular case. The ICC often considered the interests of parties that did not appear in the proceedings. The reports, additionally, make it clear that the burden of proof was on the applicants to show that the consolidation was in the public interest.[115] Often the reports turned on the amounts that subscribers would save as a result of the consolidation.[116] The ICC, in reaching its decisions, considered not only present circumstances, but also the fu-

[112] Stuart Daggert, "Telephone Consolidation under the Act of 1921," *Journal of Land and Public Utility Economics* 7 (1931): 24.

[113] Stipulation/Contention Package, episode 3, par. 200, *United States* v. *AT&T*.

[114] Daggett, "Telephone Consolidation," pp. 27, 28.

[115] See *Consolidation of Ohio Bell and Ohio State Tel. Cos*., 70 ICC 463 (1921); *Purchase of Properties by S.W. Bell Tel. Co*., 76 ICC 709 (1923); and *Purchase of Properties by C&P Tel. Co. of W.Va*., 79 ICC 37 (1923), as examples of participation, interest, and proceedings.

[116] *Purchase by Bell Co. of Property of P&A Tel. Co*., 72 ICC 829, 830 (1922).

ture plans of the consolidated company to meet population growth.[117] Finally, it should be noted that the applications for consolidation were heard by Division 4 of the ICC, consisting of three commissioners, and dissents sometimes occurred.[118] In summary, the ICC appears to have handled its consolidation work fairly and well.

Critics of ICC regulation (and those, more generally, of regulation) have faulted rate regulation—or more accurately, the lack of it. The ICC dealt with telephone rates in only eight cases from 1910 to 1934. All were instituted pursuant to complaints, but only two were resolved by order, the remainder dismissed. The ICC, according to one critic, undertook no general rate investigation, and the cases it did handle were of minor importance.[119] Although the ICC's authority over rates was not as comprehensive as the FCC's, under public service liberalism the ICC activity is measured by results rather than the number of investigations and actions brought by regulators.

First, the number of intrastate toll calls was far greater than interstate calls. A study of Bell System *toll* traffic considered during the last phases of ICC jurisdiction showed that 80.5 percent of toll calls were intrastate and 19.5 percent were interstate. Public utility commissions existed in forty-five states to handle intrastate rates.[120] Second and even more important, AT&T interstate rates in the 1920s were reduced as rapidly as they were from 1934 to 1940.[121] In short, ICC inaction is as consistent with a general satisfaction about AT&T long distance performance as it is with other explanations.

Satisfaction was certainly the overriding sentiment in the telephone industry as the 1920s unfolded. When the decade began AT&T and the independents were in harmony. The independent associations and periodicals, previously hostile to state commission regulation, now accepted it and were hostile to municipal regulation and franchising. At the 1922 convention of the National Association of Railway & Utilities Commissioners (NARUC) the presidents of AT&T and the U.S. Independent Telephone Association (USITA) outlined a formal policy of cooperation, a policy maintained until the breakup, notwithstanding a few spats.[122] In-

[117] See *Purchase of Properties of Hammonton T&T Co.*, 154 ICC 111 (1929).

[118] See, for example, *Purchase of Properties by Princeton Tel. Co.*, 76 ICC 401 (1923).

[119] Richard Gabel, "The Early Competitive Era in Telephone Communication, 1893–1920," *Law and Contemporary Problems* (1969): 357, 358. See also Stipulation/Contention Package, episode 3, par. 207, *United States* v. *AT&T*.

[120] James M. Herring and Gerald C. Gross, *Telecommunications Economics and Regulation* (New York: McGraw-Hill, 1936), pp. 213, 214.

[121] Arthur W. Page, *The Bell Telephone System* (New York: Harper & Bros., 1941), pp. 186–88. This book, written by an AT&T employee, must be used with care as, of course, must some government documents prepared by AT&T-phobes.

[122] "Telephone Men Talk to Rate Makers," *Telephony* (November 25, 1922): 12; and "Reg-

deed, in keeping with the new spirit, AT&T Vice-President E. K. Hall transmitted a letter on June 14, 1922, to the president of USITA (known as the Hall Memorandum) assuring him that AT&T did not seek to aggressively swallow independent companies. Only in special cases—when the public would be better served, state commissions requested consolidation, or it was necessary to protect the property of a Bell System company—would an acquisition be considered. Hall promised to notify USITA thirty days in advance of any formal acquisition agreement so that the association could comment on the proposal.[123]

When Theodore N. Vail retired as AT&T president on June 18, 1919, he could anticipate with satisfaction the completion of the commission–regulated network manager system to which he had made such an important contribution. The system, based on a partnership between government agencies at the state and federal levels, a giant corporation, and many smaller ones, worked very well for many decades. This basic institutional structure represented a triumph of public service liberalism. Certainly, interest group politics played a significant role in the shaping of this structure. AT&T, the independents, and others all had their private agendas. But AT&T was able to persuade others that its policy formulations would best serve the goals of public service liberalism. Its program would best promote the rapid spread of the telephone and its widespread adoption. Once the institutional structure was in place it remained for the administrative agencies, together with the private interests, to elaborate the particular rules intended to achieve the goals. And it remained for regulators to prod the private interests when they were reluctant to attain the goals.[124] In the next chapter we will examine how the completed system worked to attain public service goals.

---

ulated Telephone Competition," *Telephony* (November 25, 1922): 23–25. See also William J. Hagenah, "State vs. 'Home Rule' of Utilities," *Telephony* (October 9, 1920): 31–34.

[123] See Hall Memorandum, 14 June 1922; E. K. Hall to F. B. MacKinnon, 19 July 1927, modifying the procedure in the Hall Memorandum and other materials in folder titled "Independent Telephone Companies—Acquisition by Bell System—Hall Memorandum," in AT&T Archives, Box 1119.

[124] For example, one charge that has been made is that the Bell System and the independents ignored the telephone interests of farmers. Claude S. Fisher, "The Revolution in Rural Telephony, 1900–1920," *Journal of Social History* 21 (Fall 1987): 5–26.

7

The Administrative State and Public Service Liberalism

The Theory of the Public Utility Commission

In the aftermath of World War I public service liberalism was not dismantled. Rather, it was elaborated and refined in a few sectors such as electricity and telephones. As this chapter will show, it remained a dynamic theory of public policy in these few areas. But public service liberalism was clearly contained. Thus, as we will see in the next chapter, radio, the new communications technology of the 1920s, was not regulated according to public service liberalism's principles. After the development of public utility commissions, they were the principal institutions that carried forward public service liberalism far into the post–World War II era. The locus of decision making devolved to these authorities at the federal and state levels. And only by examining these institutions can we understand the continuation of public service liberalism in those sectors in which it continued to operate.

While regulation of telephone companies originated almost simultaneously with the recognition of the telephone's commercial possibilities, this was originally done through state statutes, municipal ordinances, and judicial decisions. This mode of regulation became increasingly cumbersome as technology and cost considerations changed rapidly. Moreover, the costs to a private individual bringing suit against a telephone company for unreasonable rates or unsatisfactory service were almost always excessively high relative to the expected gains. For these and other reasons public utility commissions (often lineal descendants of railroad commissions) assumed jurisdiction over telephone company rates and service, as well as other aspects of telephone company behavior. In 1900 only three states had public utility commissions with jurisdiction over telephone companies. By 1910 the number had risen to eighteen. And by 1920 all but three of the forty-eight states had such commissions with powers to regulate telephone companies.[1] Not all of these commissions had precisely the same powers over telephone companies, but virtually

[1] Stipulation/Contention Package, episode 3, par. 18, *United States* v. *AT&T*.

all of them asserted jurisdiction over at least one other public utility service, including gas, electricity, water, and transportation.[2]

Public utility commissions have for some time been under siege from left and right. Considering the vast number of decisions they have made, there have been very few instances of *proved* corruption or blatant disregard for the laws that these commissions are charged with enforcing. Critics often treat us to naked allegations that these commissions are "captured" by the interests that they regulate—again, without proof. Or some self-righteous crusader will claim to represent the "consumer" interest in his opposition to rate increases, for example. But these sweeping claims are nothing more, as R. H. Coase has cogently stated, than the critic's version of what the consumer interest *should* be.[3] In this chapter we will see that these institutions created under public service liberalism were carefully crafted to try to assure that public service goals would be attained.

Some observers have claimed that such agencies may have been designed to advance public service goals, but that over time this mission is forgotten, and they are subverted by regulated interests. They have advanced theories suggesting that regulatory commissions go through "life cycles" from early vigor to later inactivity (even senility). These deterministic theories are usually based on a single commission example (whose decisions are frequently misinterpreted) or on grand theorizing with no attempt at empirical application whatsoever.[4] In essence, the advocates of these theories assume, first, that regulated firms have a single and unified interest; second, that these firms are able to unduly influence legislators to protect industry interests to the exclusion of other interests by creating agencies that legitimize cartel practices; and third, that the incentives available to regulators are predominantly biased in favor of the interests of the regulated firms.

There are numerous flaws and oversimplifications in these theories. But the most fundamental criticism of the "life cycle," "capture," and similar theories is that their applicability depends on a much simpler and different regulatory world than actually exists. Legislators and regulators are assumed to be functionally the same. Regulated industries are treated

[2] Eliot Jones and Truman C. Bigham, *Principles of Public Utilities* (New York: Macmillan, 1931), pp. 170–175.

[3] R. H. Coase, "Discussion," in Warren J. Samuels and Harry M. Trebing, eds., *A Critique of Administrative Regulation of Public Utilities* (East Lansing: Institute of Public Utilities, Michigan State University, 1972); pp. 311–16.

[4] Among the leading examples are Marver H. Bernstein, *Regulating Business by Independent Commission* (Princeton: Princeton University Press, 1955); Gabriel Kolko, *Railroads and Regulation* (New York: Norton, 1965); Samuel Huntington, "The Marasmus of the I.C.C.," *Yale Law Journal* 61 (1952): 467–509; and George Stigler, "The Theory of Economic Regulation," *Bell Journal of Economics* 2 (1971): 3–21.

as monoliths. Firms and persons affected by the regulated interests (such as communications users or shippers who use railroads) are assumed to have no impact on the outcome of proceedings. Regulators are assumed to be mere tools of the regulated interests, cynical or hopeless, in contrast to the virtuous critics and concerned academics. Most critical, the entire regulatory process is assumed to be unconstrained by political, administrative, legal, and technological considerations.

Regulatory agencies were designed in terms of a set of constraints that were imposed on them, as well as a set of incentives to govern their conduct. These incentives and constraints were intended to assure compliance with the agency's missions, which were usually spelled out in statutes. Thus, these institutions were a flowering of public service liberalism. This is not to suggest that the varying abilities and regulatory philosophies of administrators have played no role in decisions. Nor would I suggest that no regulatory commissioner has ever taken a bribe or engaged in opportunistic action; of course, some of them have, just as judges, legislators, and executive branch officials at the state and federal levels of government occasionally have. But just as there is a constitutional scheme to check arbitrary action in traditional governmental institutions, there is also one that has arisen in connection with the public utility commissions (PUCs).

The constitutional scheme that arose in federal and state regulatory commissions is, of course, different from the grand plan of the U.S. Constitution, which Madison explicated in his *Federalist Paper* No. 10. Close attention to the regulatory scheme, however, reveals that, just like the U.S. Constitution, it does set up structures that subject both the altruist and the opportunist to the same constraints. For example, if a regulatory agency knows that a decision will be reversed on appeal to the courts, taking such a step would be a futile gesture. This would be so regardless of the regulator's political philosophy, degree of honesty, or opportunism. The regulator might take the step indifferent to the fact of almost certain reversal. But the constitutional system would still be working effectively, for the structural constraints (in the form of reversal) and incentives (not to engage in futile and foolish gestures) would still be operative. Even the supposed incentive of leaving government employment to work for a favored firm could not counter these. Not only is there the threat of criminal prosecution, but also because potential employers tend to respect the diligence and competence of their adversaries—their professionalism—such behavior would likely backfire.[5]

[5] See the excellent study by Paul J. Quirk, *Industry Influence in Federal Regulatory Agencies* (Princeton: Princeton University Press, 1981), passim.

The Constitutional Scheme of Public Utility Commissions

Several considerations must be borne in mind to understand the constitutional scheme: other institutions can (and do) check public utility commission performance; PUCs must undertake their actions pursuant to certain procedures and are required to set forth the rational bases for decisions; and many interests and organized groups have opportunities to effectively contest and protest decisions that might adversely affect them. In short, it is simplistic in the extreme to think that public utility commissions can do what they want just because they are ordinarily outside the purview of the general public.

The starting point in understanding the constraints imposed on PUCs is the legislative mandate. Agencies are entrusted with a set of missions usually spelled out in the preamble of their organic statutes and elsewhere in the legislation governing them. Admittedly, such language is often general or vague. But it is incorrect to treat such language as little more than a collection of meaningless words that a PUC may construe or ignore in any way it sees fit. In *New York Central Securities*, the leading federal case on the subject, the Supreme Court stated: "It is a mistaken assumption that the [public interest] criterion is a mere general reference to public welfare without any standard to guide determinations. The purpose of the Act, the requirements it imposes and the context of the provision in question show the contrary."[6] In that case (concerning ICC actions under the Transportation Act of 1920) the Court measured agency actions against such statutory goals as efficiency, adequacy of service and the best use of transportation facilities. It even used such ambiguous language as "public convenience and necessity should be interpreted so as to secure the broad aims of the statute."[7]

Precisely the same principles apply at the state level as well. Consider, as an example, the case of compulsory interconnection between telephone companies. At common law there was no public utility requirement imposed on telephone companies to interconnect with other telephone companies. Nevertheless, states increasingly imposed interconnection requirements on them in statutes. In the numerous situations that arose under these laws, courts almost always upheld interconnection and required PUCs to comply with the statutory mandate.[8] When PUCs' orders have been challenged by telephone companies under these and similar provisions, the reviewing courts usually point to

[6] *New York Central Securities Corp.* v. *United States*, 287 U.S. 12, 24 (1932).

[7] *ICC* v. *Parker*, 326 U.S. 60, 69 (1945).

[8] See, for example, *Blackledge* v. *Farmers Independent Telephone Co. and Lincoln Telephone Co.*, 105 Neb. 713 (1921), and the many cases cited therein.

statutory obligations and uphold PUC orders directing telephone companies to do something that they were resisting.[9]

Anyone who believes that such cases are unusual should think again, for many volumes could be compiled consisting of PUC orders challenged under state utility statutes. Courts have frequently reversed regulatory agency decisions on such grounds. Regulatory policymakers who have substituted their notions of "sound" public policy for those found in their statutes usually see their decisions reversed.[10] Courts impose on PUCs, under threat of reversal, the requirement of reasoning like courts. Unlike legislatures, PUCs may not bargain, compromise, or logroll as the principal means to achieve a decision, although, of course, they may use such techniques within the context of close reasoning based on the pertinent facts.

To a considerable extent public utility commissions share many of the same virtues as courts. But they also enjoy advantages that courts do not, the most important of which is that by working day after day in a field such as telecommunications they develop expertise in it. Information is a costly good, and it takes time to obtain it. Judges (except those few who are in highly specialized courts) are like dilettantes, briefly involved in a product liability question, then a banking question, then an antitrust question, and so on. Although public utility commission staffs sometimes have insufficient resources to expedite their tasks, they are still in a better position in that respect than courts or, for that matter, legislatures. A PUC staff does have continuing experience in a field and is invariably far larger than the small staffs of judges. Moreover, PUCs' staffs contain people with a variety of pertinent skills—engineering, economics, and so on—whereas a court's professional staff almost always consists of attorneys. PUCs incorporate the strengths of judicial policymaking (such as close reasoning based on facts) without its weaknesses.

Public Utility Commission–Industry Cooperation

The most important facets of PUCs are that they are, mission oriented and collaborative. The PUC is jointly blamed with the regulated utility for performance deficiencies; unlike legislatures, PUCs cannot easily shift blame onto the shoulders of anyone else. Even blaming the regulated company for a deficiency does not absolve the PUC of blame for lack of foresight or inadequate supervision of the industry. Regulator and regu-

[9] See, for example, *McCardle et al. Commissioners* v. *Akron Tel. Co.*, 156 N.E. 469 (Ind. App., 1927), rehearing denied, 160 N.E. 48 (1928).

[10] *FCC* v. *RCA Communications Inc.*, 346 U.S. 86, 97 (1952).

lated are jointly charged with fulfilling a mission. The Wisconsin public utility statute illustrates this: "Every public utility is required to furnish reasonably adequate service and facilities. . . . The charge rendered . . . shall be reasonable and just."[11] The language, although general, can be spelled out in quantitative terms so that deficiencies can readily be measured. Moreover, numerous PUC and court decisions and rules, as well as other sections of public utility statutes, elaborate in concrete terms what such general phrases mean. For example, one summary of various PUC rules and decisions concluded that the following norm was imposed on telephone companies: "The lines and equipment of each telephone utility should be so constructed and maintained as to eliminate, as far as practicable, all cross-talk and noise resulting from leakage and induction, improperly placed and defective grounds, long parallels of grounded lines."[12] Obviously, such performance norms can be measured.

As long as performance can be measured and evaluated in both long- and short-range terms, a PUC has a strong incentive to be mission oriented and cooperative at the same time that it is supervisory. Of course, one way of collaborating with an industry is to succumb to the every demand of the industry—that is, for the agency to be "captured." But on its face, this strategy would often be in conflict with the mission orientation goals. For example, if a complacent company refused to install modern or more progressive station equipment and a PUC did nothing about it, the commission's behavior would conflict with its mission. But though this could conceivably happen, incentives and structures have been designed to minimize it. Judicial review, the scrupulous requirements of administrative law, the fact that PUCs must provide cogent reasons in support of their conclusions, the doctrine of precedent that requires agencies to usually follow settled principles—all act to constrain PUC action in conflict with its mission. Unless there is a breakdown in the carefully constructed constitutional scheme that was designed for PUCs, an agency's flagrant disregard of its obligation to render decisions consistent with its mandate and the decisions spelling it out will leave the PUC vulnerable indeed.[13] PUC collaboration, in fact, is consistent with the prodding of a firm to perform better. Collaboration can be accommodated with some conflict. In this respect PUCs demonstrate one of the quintessential features of public service liberalism.

Of course, one might argue that if no one notices the transgressions of

[11] Wisconsin Statutes, chap. 196, par. .03(1).

[12] Ellsworth Nichols, *Public Utility Service and Discrimination* (Rochester, N.Y.: Public Utilities Reports, 1928), p. 640.

[13] The evolution of the constitutional scheme that governs PUCs and other independent regulatory commissions is spelled out in Robert E. Cushman, *The Independent Regulatory Commissions* (New York: Oxford University Press, 1941), chaps. 2–4, 6.

the PUCs and the regulated firms, all of the safeguards and procedures outlined above count for nothing. The fact is that there are many eyes on the results of PUC regulation, as well as on the ways in which these results have been achieved. All decisions, including regulatory ones, allocate resources so that there are winners and losers. If your rates go up and your service does not improve, *you* know it. If you suffer from a brownout, *you* know it. And sufficient criticism of this nature is invariably picked up by political actors willing to exploit it as an issue. In addition, others are deeply involved in most regulatory decisions, either as close watchers or as participants. Among those interests economically concerned about and involved in PUC telephone proceedings: residential subscribers, consumer organizations, business users, trade associations, equipment suppliers and potential suppliers, competing carriers, large telephone companies, independent telephone companies, rural cooperative telephone companies, telephone company trade associations, companies involved in other markets or technologies on which decisions may impinge, government purchasers of telecommunications services such as the Department of Defense, and labor unions.

This, of course, is not to suggest that all of these economic interests are represented in every proceeding. Rather, they hire attentive lawyers and lobbyists to closely scrutinize what PUCs and the FCC are doing or about to do so that any economic interest that may be affected by a proceeding usually has an opportunity to intervene. If that is not enough, a variety of governmental institutions claim to represent not their own interests but the "public interest" (often with varying and divergent views of what constitutes the public interest). Those "guardians" of the public interest include: other state PUCs, other regulatory agencies of state and local governments, NARUC, the association of regulatory commissioners whose predecessor dates from 1889, local governments, state attorneys general, local prosecutors, the Justice Department, the Department of Commerce and various divisions of it concerned with telecommunications, the Office of Management and Budget and parallel state agencies, various watchdog agencies such as the General Accounting Office and its state parallels, legislators who are attentive to public utility and telecommunications issues, legislative committees, other agencies (such as agriculture departments in rural telecommunications matters and the Department of State in international matters), and of course the courts. Always present, too, is the threat of a special committee appointed to critically examine the performance of PUCs or regulatory agencies generally. Finally, one should not overlook the many instances in which an agency staff member, conceiving that a PUC has taken a wrong turn, has leaked information to the press, legislators, and would-be guardians of the public interest.

There are many private guardians of the public interest. Ralph Nader was not the first "Ralph Nader"; muckrakers existed even before public utility commissions did. And though these would-be guardians of the public interest are often biased, wrong, or simplistic or have failed to investigate a question reasonably well, they nevertheless serve a useful purpose. The threat of investigation clearly provides regulators with an incentive to be able to defend each and every action they have taken and to show that the alternatives not taken were less attractive. These "political entrepreneurs," as James Q. Wilson terms them, are adept at mobilizing latent public sentiment by revealing a scandal or capitalizing on a crisis.[14] Regulators, who are thus put on the defensive, must always be ready to defend their actions. In their endeavors, the private guardians of the public interest are often surreptitiously aided in obtaining information from PUC insiders. They are also adept at leaking what they have learned to media reporters. Even industry trade journals can report unpleasant doings of PUCs.

Legislative appropriations and oversight hearings, executive branch review, adverse publicity, the threat of forced resignations, and even criminal prosecution provide strong supplementary incentives to follow the paths dictated by these constraints. Further, as Richard Posner has argued, narrow self-interest provides agency personnel from top to bottom with an incentive to achieve the statutory goals efficiently, for in that way the value of their human capital is enhanced.[15] The only reasonable counterincentive would be a bribe sufficient in value to outweigh the above-described incentives and the penalties and risks associated with them—probably a very rare situation.

In summary, considerable effort has been devoted to showing that PUCs are "captured" by the firms they regulate or that they are inherently incompetent. These views cannot withstand scrutiny. To the contrary, the constitutional scheme that was devised under public service liberalism to govern PUCs sharply directs them to review critically the behavior of firms under their jurisdiction, but to do so in a context of cooperation, looking to achieve statutory goals. Thus, a structure was designed to assure the attention of PUCs to their public service goals. Like other structures, it is not foolproof, but it has been reasonably effective.

An Overview of Public Utility Commission Regulation

Even if on a theoretical level PUCs are the appropriate governmental institution to regulate public utilities, it does not necessarily follow that

[14] James Q. Wilson, *The Politics of Regulation* (New York: Basic Books, 1980), p. 370.

[15] Richard Posner, "Theories of Economic Regulation," *Bell Journal of Economics* 5 (1974): 337–39.

the statute defining "public interest" will, on balance, be effective. Again, the costs associated with enforcement might outweigh benefits. Legislators and administrators can make mistakes in honestly carrying out their mandates. At the legislative level, interest group politics can shape statutes that administrators must carry out.[16] At the administrative level, the task of incorporating many considerations, including the condition of the network, intermodal competition, user considerations, the health of localities, and other concerns is, as I. L. Sharfman, the principal historian of the ICC, has shown, an arduous and complex one. Mistakes will be made in reconciling the trade-offs in the various goals constituting the public interest. That such mistakes will be made and judgments will differ does not undermine public service liberalism.[17]

The history of PUC telephone regulation illustrates, in contrast, the beneficial results of PUC regulation in the public interest. Whereas railroads did not flourish under the planning style of regulation imposed by the Transportation Act of 1920, telephones—and the increasingly large public they served—prospered under public service liberalism. PUCs were required to be attentive to the wider goals, best stated in section 1 of the Communications Act of 1934: "to make available as far as possible, to all the people of the United States a rapid, efficient, Nationwide and worldwide wire and radio communications service with adequate facilities at reasonable charges." Although the jurisdiction of state commissions must necessarily be more limited in geographical scope, their mandates parallel the FCC's.

Probably the most important problem PUCs face is obtaining information. Under public service regulation PUC commissioners are supervisors of an industry's performance; they are not the industry's managers. PUC administrators obtain a great deal of information about the industries they regulate, but this perspective is different from one obtained in planning a corporation's short- and long-term goals and operating the company. Long before this century's dreary experience with nationalized industries, Theodore Vail (obviously out of self-interest) warned: "Experience also has demonstrated that this 'supervision' should stop at 'control' and 'regulation' and not 'manage', 'operate' nor dictate what the management or operation should be beyond the requirements of the greatest efficiency and economy. . . . State control or regulation should be of such character as to encourage the highest possible standard in

[16] Shipper interests, for example, dominated railroad legislation from 1897 to 1917 to the detriment of the railroad network. See Albro Martin, *Enterprise Denied* (New York: Columbia University Press, 1971), passim.

[17] See I. L. Sharfman, *The Interstate Commerce Commission* (New York: Commonwealth Fund, 1936), 3B:33–97; see also pp. 290–308. On the ICC's attention to shipper interests—one of the many considerations to which it paid attention—see William Z. Ripley, *Railroads: Rates and Regulation* (New York: Longmans, 1913).

plant, utmost extension of facilities, highest efficiency in service . . . rigid economy in operation."[18]

Of course, the other major problem PUCs face is being hoodwinked by telephone companies so that the agency becomes a mere rubber stamp of the regulated firms' desires. As we saw earlier, the large number of interests hostile to the telephone companies' demands who are only too willing to bring information to the attention of regulators makes hoodwinking an unlikely, but not impossible, outcome. The outcome hoped for, avoiding the two principal pitfalls, is accomplished by scrutinizing carefully telephone company requests and the complaints of other parties and prodding the telephone companies to move in desired directions. But at the same time PUCs must appreciate that telephone companies have to compete in capital markets, expend funds for research and other activities, and make a reasonable profit. Clearly, balancing competing considerations in concrete cases is not always easy. Moreover, given the tens of thousands of individual decisions that PUCs have made since they began regulating telephone companies, it would be foolish to suggest that *every* decision has been a wise one. Nevertheless, measured by the aggregate data, the system has worked very well.

PUCs have also been criticized on the ground that they lack the resources necessary to undertake continuous and vigorous supervision. In this connection, agency officials sometimes join the criticism, complaining that they lack adequate resources in the form of money or personnel, or legal powers to do their tasks appropriately.[19] Although in some instances these charges might have some merit, one must also maintain a degree of skepticism. Virtually every law enforcement agency—perhaps every governmental agency—annually seeks more money, more personnel, and more powers. Invariably, agencies point to things that they have not accomplished to justify the demand. Thus, police departments will use the fact that they solve few crimes to request bigger budgets, more police, and greater powers with respect to citizens.

For this reason it is illogical to accept at face value an administrator's statement that he or she has been unable to do a good enough job. Logically, such a statement should be treated as self-serving if it is used to request more resources. The evidence, to the contrary, demonstrates that state PUC regulation and utility performance was until recent times excellent. Even though the period covered occurred much later than 1934, the following conclusion is still applicable, since the industries mentioned were then governed by public service liberalism:

[18] AT&T, 1910 Annual Report, pp. 32, 33.

[19] On these complaints, see generally Charles F. Phillips, Jr., "The Effectiveness of State Commission Regulation," in Samuels and Trebing, *Administrative Regulation*, pp. 71–87.

> Thus, the growth of public utilities in the postwar period has far exceeded that of unregulated industries; the productivity of public utilities during the last decade [the 1960s] has increased faster than for any other industrial sector; and public utility rates have risen far less than prices generally throughout the economy. Moreover, the rate of return earned by public utilities has been considerably below that earned by other industries, while the public utility sector has accounted for a significant percentage of the nation's capital formation.[20]

Information and Procedure

As we observed earlier, one of the most important underlying problems for PUCs is obtaining information. Thus, in an important sense money and personnel are secondary problems if there is no good way to obtain the information necessary to perform a job properly. Viewed in this light PUCs are provided with a number of valuable resources and opportunities to perform their tasks well with relatively modest resources. First, unlike the police, for example, who have jurisdiction over large numbers of people, PUCs regulate relatively few companies. Second, whereas crimes are often unobserved or seen only by a person who cannot readily identify the perpetrator, knowledgeable persons and organizations closely follow the affairs of public utilities. Many of these persons are eager to bring the most trivial transgression to the attention of the PUC. Third and most important, a variety of procedures and techniques exist to make the investigative problems of PUCs relatively easy. Although statutes and powers vary from state to state, the Wisconsin statute provides an example of some of these procedures and techniques.

The Wisconsin PUC is empowered to make reasonable rules and regulations for the inspection, investigation, and auditing of utilities. It may also engage in tests to determine the level of public utility functioning.[21] The agency is more generally empowered to obtain any relevant information about a utility's management practices, payouts, and management structure. In short, the PUC can easily shift the costs of preparing information onto the utility, using spot checks and other sampling techniques to test for veracity and accuracy. If the public utility balks at producing information, the agency is empowered to subpoena any pertinent "books, accounts, papers or records" at the time and place the PUC designates.[22] Additionally, the PUC may independently value and revalue a utility's property for rate-making purposes. Further, the PUC is empowered to

[20] Ibid., p. 80.

[21] Wisconsin Statutes chap. 196, par. .02(3).

[22] Ibid., par. .02(6).

require the utility to maintain accounting practices in a manner that will facilitate the agency's ability to understand the utility's business. Moreover, utilities are required to publish and make publicly available their balance sheets and income accounts so that independent parties may scrutinize them.[23]

If the commission wishes to probe information supplied by the utility further, it may require the utility to supply itemized accounts of every detail, including wages, depreciation, receipts, expenses, quantity and value of materials, and dividends. Since this data is filed with the PUC for the purpose of changing a rate or practice, it is important that the agency should have some powers to enjoin a proposed rate or practice from going into effect pending a determination as to its lawfulness. The statute accordingly provides: "No change in any utility rule which purports to curtail the obligation or undertaking of service of such utility shall be effective without the written approval of the commission after hearing." The PUC may make emergency provisions pending its final decision. Moreover, it is permitted under certain circumstances to suspend rates for four months.[24] In the event that the PUC certifies an unlawful rate (one, for example, yielding an unreasonably high rate of return), a court may set it aside on the complaint of any adversely affected user.[25] Industrial subscribers and municipalities often have the interest and means to institute such a proceeding.

Many kinds of organizations, municipalities, or any group of twenty-five persons (the number designed to discourage frivolous use of public proceedings) may bring a charge concerning rates or practices (including inadequate service) before the PUC. The PUC *must* conduct an investigation and hold a hearing under these circumstances. Additionally, any individual can bring an informal complaint to the PUC's attention and the agency may conduct an investigation on that basis. It may, of course, also initiate investigations.

Since PUCs usually consist of three or more persons appointed at different times for fixed terms, the possibility of collusion among its members is further reduced. Most important, as we observed earlier, agencies must present reasons for what they do, usually in the forms of findings of fact and a discussion of the applicable legal principles.[26]

As this overview of PUC operations indicates, these agencies are constrained to behave in a very formalized manner. A variety of protections have been built into their structures, and procedures exist to guard

[23] Ibid., par. .06–.08.

[24] Ibid., par. .20(1).

[25] *City of Eau Claire* v. *Railroad Commission*, 178 Wis. 207 (1922).

[26] Wisconsin Statutes, chap. 196, par. .26. See also *Wisconsin Tel. Co.* v. *Public Service Comm.*, 287 N.W. 122 (1939), on fundamental procedural rules in PUC proceedings.

against undue influence and to minimize the intercession of political considerations into their affairs. This does not mean that undue influence *never* exists or that political considerations never play a role in influencing events. Of course they do. But one must be careful not to generalize from the occasional instance of malfunctioning. The rare scandal invariably erupts into headlines, but the innumerable daily proceedings that constitute the overwhelming bulk of PUC work never do.

Finally, one should not confuse differences of opinion among commissioners with the intervention of "politics" in the crass sense. In most cases PUC commissioners operate within the constitutional scheme that has been delineated. Within such conceptions as a "reasonable rate" there is obviously room to differ, just as there is room to differ on many judicial questions that are outside the framework of public utility law. That there is a zone of reasonableness governed by general guidelines within which reasonable persons may disagree should not be misconstrued to mean that there are no limits. Obviously there is room for discretion, but just as the common law system has evolved over the centuries to establish vast numbers of set principles, so also has public utility regulation. Similarly, in both systems there are points about which people may differ.

How Public Utility Commission Principles Are Established

Public utility law existed before utility commissions came into existence. As we saw in chapter 2, the roots of public utility law can be traced back to the early days of the common law. Nevertheless, the rapid spread of PUCs in the second decade of the twentieth century led to a veritable explosion of new principles as well as to statutory changes of some earlier common law principles, such as that requiring compulsory telephone company interconnection. Since PUCs have to reduce their opinions to writing in most cases, this led to the establishment of a series of PUC reports in each state. Because all of the PUCs faced similar issues, they felt a need for a centralized reporting system. The problem was met in 1915 with the initial publication of *Public Utilities Reports* (*PUR*).

Published periodically and patterned on the reporting of court decisions, the *PUR*s print verbatim many of the major decisions in the public utility field. Early on, a headnote system, again patterned on judicial reporting systems, was begun, and periodically a digest grouping together the cases and principles under each subject is published. (The digests also include summaries of cases that were not included in the *PUR* series.) Thus, the *PUR*s supplemented by state reports, constitute the basic raw materials that allow PUC commissioners and staff to learn evolving principles. In this way the *PUR* system established something approaching a national system of public utility principles.

Since 1915 the *PUR*s have filled many volumes containing thousands of individual decisions. The cases, of course, have concerned many topics, including esoteric and complex accounting matters and technical jurisdictional questions, but three topics above all concern us here—structure, service, and rates—and they provide a sketch of some of the most important principles in telephone matters developed by the PUCs. Many of the most important principles were established before 1930, including the central structural one—monopoly. Public service liberalism, as these principles show, was capable of logical, incremental growth to confront novel situations and issues.

Public Utility Commissions and the Monopoly Privilege

As we saw earlier, aversion to telephone competition was very much in vogue by the second decade of the twentieth century. In a series of 1915 cases involving proposed telephone consolidations of two competing companies into local monopolies, the California Railroad Commission summarized the dominant view on the benefits of single telephone companies in each locale: "It may be said, however, that after a number of years of experience with two telephone systems in these communities, the subscribers almost unanimously demand a consolidation into one system. . . . The Telephone being a natural monopoly there should be one universal service . . . in addition to the usual advantages . . . resulting from the elimination of duplicate property and duplicate operating expenses."[27]

In a case involving the sale of the Pacific Telephone and Telegraph property in Redondo, California, to an independent, the commission observed: "This is not a field within which two competing companies can operate with financial success."[28] In the San Diego telephone consolidation case the California commission reported that, based on the public's experience with dual systems, "There is a widespread demand in San Diego County for the elimination of a double telephone service. . . . Almost without exception there is serious and widespread objection to the operation of two telephone systems in competition."[29]

The same *general* attitude prevailed throughout the nation; from

[27] *Pacific Teleph. & Teleg. Co.*, 15 Cal. R.C.R. 993, 994 (1918).

[28] Ibid., p. 991.

[29] *San Diego Home Teleph. Co.*, 15 Cal. R.C.R. 995 (1918).

Maine to California competition was discouraged, consolidations approved, and certificates of convenience and necessity denied to applicants when existing companies could adequately serve the demand in their areas. Commissions, however, did not simply incant the virtues of monopoly. They required a showing in the forms of testimony and exhibits that local monopoly would benefit consumers in such ways as rate savings or superior service.[30] The same principles were applied to toll lines as they were to local exchanges.[31]

Because a showing was required in each case, however, there were exceptions and caveats to the general rule favoring monopoly provision of local service and toll service. Like courts, PUCs are capable of making acute distinctions to the end of serving the higher public service standards. That standard, articulated by the Indiana public service commission (PSC), was: "A public utility has a high responsibility to render adequate public service in return for the monopolistic privileges granted to it. . . . When such utility fails, refuses to, or is unable to render such service, it should forfeit its exclusive right to operate in such territory. The public interest is paramount."[32]

Thus, where telephone companies were unable or unwilling to provide adequate service, their franchises were forfeited and other firms were permitted to compete against or replace the laggards.[33] For the same fundamental reasons telephone companies whose service was temporarily inadequate were permitted to continue for trial periods their local monopolies against the entry requests of other companies when the monopolists were making good faith efforts to improve the system and provide proper service.[34] A PUC must, in short, consider not only service to consumers, but also that a firm will be reluctant to undertake long-term investments to improve service if its franchise could be lifted for temporary difficulties that the company is trying in good faith to remedy.

Obligations that accompany monopoly privileges also include the duty not to abandon a part of the telephone company's assigned territory, even if it is unprofitable or difficult to serve.[35] Again, the subscribers' interests

[30] See the decision of the Illinois commission, for example, in *Illinois Bell Teleph. Co.*, PUR 1926C, 1; and the Oregon PSC in *Troutdale* v. *Pacific Teleph. & Teleg. Co.*, PUR 1926A, 310.

[31] See, for example, *Bevier Teleph Co.* v. *Macon Teleph Co.*, PUR 1926A, 545, 547 and *Lake Shore Tel. Co.*, PUR 1926A, 137, 138–39.

[32] *Sharpsville Telephone Co.*, 75 PUR (N.S.) 189, 192 (1948).

[33] See, for example, *In Re Cole*, PUR 1921C, 385, 392; *Lanagan Teleph Co.*, PUR 1919F, 596, 600; *Gasconade Central Teleph. Co.*, PUR 1926A, 573, 579–81 and the Indiana Supreme Court's opinion in *Farmers & M Co-op Teleph. Co.* v. *Boswell Teleph. Co.*, 187 Ind. 371 (1918).

[34] *United Teleph. Co.*, PUR 1926E, 777.

[35] For example, *Steuben County Teleph. Co.*, PUR 1926C, 367.

are paramount and the telephone company is expected to cross-subsidize to make up for such losses. Nevertheless, PUCs have flexibly sought to devise other arrangements to mitigate telephone company losses, while guaranteeing reasonable service. For example, the California PUC recommended seasonal abandonment of service in a resort area, while the Kansas commission prodded a telephone company to increase its revenues by undertaking renewed efforts to obtain additional subscribers.[36] Nevertheless, there are limits and PUCs have permitted small companies to abandon service when the losses incurred in serving very few customers (for example, twenty-two) are substantial and efforts had been made to secure alternative service.[37]

PUCs, in summary, had adopted a rule of reason in cases involving structural questions such as competition or abandonment. The consumer's interest is paramount, but not to the point that public utilities were required to incur substantial losses for very sparse service. Such a burden on the utility would inexorably be passed on to the other subscribers who were expected to contribute to cross-subsidy, but not unreasonably so.

For similar reasons, as the California Railroad Commission summarized in a case involving ferry service, a new utility will not be permitted to enter an area already served by another firm if five conditions are met: (1) the occupant pioneered the area, (2) it is rendering efficient service and (3) cheap service, (4) it is otherwise fulfilling its service obligations, and (5) it is generally serving its assigned area and is reasonably attempting to expand its coverage.[38] It follows too that a PUC may order a utility to extend its lines.[39] Even then, however, if a potential entrant can *clearly* show (and not just assert) that it can supply the service more cheaply and efficiently than an existing utility, it may be permitted entry or substitution for the existing utility.[40] Of course, if such a showing of a promised lower rate is based on specious or questionable accounting, the potential entrant will be turned down.[41]

End-to-End Responsibility

A rule of reason applies in the service area as well. Telephone companies have not been required to do the impossible, such as supplying everyone

[36] *Raymond Teleph. Co.*, PUR 1927E, 588; and *Haven Teleph. Co.*, PUR 1915A, 1014.

[37] *Monticello Telephone Co.*, PUR 1919C, 473.

[38] *Golden Gate Ferry Co.*, 27 Cal. R.C.R. 499 (1926).

[39] *Young* v. *Hyde County Teleph. Co.*, PUR 1919C, 559.

[40] See *Great Barrington Appeal*, 10 Ann. Rep. Mass. G.&E.L.C. 18 (1895); and *Easthampton Appeals*, 23 Ann. Rep. Mass. G.&E.L.C. 28 (1908).

[41] *Greaves Teleph. Co.* v. *Lewis Mut. Teleph. Co.*, PUR 1923A, 804.

with picturephone service at low cost. Nevertheless, PUCs have monitored utilities to make sure that their service standards are as high as reasonably possible. By 1920 thirty-eight statutes required their PUCs to oversee telephone service quality. The 1911 New Jersey statute, for example, empowered the PUC "after hearing, upon notice . . . to require every public utility . . . to furnish safe, adequate and proper service and to keep and maintain its property and equipment in such condition as to enable it to do so." Similarly, the 1913 Missouri statute required "every telephone corporation [to] . . . furnish and provide with respect to its business such instrumentalities and facilities as shall be adequate and in all respects just and reasonable."[42] In turn, these statutes led PUCs to adopt what AT&T has termed the end-to-end service concept, also known as end-to-end responsibility. Under end-to-end the public utility is responsible for the quality, safety, and effectiveness of each component of its network. Further, that responsibility is a continuous one and is not to be discharged sporadically.[43]

In *United States* v. *AT&T*, end-to-end was a significant issue. The Department of Justice contended that the concept had no basis in historical fact and was not an established regulatory policy, notwithstanding AT&T's citation of innumerable cases supporting the doctrine. The reason for the heated dispute over what might appear to be just a historical issue is that it had (and has) important contemporary implications. End-to-end in the telephone industry implies that in local service a single company should have the responsibility for customer premises equipment, transmission, and switching from one end of the local loop to the other. Toll service is different only in that the toll company had full responsibility from the moment a message was switched from the local loop until it was switched back into the receiving party's local loop. Moreover, in long distance, proper regulation precluded a no-man's-land in which neither the long distance nor local companies had responsibility. Accordingly, under end-to-end the network required a manager—AT&T—that would have the final word on the problems of coordination and compatibility.

In short, end-to-end justified the system that existed before the AT&T breakup.[44] Many of the end-to-end cases involved smaller companies with inferior equipment. Nevertheless, in keeping with a fundamental principle, the same rule has been applied to small firms and large ones.

Within that context the leading public utilities text, which was dispassionately prepared so that legal practitioners would know applicable prin-

[42] Cited in Stipulation/Contention Package, episode 4, par.. 7, *United States* v. *AT&T*.

[43] Ibid., par.. 6, *United States* v. *AT&T*.

[44] Ibid.

ciples, stated as early as 1928 (citing cases): "A telephone company should assume responsibility for the maintenance of all lines and equipment according to the prevailing opinion. . . . The burden of properly maintaining telephone service rests with a telephone utility and not with its subscribers and it is the business of the utility to inspect its lines and keep them in such condition that good and continuous service will result; and that telephone companies should not permit subscribers to maintain parts of their line."[45] In brief, end-to-end became an established principle early in the history of PUCs. The initial cases arose in connection with poor maintenance practices of rural and mutual telephone companies. These firms allowed subscribers to fix lines or install their own equipment. Since bad practice in one part of the network can adversely affect the quality—and sometimes the safety—of other subscribers, the South Dakota Board of Railroad Commissioners said in 1916: "Those companies which permit, and in some cases compel, subscribers to purchase equipment, including telephone instruments . . . are, to say the least, engaging in improper telephone practices."[46]

In 1924, with more experience in the matter, the South Dakota commission, considering the complaint of forty subscribers of a rural telephone company about the quality and cost of service, noted that subscribers were required to maintain branch lines. The commission found that "such practice appears to have intensified the conditions complained of and to a large extent the failure to properly maintain such branch lines has been very largely a contributing cause to the service conditions complained against."[47] Accordingly, the company was required to improve and maintain branch lines as well as all equipment used in the telephone service. Similarly, over the objection of a rural telephone company, the Wisconsin PUC held in 1922 that the only way a telephone system could meet the the commission's "Standards of Service for Telephone Companies" concerning loading of lines was for the company to have full local responsibility for equipment and service. The PUC stated, "We believe that the subscribers are correct in their contentions that the company should maintain all property used and useful in serving them. The practice advocated by the company cannot be too severely condemned for we have found in practically all cases where there is a division of responsibility in the operation and maintenance of a utility property, that the result is poor service."[48]

For the same set of reasons the end-to-end principle required a telephone company to maintain service accessories and batteries for tele-

[45] Nichols, *Service and Discrimination*, pp. 473, 474.

[46] *Letcher Teleph. Co.*, PUR 1916E, 486, 494.

[47] *Jones* v. *Rapid Valley Telephone Co.*, PUR 1924D, 321, 323.

[48] *Mt. Vernon Teleph. Co.*, PUR 1922D, 139.

phones. The telephone company should exclusively maintain customers' premises equipment and it would be required to furnish new instruments to replace subscriber-owned equipment. Divided maintenance responsibilities were held to be impractical and unlawful.[49] Even when subscribers used batteries for purposes in addition to telephony, commissions, over the objections of telephone companies, required end-to-end responsibility.[50] Nor were telephone companies permitted to offer a lower class of service or a discount based on customers' maintenance or supply of equipment.[51]

In summary, end-to-end was the single most important service obligation of public service companies. It arose not from the connivance of telephone companies, but frequently over their objections. The principle arose because PUCs saw it as essential to fulfilling their statutory obligations to assure safe and efficient telephone service as well as to achieve other goals under the public service conception.

The Consequences of End-to-End

Once end-to-end is understood as an obligation of telephone companies and not as a conspiratorial move by them to deny entrance to rival firms, many of the other service obligations follow. For example, a telephone company has the right to enter a subscriber's premises at a reasonable time to do repair and maintenance work.[52] When one weds end-to-end to the network idea, another important consequence follows: customer premises equipment should be leased and not sold. As the New York PSC, Second District, said:

> Every subscriber is interested in the entire service, including the equipment of every other subscriber with whom he has occasion to communicate. The best equipment at one station is of no avail if communication is prevented by defective equipment at the end of the line. . . . The public has a right to generally efficient telephone service, and it has a right to look to the telephone company for such service and to hold it responsible therefor. Generally efficient telephone service would, we believe, be impossible if subscribers generally owned their own station equipment buying such as they saw fit and maintaining it as

[49] *Farmers Fountain Teleph. Co.*, PUR 1926C, 363; *Spring Green Teleph. Exch. Co.*, PUR 1921E, 184; *Jordon* v. *Peoples Teleph. & Teleg Co.*, PUR 1919C, 226; *Littlepage* v. *Mosier Valley Teleph. Co.*, PUR 1918E, 425; *Bluffs & W. Teleph. Co.*, PUR 1915A, 928; and *In Re Green*, 17 Cal. R.C.R. 744 (1920).

[50] *Cadott Teleph. Co.*, 21 Wis. R.C.R. 645 (1918).

[51] *Riverview Teleph. Co.*, PUR 1916B, 442; and *Nelson Farmers Teleph. Co.*, 24 Wis. R.C.R. 648 (1920).

[52] *Lafayette Teleph. Co.*, PUR 1915A, 930.

> best they could. A system of private ownership or a practice permitting subscribers at their will to install private systems would certainly lead to great deterioration if not demoralization of the service.[53]

Nevertheless, commissions did not apply the lease-only rule in a mechanical manner. Like common law judges, PUCs have been capable of making significant distinctions. Thus, a telephone company was not allowed to refuse service when its predecessor agreed to allow a hotel to own its private branch exchange (PBX). The New York PSC, Second District, noted that "such equipment was installed by invitation or requirement of the [telephone] company. . . . Moreover, the actual installation and the selection of the equipment was by the telephone company itself."[54] It was the standard equipment used by New York Telephone (a Bell System company), and the hotel gave the telephone company complete control over the equipment and its connections into the network.

From such cases an important principle emerged that was to have major consequences in the post–World War II era. A telephone company is obligated to serve not only all customers willing to pay reasonable charges for customary uses, but it is also expected to make arrangements often for untraditional or novel uses. If the telephone company chose not to install its equipment under such circumstances, customers might lease or buy their own. But—and this is a big but—the telephone company was required to assure that the equipment chosen was compatible with the telephone company's, and it was obligated to have control and supervision over all connections and equipment. Only in this way could the integrity of the network be guaranteed.

Consistent with this principle, the Washington public utility regulator approved an arrangement allowing several lumber companies to own and erect subsidiary lines within their camps. The telephone company was, however, required to install the main lines, PBX, and all other equipment. And the lumber companies agreed to bring all equipment in need of repair to the exchange office. Approving the agreement as an exceptional one, the PUC said, "It is always desirable that the utility should own the instrumentalities, unless there are exceptional reasons for doing otherwise. Dividing responsibility and authority with respect to operation and maintenance is apt to result in confusion and trouble and oftentimes poor service."[55]

[53] *State Agricultural & Industrial School* v. *New York Teleph. Co.*, 4 P.S.C.R. 206, 219 (2d Dist., N.Y.), quoted in Nichols, *Service and Discrimination*, p. 479. See also *Industry Teleph. Co.*, PUR 1928A, 435; *Franksville Teleph. Co.*, PUR 1917A, 270; and *Pulaski Merchants & Farmers Teleph. Co.*, PUR 1925E, 674.

[54] *New York Teleph. Co.*, PUR 1916D, 688, 690.

[55] *Department of Pub. Works* v. *Montesano Teleph. Co.*, PUR 1925A, 676, 682.

The end-to-end principle was also followed by the Massachusetts Supreme Judicial Court when, in a complex labor dispute, a hotel was forced to permit union employees not associated with the telephone company to install telephone wiring. The telephone company was not permitted to install or supervise wiring and equipment and consequently refused to provide telephone service. Overruling the Massachusetts PUC, which required the telephone company to provide service, the court held: "The determination whether certain wires are suitable and are properly installed is a detail of management in the administration of the business of the Telephone Company. To substitute the judgment of others for that of the Telephone Company in that matter is an interference with the right of management which goes beyond the reasonable limit of public control."[56] Again, we see that end-to-end was conceived as a public service responsibility and not, as later fiction would have it, as a monopolistic justification for AT&T to control every phase of the telephone business.

The Quality and Quantity of Service

The responsibilities discussed were coupled with the universal service conception early in the history of PUC regulation. Recognizing the rapid progress of telephone technology, PUCs often demanded that telephone companies upgrade or increase the service they provided. In 1919 the Vermont PSC chastised New England Telephone & Telegraph, an AT&T licensee, for not making efforts to increase the use of its service. Disapproving part of the telephone company's plan that, in the judgment of the PSC, would have discouraged additional use, it said: "The use of the telephone is a business and social necessity, and we believe that the efforts of a great telephone company . . . ought to be directed to some extent toward the idea of encouraging the increase of the use of the telephone rather than discouraging that use."[57] This was done even though New England Telephone had operated in the relevant area at a loss for several years.

Indeed, so entrenched has the cross-subsidy idea been that, within a very broad range, telephone companies, like other utilities, could not plead service at a loss as an excuse for not providing service or for providing inferior service. As the Missouri PUC said in a water case that is equally applicable to telephones: "The fact that some water main extensions may not furnish a profitable income will not prevent the Commis-

[56] *New England T&T Co.* v. *Dept. of Pub. Util.*, PUR 1928B 396, 406.
[57] *St. Johnsbury* v. *New England Teleph. & Teleg. Co.*, PUR 1919C, 964, 967.

sion from ordering them if necessary for adequate service. Rates are not based on the cost of service to each individual consumer or group of consumers, but are based upon the cost of serving all customers."[58]

During the infancy of PUC telephone regulation, service matters were decided on a case-by-case basis. As experience accumulated, the PUCs more actively prodded telephone companies to improve service. Although there were very early attempts to develop explicit quantitative standards in some telephone service areas, it was not until the development of computer-controlled monitoring in the post–World War II era that very sophisticated metering and measuring of telephone service were possible. Nevertheless, as we shall see, within the limits of available techniques, PUCs were sometimes far from satisfied with the quality (and occasionally quantity) of service provided by some telephone companies. The cases reveal that consumers and other groups have been critical of telephone service—sometimes unjustly so—since PUC regulation began. And while they sometimes complained to mayors and legislators instead of the PUC, the message did get through, and watchful governmental eyes were focused on the accused telephone companies. That most of the cases involved small independent companies, and not Bell licensees, only demonstrates the substantially higher quality of service AT&T and its affiliates rendered. Service crises have occasionally erupted in the Bell System (most notably in the 1970s), but this was not the case when PUCs established basic service principles.

Although no longer a significant service issue, the number of parties on a single line was one of the most important ones until the late 1960s. In 1950 approximately 75 percent of residential telephones were on party lines. But by 1960 the figure had been reduced to 40 percent and by 1965 to 27 percent. Gains continued to be made thereafter. Party lines were introduced in New York in 1891, but their main use was in smaller systems, especially in rural areas.[59] Restricting the number of lines through party-line arrangements reduced the size and complexity of switching equipment and the number of trunks required, and it yielded other cost savings as well. Thus, until technological advances reduced costs sufficiently, party-line arrangements were an economical way to extend coverage and more closely approach the universal service goal. But at the same time, party-line arrangements were a built-in source of conflict between subscriber and telephone company. These controversies, therefore, provide an interesting examination of PUC behavior. Did they consistently favor telephone companies or subscribers? Or did the PUCs

[58] *Webb City* v. *Missouri Public Utilities Co*., PUR 1919E, 285. See also, for the same principle applied to railroads, *Maine C.R.Co.F.C. 242*, PUR 1919E, 282, 283.

[59] John Brooks, *Telephone* (New York: Harper & Row, 1976), pp. 100, 267.

seek to strike a balance between cost considerations and subscriber convenience?

The short answer is that PUCs often required telephone companies to reduce the number of parties on a line but adopted this tack within reason. Thus, in 1923 the Alabama PSC encouraged Southern Bell to adopt four-party service where only two-party service had been available because the new service would make telephone service available to many who otherwise would not have been able to afford it.[60] On the other hand, where no demand (or very little) existed for four-party service, it was permitted to be discontinued. But where demand persisted, a telephone company was not permitted to abandon four-party service.[61] PUCs required telephone companies generally to offer the variety of services that subscribers or potential subscribers required within reasonable cost limits. For this reason the Wisconsin commission required a rural telephone company not to abandon the higher-grade two-party system.[62]

While PUCs did not at first articulate explicit standards with respect to party lines, instead largely using the common law method of inclusion and exclusion, they did establish almost from the outset certain minimum standards. The most important of these concerned overloading and ringing. In general two forms of ringing were available on party lines: code ringing and selective ringing. The latter was far better—but also more costly to operate—since the telephone only rang in the premises where it was intended to ring. Whenever cost justified, PUCs ordered, often over the objection of telephone companies, the substitution of selective ringing for code ringing.[63] Many cases concerned the practice of overloading—placing too many subscribers on a party line. Almost without exception PUCs upheld subscriber complaints of overloading. They either refused to grant a telephone company's application for a rate increase until the condition was remedied or required the condition to be remedied within a designated time.[64]

PUCs continuously reviewed their older orders in the light of advancing technology and changing economic conditions. Thus, in 1937 the Il-

[60] *Public Service Commission* v. *Southern Bell Teleph. & Teleg. Co.*, 34 Ann. Rep. Ala. PSC 38 (1923). See also *St. Johnsbury* v. *New England Teleph. & Teleg. Co.*, PUR 1923C, 365.

[61] *Couer D'Alene* v. *Interstate Utilities Co.*, PUR 1916C, 438; and *New York Teleph. Co.*, PUR 1930C, 325.

[62] *Barron County Teleph. Co.*, 28 Wis. R.C.R. 724 (1925); and *St. Johnsbury* v. *New England Teleph. & Teleg. Co.*, 75 PUR 1919C, 964.

[63] *Kirkman*, 20 Cal. R.C.R. 966 (1921); *Wisconsin Teleph. Co.*, PUR 1922B, 91; and *Pacific Teleph. & Teleg. Co.*, 75 PUR (N.S.) 379 (1948).

[64] See, for example, *San Antonio Home Teleph. Co.*, PUR 1922A, 45; *New London Co-op Teleph. Co.*, PUR 1925E, 368; and, more recently, *Midwest Teleph. Co.*, 23 PUR 3d 26 (1958).

linois Commerce Commission noted that technical improvements had greatly reduced the cost margin between four-party and two-party service, and that the growth in traffic in Chicago was causing lengthy delays, increased busy signals, and more. The commission concluded that the need for four-party service had disappeared and ordered a tariff revision accordingly.[65] Similarly, the New Jersey commission permitted New Jersey Bell to discontinue four-party service in 1961 on the ground that such service had become obsolete. When the number of subscribers to a service had become very small and a superior alternative was available at a small additional charge, the older service could be abandoned.[66]

The party-line cases typify PUC action in other service areas as well. Commissioners more often sided with complainants than with telephone companies, which were held to stringent service obligations. Moreover, even when telephone companies were not required to meet subscriber demands, it was because the costs imposed on the telephone company were excessive or other reasonable alternatives were available to complaining subscribers. Consider in this light the cases in which subscribers (sometimes only one) sought the extension of a line over the objection of a telephone company complaining that the costs of serving such marginal customers were in excess of the revenues that would be received. An early case involved the complaint of one person living in a remote part of South Dakota. The commission, siding with the complainant, stated that it "has held in numerous cases that a company might be compelled to construct lines a reasonable distance. . . . If the party desiring service resides at more than a reasonable distance from the line," other arrangements might be made, such as the prospective customer incurring some of the telephone company's costs.[67] Far from constituting capture by telephone companies, PUC telephone cases might almost be caricatured as examples of the adage "the customer is always right."

When telephone companies have complained that the cost of serving distant customers will forever be undertaken at a loss with no hope of breaking even, PUCs have responded that as long as the service *as a whole* is profitable and other subscribers are not unduly burdened, a telephone company must still continuously serve unprofitable customers. Cross-subsidy, in a word, was an accepted PUC doctrine, especially when the matters involved Bell licensees.[68] While several theories were

[65] *Illinois Commerce Comm.* v. *Illinois Bell Teleph. Co.*, 21 PUR (N.S.) 273, 276 (1937).

[66] *New Jersey Bell Teleph. Co.*, Docket 615–373 (1961) and *Chesapeake & P. Teleph. Co.*, 73 PUR (N.S.) 12 (1947).

[67] *Latta* v. *Medicine Valley Teleph. Co.*, PUR 1917E, 950, 952. See also *Young* v. *Hyde County Teleph. Co.*, PUR 1919C, 561.

[68] *Branch* v. *Southwestern Bell Teleph. Co.*, PUR 1922B 631; and *Castner* v. *New Jersey Teleph. Co.*, PUR 1923D, 517.

employed by PUCs to justify ordering telephone companies to extend their facilities into revenue-losing territories, two were paramount. First, since a public utility traditionally was required to treat all of its customers without discrimination, it cannot deny service to some, and moreover, must charge the same rates to all.[69] Of course, there are limits to this idea, and where the costs of service to relatively few were extremely high, an exception was made.[70] Second, both independents and Bell licensees, being public service companies, were required to satisfy the developmental needs of the communities they served and were, therefore, required to expand into areas unprofitable for them but important to the wider community. This reasoning applied even in the case of a seasonal resort.[71]

In virtually every other service quality area the same PUC approach of requiring, within reason, better telephone company performance prevailed. Moreover, PUCs used telephone company applications for rate increases as opportunities to scrutinize carefully aspects of telephone company service. Under such circumstances even vague and ambiguous dissatisfaction was treated seriously. In one Oklahoma case involving Southwestern Bell, complainants, thirty-five citizens and business firms in Shawnee, listed eleven complaints, including dilatory connections, frequent interruptions, and company indifference to complaints. Finding that the complaints were true, the Oklahoma commission ordered Southwestern Bell to correct all of the service defects promptly.[72] In other cases customer complaints concerned specific deficiencies. These involved repairing poles, testing and repairing telephones, replacing switchboards and coils, employing more personnel, adding new trunk lines, and installing automatic equipment in place of manual equipment.[73]

Standards

In the earlier days of commission regulation decisions were usually made with reference to a qualitative standard on what constituted adequate ser-

[69] Typical of the many cases advancing this theory is *Hyde* v. *Vincent-Bethel Teleph. Co.*, PUR 1919E, 655, 658.

[70] As in *Chesapeake & P. Teleph. Co.*, 10 Ann. Rept. Dist. Col. PUC 59.

[71] *Hanson* v. *Mountain States Teleph. & Teleg Co.*, PUR 1920B, 547; and *High Hill Beach Impro. Assoc.* v. *New York Teleph. Co.*, PUR 1916E, 1043.

[72] *Roebuck* v. *Southwestern Bell Teleph. Co.*, PUR 1918D, 210.

[73] *Hizor* v. *Coshocton Co. Teleph. Co.*, 4 Ann. Rep. Ohio PUC (1916); *Southern California Teleph. Co.*, PUR 1925C, 627; *Grantman* v. *Theresa Union Teleph. Co.*, PUR 1915A, 103; *Coady* v. *La Crosse Teleph. Co.*, PUR 1915A, 565; *Michigan State Teleph. Co.*, PUR 1921C, 545; *Madison Rural Teleph. Co.*, 18 PUR 3d 273 (1957).

vice and equipment. Even then, however, a few PUCs attempted to set forth, as far as possible, quantitative standards of good service. Thus, Wisconsin in 1914 was one of the first states in which the PUCs devised a set of quantitative standards. This move stemmed not only from a desire to upgrade service, but also from an attempt to assure consistent service and a coordinated network throughout the state. Incompatible practices and equipment were perceived as major impediments to Wisconsin-wide communication.[74]

Early quantitative rules covered such matters as switchboard capacity, number of subscribers on any single line, operating force, and promptness in handling calls. The rules, while sometimes flexible in view of the differences between rural and urban services, were also at times inflexible. Wisconsin PUC rule 7, for example, stated that at exchanges serving five hundred or more subscribers, 94 percent of the calls should be answerable within ten seconds or less; at other exchanges, 90 percent of the calls should be answerable within ten seconds or less. In the early twentieth century *exact* telephone standards were harder to devise than for other utility services such as gas, where exact BTU standards were specified.[75] Nevertheless, some standards, such as mandatory interconnection between telephone companies, were explicitly set forth in statutes and consistently enforced when complaints were made to PUCs.[76]

Generally, however, early PUC standards should be understood in the sense of cajoling telephone companies toward greater progressiveness in the light of technological and economic advances. Usually more progressive telephone service elsewhere provided the standard employed by the PUC. The multiparty cases that we have looked at provide a good example because this was a paramount issue. But another early issue of importance, especially in rural areas, was twenty-four-hour service. For obvious reasons, some telephone companies would resist employing operators during periods of very light traffic. Once again, however, PUCs employed a rule of reason that tended to require twenty-four-hour service. Again, PUCs tended to apply commonsense solutions in appropriate cases so that only emergency night service was required in some exceptionally light traffic situations.[77]

Subscriber dissatisfaction can also be traced to the far more complex

[74] *Fixing Standards of Telephone Service*, 15 W.R.C.R.1 (1914); and Fred L. Holmes, *Regulation of Railroads and Public Utilities in Wisconsin* (New York: D. Appleton, 1915), chap. 5.

[75] See, for example, *New York & Q Gas Co.* v. *Prendergast*, PUR 1924B, 138.

[76] *Railroad Commission* v. *Northern Kentucky Teleph. Co.*, 236 Ky. 747 (1931).

[77] Typical hours-of-service cases include *Raymond Teleph. Co.*, PUR 1927E, 588; *Troy Teleph. Co.*, PUR 1926D, 581; *Oakdale Teleph. Co.*, PUR 1926D, 59; and *La Crosse Teleph. Corp.*, PUR 1931C, 81.

area of central office equipment. Here, of course, we are in the difficult border between regulatory supervision and management responsibility. Regulatory agencies ordinarily did not tell telephone companies precisely what kind of switches to install. Nevertheless, PUCs had to have some familiarity with switches in order to assure high-quality service. Faced with this quandary, as early as 1917 the New Hampshire commission established the appropriate standards principle. Without telling telephone companies what equipment to install, companies were required to make frequent traffic studies showing that sufficient equipment of a satisfactory type was available to handle the desired volume of calls. The commission had developed information showing the average number of calls that were requested in a typical month. Comparing that data to the number handled, a wide discrepancy would indicate equipment or personnel difficulties that the telephone company would be required to clear up.[78] It should be noted that such rules were applied to Bell licensees and other large companies, as well as to rural and small independent companies.[79]

As commissions learned more about telephony, the accumulated engineering knowledge allowed them to become more specific in their standards. As early as 1915 the South Dakota commission was bold enough to order a telephone company to use no. 12 wire instead of no. 9 to relieve overloading of the lines and solve other transmission difficulties.[80] Similarly, in 1930 the Louisiana regulators directed the installation of a common battery system in place of a magneto system.[81] Monitoring requirements grew more comprehensive and sophisticated. In 1930, for example, the California PUC required an independent that failed to meet the commission's service standards to furnish periodic progress and operating reports. In another matter, telephone companies were required to submit reports on customer requests for service and to explain any request unfilled after fifteen days.[82] Again, based on experience, the Georgia commission found that a Continental Telephone affiliate's central office had too many connector switches not in operation at a single time.[83] Finally, the Washington Department of Public Service was able, in a rate investigation of Pacific Telephone and Telegraph, to require more rapid substitution of dial service for operator-assisted service.[84]

[78] *Standards for Telephone Service*, PUR 1917B, 676.

[79] *Southern California Teleph. Co.*, PUR 1925C, 627.

[80] *Cheyeene Valley Electric Teleph. Co.*, PUR 1915F, 932.

[81] *Public Service Commission* v. *Southern Bell Teleph. & Teleg. Co.*, PUR 1930E, 491.

[82] *Doggett* v. *California Water & Teleph. Co.*, 65 PUR (N.S.) 502 (1946); and *Telephone Service*, Cal. PUC decision no. 53312, case no. 5537 (1956).

[83] *Monroe* v. *Georgia Continental Teleph. Co.*, 25 PUR (N.S.) 95 (1938).

[84] *Dept. of Pub. Serv. of Wash.* v. *Pacific Teleph. & Teleg. Co.*, 34 PUR (N.S.) 193, 214–16 (1940). See also the cases cited in Stipulation/ Contention Package, episode 4, par.. 19, in *United States* v. *AT&T*.

By the late 1960s and early seventies, the increased technical capabilities of PUCs and the FCC, coupled with inferior telephone service in certain localities, led the FCC to institute a formal reporting procedure on service standards. The FCC, with the cooperation of the Bell System and a NARUC committee on telephone service, set out a standard consisting of ten service quality areas and defined "bench marks" for seven of these.[85] By the early 1980s, thirty-three state commissions had developed highly detailed service quality rules together with means to obtain the pertinent information. Of course, we must always bear in mind that there have been many persons and organizations more than happy to monitor telephone company behavior and accuse the company before a PUC of every known transgression. Thus, ordinary subscribers can ascertain whether a Maryland telephone company is in *possible* violation of a rule requiring 98.5 percent of calls to receive a dial tone within three seconds. Any Oregon subscriber (or any busybody) can ascertain whether a telephone company is violating that state's service standard requiring installation within three days of an order if the subscriber lives within fifteen miles of the telephone company's plant.[86]

End-to-end responsibility worked. Service standards worked. The American telephone system, notwithstanding some rough spots and downturns, got better and better in the long run. To take a single example, before World War II the time required to establish a long distance telephone call was approximately three minutes. At the beginning of the 1960s the time was down to one and a half minutes, and by 1973 it was further reduced to under forty seconds.[87]

Rates

Few subjects are more shrouded in mystery than public utility rate making. The elemental principles are relatively easy and appeal to common sense. The details that are worked out by cost accountants are tedious and complex. The formulas determining allocations are extraordinarily complex and fully understood by the economists and cost accountants who specialize in the subject. Fortunately, the state PUC's job in this area is really not about rates. As law professor Neil N. Bernstein observed, "Rate regulation is a misnomer. Public utility 'rate cases' are not cases about rates; instead they mainly deal with the amount of revenues that the utility ought to earn. The typical revenue measure provides that

[85] Stipulation/Contention Package, episode 4, pars. 27–30, *United States* v. *AT&T*.

[86] Ibid., par.. 32.

[87] Ibid., par.. 39.

the utility is entitled to collect sufficient revenues from its customers to cover proper operating expenses, depreciation expense, taxes and a reasonable return on the net valuation of property."[88] In short, regulators are not concerned about each and every item in a utility's documentation supporting a rate change. Their principal focus is on the end result. Only if revenues are too great or too small are rates adjusted.

Critics have attacked even this aspect of rate regulation. Again, as Bernstein, summarizing many critics of the process, states: "The public should be primarily interested in the quality of service it receives from the utility and the price it must pay therefor. If quality is high and prices are modest, we should have no concern even if the supplier's investors are reaping rewards beyond their wildest dreams. On the other hand, if service is poor and rates are exorbitant, it is small comfort to learn that the utility's profit picture is gloomy."[89] If one adopts this very sensible perspective, one must conclude on the basis of what we have seen that the state PUCs have done an excellent job of regulating telephone rates. Until the AT&T breakup, service had been good and improving, the system got closer and closer to universal service, and virtually every American had been able to obtain telephone service at reasonable rates. And even if resentment at high returns is allowed to enter the picture, AT&T and other public service telephone companies had always earned a reasonable, and never an exorbitant, profit. PUCs were able to avoid a conflict between traditional rate-making theory and the more modern view advanced by Bernstein. Figuratively, the PUCs have kept one eye on sensible regulatory practice and the other eye on the legal constraints imposed on their rate-making actions by court decisions.

PUCs must be concerned not only with rate levels but with rate structures as well. Rate structures can have an important impact on a firm's efficiency and, therefore, on its rate level. Consider, for example, an issue that has raged almost since the initial commercial exploitation of the telephone—flat rates versus measured service (on the basis of duration, distance, and time of day). As Alfred Kahn explains:

> In a sense the flat charge is inefficient because it involves a zero price for additional phone calls, whereas their MC [marginal cost] is certainly not zero. . . . On the other hand (1) most customers apparently prefer the flat charge because it gives them the freedom to make local calls without having to worry about their bills; (2) a flat rate is also preferred by the phone companies because when they charge on a per call basis they incur additional costs—of equipment to count the calls, employees to handle customer complaints about

[88] Neil Bernstein, "Utility Rate Regulation: The Little Locomotive That Couldn't," *Washington University Law Quarterly* 1970 (Summer 1970): 223.

[89] Ibid., p. 226.

their bills, and so forth; (3) also, the incremental costs of local calls are well below average costs because of economies of scale when the dimension along which output is expanded is the number of calls made per subscriber.[90]

In short, rates and rate structures are functions of more than economic and accounting principles, notwithstanding the language of some regulatory decisions. They are usually made with careful consideration of service obligations and distributional considerations. For example, flat rates inexorably redistribute resources from those who rarely use their telephones to those who jabber constantly. But, as we have noted, the concept of cross-subsidization is one of the underlying aspects of the public service idea. That is why electric companies grant discounts unrelated to cost savings to hospitals, why urban transportation systems use flat rates, and why public utilities generally provide many services that do not pay their way in the marketplace. Such "taxation by regulation," to use Richard Posner's term, is rampant.[91] But cross-subsidies are intended to serve values other than efficiency. For this reason free water is provided to fire companies and lower-cost telephone service is provided to persons otherwise unable to afford it.

When we avoid the *general* issue of "fairness"—something that in rate matters, as economist James C. Bonbright shows, is rife with ambiguity and multiple meanings—and instead focus on the attainment of social goals at reasonable levels of efficiency, issues become clearer.[92] We may argue endlessly about the meaning of "fairness," but few persons at our level of civilization would argue that the mentally retarded should starve to death because they do not pay their way or produce anything of value for society. From the foregoing it follows that if public service goals (such as universal service) are being met and the firms doing so are technologically progressive and reasonably efficient, the joint rate-making activities of regulators and telephone companies should be considered successful. Public service liberalism is thereby served.

By the time public utility commissions began to regulate the telephone companies, the Bell companies and others had already developed certain rate principles, local governments had instituted telephone rate structures, and the courts had shaped a set of regulatory and constitutional principles that the PUCs were required to employ in making rate evaluations. One of the earliest problems arose in connection with value-of-

[90] Alfred E. Kahn, *The Economics of Regulation*, 2 vols. (New York: John Wiley, 1970), 1:182.

[91] Richard A. Posner, "Taxation by Regulation," *Bell Journal of Economics* 2 (1971): 22–50.

[92] See James C. Bonbright, *Principles of Public Utilities Rates* (New York: Columbia University Press, 1961), pp. 124–27.

service pricing. In the nineteenth century AT&T had instituted a complex set of pricing principles of which value of service was one of the most important. In essence, this principle allows rates to be set on the basis of their value to consumers rather than the cost to provide the service. Thus, value of service provides the telephone company with an incentive to make service more valuable. This translates into enlarging the system by adding more subscribers and by providing new uses for the telephone. Value-of-service pricing provides an incentive to technological progressiveness and universal service. It also provides a justification for charging business firms higher rates than residential subscribers—a Bell practice from its inception. In its 1907 Annual Report, AT&T stated: "Business rates are higher for the reason that presumably the business subscriber connects with the greatest number of other subscribers, and consequently makes use of the greatest number of circuits and operating facilities in an exchange. . . . The value of an exchange depends on the area covered and the maximum number of desired individuals that can be reached."[93]

Value-of-service pricing has been only one of a complex mix of rate principles developed and employed by AT&T. But it has been a very important one. When PUCs assumed jurisdiction over telephone rates, they were called on to accept, reject, or modify the principle. Subject to modification and the application of other principles, PUCs have, in James Bonbright's term, "acquiesced" in value of service. In so doing the PUCs have acquiesced in "the levy of higher charges on subscribers in large cities than on subscribers for comparable service in smaller communities. In any given city, moreover, higher rates are quoted for business use than for residential use."[94] In short, the PUCs acquiesced in AT&T playing Robin Hood—a system that could operate effectively only as long as competitors were not legally able to undercut the rates of the subsidizing traffic. The approving attitude of most PUCs in these arrangements is best stated in a Michigan PSC decision.

> Whether the Commission or the Bell System initiated the effort, condoned by the other, is unimportant. The result is that cross subsidization exists in the rate structure. Some services are priced far above cost to serve and some far below costs to serve. This structure fostered a legitimate purpose—universal availability of basic service. It placed residential service below the cost to serve—to be subsidized by other services priced well above cost.[95]

[93] AT&T, 1907 Annual Report, p. 21.

[94] Bonbright, *Principles*, p. 83.

[95] *Grand Trunk Radio Communications* v. *Michigan Bell Tel. Co.*, Case No. U-5082 (Mich. PSC, March 8, 1978), p. 8. See also the numerous cases in Stipulation/Contention Package, episode 4, par. 114, *United States* v. *AT&T*.

As a result of this basic value-of-service concept PUCs could justify ordering extensions into remote unprofitable territory in many cases (although rejecting others as inordinately costly relative to benefits). As the Missouri PSC stated in 1912, "The losses thus brought about are to be recovered from the business of the company in its entirety."[96] For the same value-of-service reasons PUCs justified, in some cases, higher rates for larger cities than smaller ones; the telephoning opportunities are greater in the former than in the latter, as shown by higher calling rates.[97] Similarly, PUCs justified low rates for such services as local coin box service, 911 emergency service, and so on.[98] Luxury items, like decorator phones, were also priced to support other services, such as those that were intended to serve social needs and attain universal service.[99]

The result of all of these subsidy flows has been, of course, to make it difficult to specifically allocate costs to particular services. But we must recall where we began. The purpose of PUC rate regulation is not to demonstrate that cost accounting is a science. Under public service liberalism the point has been to achieve a set of service goals. As the Colorado PUC concluded in a 1974 proceeding, rate setting is as much an art as it is a science. Although there are limits, allocations of plant and equipment can be made within a broad range. Nevertheless, PUCs consciously approve rates and structures that include value-of-service concepts, actual costs, reasonable rates of return, usage of telephone service, and social goals. In so doing they deliberately have attempted to approve subsidy flows from business to residential users and urban to rural users.[100] Companies design the basic structure and PUCs scrutinize them, judging by probable results as well as accounting data. While the system may not accord with textbook rate making, it has obviously worked well measured by results.

One reason that it has worked well is the constant attention that numerous interests pay to state telephone rates, forcing telephone companies, especially AT&T and the large operating companies, to be on the defensive. During the supposedly complacent 1920s—long before Ralph Nader—the *New York Times* reported that several large cities and the Ohio PUC bitterly complained about alleged high rates to the ICC and state bodies. Congressmen demanded that the telephone "trust" be bro-

[96] *Branch* v. *Southwestern Bell Tel Co*., 11 Mo. PSC 675, 678 (1912).

[97] See *Southwestern Bell Tel Co*., 95 PUR (N.S.) 1, 9 (1952).

[98] See the cases cited in Stipulation/Contention Package, episode 4, Defendants' Contention, par. 109 *United States* v. *AT&T*.

[99] See *General Tel. Co*., 80 PUR 3d 2, 74–76 (1969).

[100] *Mountain States Tel. & Tel. Co*., docket no. 867, decision no. 86103 (Colo. PUC, 1974), p. 29.

ken up.[101] Moreover, numerous interests, including cities, chambers of commerce, users, and so-called public advocates, frequently appeared at rate proceedings. For these reasons and to withstand a possible court challenge, large telephone companies expend a considerable effort in rate proceedings to prove the numerous particulars that go into the rate request.[102] They have rarely fully persuaded PUCs to grant all that they have requested. As a detailed study of Michigan PUC actions found, there is a low ratio of commission approvals of telephone rate requests.[103]

Public Service Liberalism and Public Utility Commissions

The public utility commission represented public service liberalism's response to the increasingly complex world of the early twentieth century. The PUC was created not only because the technical information necessary to make judgments had become increasingly complex and detailed, but because public service goals had become more complex and the trade-offs between them more difficult to assess. Regulatory agencies have not been pluralist institutions seeking to achieve compromises between groups. Rather—and this is far more than a nuance—they are constrained to focus on their public service goals, which sometimes involves compromise and sometimes declares clear winners and losers. But under their constitutional scheme the various interests have opportunities to express demands in informal ways, formal proceedings, or on appeal.

PUC deliberations also involve judgments. And these may sometimes be wrong. Nevertheless, the performance of the industries closely regulated by PUCs has been remarkably good measured by such standards as productivity. The telephone, gas, electric, and water services in the United States have been extremely competent; go to almost any other country and compare. In no small part this is attributable to the ingenious PUC institutional structure that combines industry-government collaboration with a mechanism to resolve disputes and differences and prod industry to achieve still higher standards. Public service liberalism has worked well. It contrasts with the case of radio, an example of modern liberalism.

[101] "Cities Join to Fight Higher Phone Rates," *New York Times*, September 4, 1926, p. 1; "Gifford Again Denies Illegal Monopoly," *New York Times*, Sept. 8, 1926, p. 29; and "Again Asks House for Phone Inquiry," *New York Times*, May 28, 1924: p. 36.

[102] A discussion of the preparation required is contained in Edward Devereux Smith, *A Telephone Rate Case* (Washington, D.C.: Public Utilities Reports, 1941), chap. 9.

[103] C. Emery Troxel, "Telephone Regulation in Michigan," in William G. Shepherd and Thomas G. Gies, eds., *Utility Regulation* (New York: Random House, 1967), pp. 141–86.

8

The Contraction of the World

An Interlocking America

LATE IN 1921 Congress enacted the Federal Highway Act, creating a Bureau of Public Roads. Careful highway maps were prepared and a ten-year road-building program was adopted. The unparalleled road-building program both stimulated and was a reaction to the great new industry of automobile manufacturing. In less noticed developments, new methods of mixing concrete for the road-building program, new road-building machinery, powerful new explosives for blasting rock, new ways of spreading asphalt, and a host of other dramatic new activities responded to the needs of automobile and truck transportation. Aviation made even more remarkable strides.[1]

Parallel to the developments in transportation during the 1920s were those in communications. Motion pictures experienced a boom, and in the last part of the decade they began to talk. Even such older activities as printing and papermaking experienced innovation. For example, four-color rotogravure came about after World War I, and new dyes opened up vast possibilities of color design. More dramatically, the transmission of pictures by wire became possible. Photoelectric typesetting came into being. But most important, radio emerged as a booming industry in the 1920s. All of these technologies in transportation and communication had their roots in prewar developments. Most could be traced to events that occurred before the turn of the century. Prescient people grasped the potential of the interactions that these innovations and further developments in older technologies, such as the telephone, opened up. The stakes in communication, already high, rose spectacularly.

The dramatic changes that took place typified the vast expansion in the variety of merchandise and services that became widely available, especially those associated with electricity and chemical processes. Automobiles, radios, motion pictures, refrigerators, toasters, rayon, other synthetic fibers, plastics, and cellophane typify this vast outpouring. At the same time, other developments encouraged the widespread consumption of these products and services, the most important being a vast expansion

[1] John W. Oliver, *History of American Technology* (New York: Ronald Press, 1956), chap. 38.

and increasing sophistication of advertising and a wider availability of installment loans and other new forms of consumer credit.

The cult of scientific management, the overall rise in average real income, and general price stability combined with these trends to create a subtle shift in public philosophy—modern liberalism. Business leaders, trade association officers, and government administrators holding parallel beliefs were, in historian Arthur Link's words, building a political economy based on "a whole new set of business values—mass production and consumption, short hours and high wages, full employment, welfare capitalism."[2] Increasingly, the public service questions were no longer asked. When an industry's performance was satisfactory in these terms, questions were not raised about other goals. Industry planning through professional management and trade association problem solving was sufficient. But when special problems arose, either because an industry was "sick," as in the case of agriculture, or because the business professionals and trade association figures could not solve problems without government intervention, as in the case of radio, public action was required. State intervention was not of the public service kind, but rather an attempt to use professional expertise to plan an industry's structure and performance. But as the case of radio demonstrates, this was not done to achieve public service goals (even if there was the facade of this). Rather, as the radio case shows, policies resulted from attempts to plan a structure that would resolve economic and political disputes. Because of this, radio and its successors— over-the-air television and cable television—were on their way to becoming vast profit centers that disregarded traditional public service goals, although the symbolic language of public service remained.

Traditional public service principles were increasingly confined to the sectors of industries in which they had been developed. PUCs and some other administrative agencies continued as islands using the public service principles, even as they were abandoned in other parts of the political economy. And when much of the confidence in the consumer society was shattered by the Great Depression, the public service philosophy was not revived.

In the course of devising public policies that respond to new technologies, it is important to bear in mind the critical issue of precisely what public values are expected to be served. In the case of the new transportation and communication technologies, certainly the widest interaction economically feasible is a paramount one. As telephone transmission expanded, so did the range of human contact. At first you could only call local people who subscribed to the same company's service. With the

[2] Arthur S. Link, *American Epoch*, 2d ed. (New York: Alfred A. Knopf, 1963), p. 833.

demise of telephone competition this expanded to all other local subscribers. Soon this also included national and then international contacts. Moreover, in the telephone system you can both transmit *and* receive information. But now consider radio (and by extension television) technology. Under policies that can be traced back to the 1920s, many can receive, but very few can originate. In short, the range of interaction has been considerably less in radio and television than in telephony. As this chapter will show, it need not have been that way.

The advent of cable, direct broadcast satellites, interactive television, and other technologies in the 1970s was expected to free television from dominance by three major networks and a blandness (with few exceptions) in entertainment and news presentations. To date the promise of wide diversity has been largely unfulfilled. AT&T had a vision of broadcasting that could have had different results. In the conception of toll broadcasting that it developed, broadcasting stations would be akin to telephone booths that anyone could use upon payment of a reasonable fee. As demand increased, the number of "telephone booths" would be increased. The common carrier (that is, AT&T) would be required to distribute all such messages, as it did in telephony, subject only to the bounds of decency. Toll broadcasting, as we shall see, had a brief but interesting career.

Invading the Ether

In October 1864 James Clerk Maxwell, the greatest nineteenth-century physicist, submitted a theoretical paper to the Royal Society titled "A Dynamical Theory of the Electromagnetic Field." Building on and synthesizing work in electricity and other branches of physics, Maxwell formulated a system of equations theoretically showing, among other things, that "electromagnetic fields could be propagated through space as well as through conductors; that if so propagated, they would travel as waves with the velocity of light; and that light itself was electromagnetic radiation, within a certain narrow range of wavelengths."[3] In short, Maxwell's extraordinary genius had opened the theoretical door for the technological development of new ways to transmit radio waves. Even light wave transmission was feasible. Indeed, Alexander Graham Bell had some success in transmitting speech over short distances through light waves as early as 1880.[4]

[3] Hugh G. Aitken, *Syntony and Spark—The Origins of Radio* (New York: John Wiley, 1976), p. 22. Throughout this chapter I have relied for much historical information (although I differ in interpretation) on Susan J. Douglas, *Inventing American Broadcasting: 1899–1922* (Baltimore: Johns Hopkins University Press, 1987).

[4] Robert V. Bruce, *Bell* (Boston: Little, Brown, 1973), chap. 26.

Nevertheless, the first major technological breakthroughs were made in the use of radio waves, not light waves. Heinrich Hertz, a German physicist, announced in 1888 that he had generated, detected, and measured electromagnetic waves. Hertz's concerns were entirely scientific; he eschewed any technological or economic interest in his research. But based on the apparatus that Hertz used, the English physicist Oliver Lodge demonstrated an effective and practical wireless system of transmission and reception in 1894. At first Lodge saw no economic possibilities for wireless transmission, although he later became involved in commercial ventures. Notwithstanding the innumerable improvements that came thereafter, the basic technology for radio was then in place.

It remained for Guglielmo Marconi to see the commercial communications possibilities of the new technology. In 1896, at the age of twenty-two, he was granted the first radio patent and soon demonstrated that the device had a practical application as a navigation aid for ships. But a quantum leap in radio occurred when Marconi transmitted the letter S in Morse Code from Cornwall at the western tip of England across the Atlantic Ocean to Newfoundland on December 20, 1901.[5] Events followed rapidly in the next few years, but from AT&T's perspective a major breakthrough in the transmission of speech was the development of the audion vacuum tube by Lee de Forest in 1906, one year before Vail's return to AT&T.

Before Vail's return, Hammond Hayes, the company's chief engineer, wrote to AT&T President Fish that there was "a reasonable probability of wireless telegraph and telephony being of commercial value to our company." Thomas D. Lockwood, the company's chief patent attorney, wrote, "For a telephone company the possibility of substituting a wireless system for a system of toll lines is the most attractive feature of the proposition."[6] The organization of small companies to exploit the new technology, even though they initially met with little success, had already begun, requiring AT&T to pay close attention to new competitors as well as to its traditional rivals in equipment and transmission.

The possibilities of wireless transmission seduced many persons seeking to become rich by achieving major breakthroughs. Among these talented people was Lee de Forest, who in 1907 was granted a patent for a "Device for Amplifying Feeble Electric Currents"—a vacuum tube.[7] A major improvement on earlier vacuum tubes, that patent and a subsequent one taken out by de Forest in 1908 went a long way toward solving

[5] Material on Marconi's role in early radio is drawn from W. J. Baker, *A History of the Marconi Company* (New York: St. Martin's, 1971), and Douglas, *Inventing American Broadcasting*, chaps. 1–3.

[6] Quoted in M. D. Fagen, ed., *A History of Engineering and Science in the Bell System* (New York: Bell Telephone Laboratories, 1975), pp. 362, 363.

[7] Quoted in ibid., p. 258.

the problem of line attenuation that plagued wire transmission. The Pupin coil had extended long distance by wire as far west as Denver. But without some additional means, it would be impossible for a voice on the East Coast to be heard intelligibly on the West Coast. Moreover, the device that solved that problem would simultaneously have to solve the problem of excessive distortion.[8]

On January 25, 1915, the first transcontinental telephone line, New York to San Francisco, was opened. Appropriate to the occasion, Alexander Graham Bell, seated in New York, uttered to Thomas Watson over a replica of an early telephone the famous line, "Mr. Watson, come here. I want you." Two weeks later regular service was opened between the two cities, and in May regular service between New York and Los Angeles began. By late October the first transatlantic telephone call, between Arlington, Virginia, and the Eiffel Tower in Paris, was made.

When Bell spoke to Watson across the continent on January 25, 1915, the necessary amplification was in part achieved through the employment of a modified version of de Forest's vacuum tube and the principles of wireless transmission. Before that conversation took place, important AT&T policymakers struggled with the problem of what the company's policy and attitude should be toward wireless and its associated technologies and inventions. After the Bell-Watson conversation, AT&T's policies and attitudes would change several times. In retrospect this is not surprising, for like many groundbreaking technologies, virtually everyone was surprised at the novel directions that radio took in a short time. In 1919, for example, few knowledgeable observers considered that the broadcasting aspect of radio would boom as it did only a few years later. Moreover, like so many new technologies, wireless transmission constituted both a threat and an opportunity to existing firms (most importantly, AT&T). At the very least, it threatened to render obsolete the vast amount of work already accomplished on wire transmission. But at the same time, wireless promised to reduce transmission costs and expand the network.

Radio and Public Policy before 1920

Like so many novel technologies, radio was viewed by many in its earlier days as impractical or a field that would never be perfected beyond a crude stage. Wireless transmission was considered by many as having very few possible uses, the most important one of which was ship-to-

[8] Leonard S. Reich, *The Making of American Industrial Research* (Cambridge: Cambridge University Press, 1985), pp. 160–64.

shore transmission. In that respect it was conceived as supplementary to wire transmission of messages. Although there were probably few technological reactionaries of the "If God intended us to fly, we would have been born with wings" stripe in important positions, very few people before 1920 had an accurate sense of the major directions that wireless would take in the boom years of the 1920s. As one of the leading histories of radio put it:

> Implicit was the possibility not only of point-to-point communication . . . but also of "broadcasting" information and entertainment to anyone with a suitable receiver. In 1900, to be sure, no one was thinking of broadcasting in those terms. The fact that radio communications were "broadcast"—that they could not be kept secret unless coded or encrypted—was generally thought of as a serious limitation of the new technology, as compared with wired systems.[9]

Because of this broadcasting "problem," AT&T considered wireless transmission an inferior form of voice transmission since it sacrificed the valuable attribute of privacy. But the privacy problem was not viewed as insurmountable. Indeed, the 1920 AT&T Annual Report noted important progress in solving the problem of privacy, stating that the company's engineers "have carried on conversations by radio telephony according to a method which they devised whereby ordinary receiving stations can hear nothing but unintelligible sounds; yet at all stations equipped with the necessary special apparatus, and in possession of the requisite operating information, the spoken words can be heard and understood."[10] The AT&T method, then in a primitive stage, consisted of dividing speech into bands that were reordered before transmission and transformed into the normal order by parallel equipment at the receiving end.

Privacy was hardly the only technological problem raised by wireless transmission as a method of telephony in the early years. The power problem presented equal difficulties. Long distance radio imposed enormous power requirements compared to wire transmission: hundreds of kilowatts compared to a few milliwatts. Aside from the economic problem this raised, there was a major technological one as well—modulating radio's great power with the relatively feeble telephone current. Fortunately, those problems were on the way to solution with de Forest's 1907 patent for the vacuum tube. Nevertheless, not until 1912–1913 did AT&T engineers and scientists begin to improve the de Forest tube, called the audion, sufficiently to use vacuum tubes as voice frequency amplifiers in addition to their original function as radio wave detectors.[11] But many

[9] Hugh G. J. Aitken, *The Continuous Wave: Technology and American Radio, 1900–1932* (Princeton: Princeton University Press, 1985), p. 12.

[10] Quoted in Fagen, *Engineering and Science*, p. 420.

[11] Ibid., pp. 364, 365; and Reich, *American Industrial Research*, pp. 160–64.

firms in addition to AT&T were simultaneously involved in such radio developments, which had important implications for the structure of the emerging radio industry. Numerous patents blocked the holders of other radio patents from commercially exploiting them. This stalemate made patent pooling virtually inevitable.

AT&T's Early Attitude toward Wireless

In early 1922 AT&T spokesman A. H. Griswold summarized the company's views on wireless's limitations in telephony. He concluded that radio telephony could never replace universal wire service but could be a valuable supplement to reach places inaccessible to wire transmission. Given the technology then available and reasonably foreseeable, there was no way that radio could conceivably handle the volume of traffic that the wire system then did or would in the next few years. Additionally, the station equipment cost was so expensive relative to that of ordinary telephones and other customer premises equipment then in use that a major trend toward radio telephony would reverse the movement toward universal service. Further, the switching problems that radio telephony presented were nowhere near solution. Even in the case of long distance, wire transmission was considerably less costly between most points. In this respect it should be noted that AT&T had been making major strides in multiplexing five conversations simultaneously over the same wires. The primary uses of wireless transmission were thus seen in ship-to-shore service, transoceanic telephony, links to islands or other places that wires could not reach or could reach only with great difficulty, and broadcasting. In 1922, AT&T was clearly interested in broadcasting (as were many others), but it was not a major concern.[12]

For these and still other reasons, before the broadcasting boom became apparent in the 1920s, AT&T adopted an ambiguous attitude toward wireless transmission that reflected the differing points of view of the company officials who considered the issue. Many in the company, most important Theodore Vail, could recall the cavalier attitude that Western Union initially had toward the telephone—until it was too late. Certainly, they did not want to repeat that mistake and realized that they could be wrong about wireless. Such considerations led AT&T to adopt early on a cautious yet defensive attitude. It was an attitude that combined skepticism about the telephonic utility of wireless transmission, but at the same time appreciated that something (although it was not

[12] A. H. Griswold, "The Radio Telephone Situation," *Bell Telephone Quarterly* 1 (April 1922): 6–8 and AT&T, 1918 Annual Report, p. 28.

clear what) would probably come of it. Although AT&T did not know what that "something" would be until the radio boom of the 1920s, it wanted to be in a position to be a participant. Most important, it wanted to head off any threat to its supremacy in point-to-point voice transmission. These considerations must be borne in mind when considering the twists and turns in AT&T's behavior.

AT&T's behavior toward new inventions associated with wireless communication illustrates this complexity. Reginald Fessenden of the University of Pittsburgh was one of the most important early American wireless experimenters. In 1906 he developed a continuous wave generator in which voice currents were superimposed on a carrier wave. (Continuous waves are those generated by pure tones.) Thus, demodulation would permit the sending voice to be received and understood. On Christmas Eve of 1906 Fessenden broadcast a program of music and speech beamed over the Atlantic Ocean. This first voice broadcast was picked up by several ships in the North Atlantic. The achievement was pathbreaking, but not until 1912 was a machine that could *regularly* generate alternating current electricity in the requisite wavelengths developed by Ernst F. Alexanderson of the General Electric Company (GE).

Nevertheless, on the basis of his startling success and follow-up demonstrations in 1907, Fessenden (and his financial backers) sought to interest one of the major communications companies—AT&T, Postal Telegraph, or Western Union—in his patents. Based on a favorable report by E. H. Colpitts, AT&T's employee in charge of investigating Fessenden's patents, Hammond Hayes, AT&T's chief engineer, recommended to President Fish in April 1907: "There is such a reasonable probability of wireless telegraphy and telephony being of commercial value to our company that I would advise taking steps to associate ourselves with Mr. Fessenden if some satisfactory arrangements can be made."[13]

Although Fish had agreed to buy, the moment chosen could not have been less propitious. As we saw, 1907 was the fateful year in which AT&T was in the throes of a financial crisis and a dramatic management shakeup that would bring Vail back to head the company. Since the company was strapped for funds, it had to cut back on far more pressing projects than radio. Its financial plight and the Panic of 1907 compelled AT&T to reverse its earlier decision and, therefore, not buy the Fessenden patents. On July 9 Thomas D. Lockwood expressed the new attitude. Although "the possibility of substituting a wireless system for a system of toll lines is the most attractive feature of the proposition . . . I have a strong conviction that this feature cannot and will not reach any practical realization within the term of years yet remaining to Fessenden's fundamental pat-

[13] Quoted in Fagen, *Engineering and Science*, pp. 362, 363.

ents."[14] Vail at that time was skeptical about the prospects for wireless telephony. Noting that wireless telegraphy had made few inroads on wire transmission, he wrote: "The difficulties of the wireless telegraph are as nothing compared with the difficulties in the way of the wireless telephone."[15]

Vail's assessment was probably a correct one *at the time*. But as we have seen, this did not mean that AT&T would turn its back on wireless. The vacuum tube's ability to act as an amplifier was brought to the attention of the AT&T high command by John Stone Stone, one of its principal researchers, in 1912 and led to AT&T's acquisition of the de Forest triode vacuum tube rights in 1913. De Forest's audion fell short of commercial viability, but it provided AT&T researchers with the foundation from which future developments would follow. AT&T's most immediate aim, as we have seen, was to produce an amplifier that would permit the transcontinental transmission of intelligible speech—a goal that was attained in 1915. Vail expressed the company's enthusiasm for wireless's supplementary role to wire communication in a September 30, 1915, telegram to chief engineer John J. Carty: "To throw your voice directly without aid of wires from Washington to Hawaii was wonderful, but to send the recognized voice part way over wire and part through the air was still more wonderful and was the demonstration of the chiefest [*sic*] use that will probably attach to the wireless, as amplifying and supplementing, not substituting, with wire system. . . . Your work has indeed brought us one long step nearer our ideal—a 'universal system.' "[16]

Even as a supplement to wire transmission AT&T had strong incentives in 1915 to deploy resources in wireless development. Much remained to be done in order to improve the de Forest audion so that it would be commercially viable and technically effective in long distance transmission. Further, improvements in wireless also carried a threat—that a potential long distance competitor would utilize the improvements and compete against AT&T with a technologically superior and/or more efficient service. The American Marconi Company, a subsidiary of the Marconi Company, a British firm and then the world leader in radio, had established transatlantic and transpacific service to supplement its shore-to-ship efforts by August 1915.[17] A national link would be the next logical step. For these reasons AT&T hired a wealth of research talent during this period. Since the problems of vacuum tube technology required fun-

[14] Ibid.

[15] Quoted in W. Rupert MacLaurin, *Invention to Innovation in the Radio Industry* (New York: Macmillan, 1949), p. 210.

[16] Quoted in Reich, *American Industrial Research*, p. 175, from the September 30, 1915, *New York Evening Post*.

[17] W. J. Baker, *Marconi Company*, pp. 154, 155.

damental scientific and mathematical research, Ph.D.s in those fields were employed. In a fairly short time the company built large triodes for transmission purposes and had developed a system of vacuum tube modulation, among its numerous accomplishments.[18]

AT&T was, however, not alone in engaging in significant radio research. In addition to independent inventors anxious to set up their own firms or sell their inventions to existing communications companies, other large companies, the most important of which were GE and Westinghouse, were also working on radio-related activities. GE, it should be noted, was the clear leader in that blend of scientific and engineering research that Bell Laboratories later came to typify. The convergence of these firms in the new field of radio and the navy's need for wartime research and development, as well as the navy's active role in determining the future industrial organization of radio, would contribute importantly to the birth of a powerful new company, the Radio Corporation of America (RCA), and the particular structure of broadcasting and telecasting—the network system—that still persists to the present (although in modified form). Wartime planning and direction under modern liberalism profoundly shaped the development of radio.

The Patent Pool and RCA

Government involvement in wireless began early. Aside from the inevitable patent involvement that occurs in most novel technologies, an interdepartmental committee as early as 1904 recommended that the navy play the primary planning role in the new technology because of its potential for national defense. The report

> admitted, however, that there may be special cases where private stations can serve a useful purpose and the Board believes that the Department of Commerce and Labor should have the duty of issuing licenses in such cases under such regulations as will prevent interference with stations necessary to the national defense. . . . This method of placing private stations under *full* government supervision is desirable. . . . To prevent the control of wireless telegraphy by monopolies or trusts, the Board deems it essential that any legislation on this subject should place the supervision of it in the Department of Commerce and Labor [emphasis supplied].[19]

[18] MacLaurin, *Invention to Innovation*, p. 160; and Fagen, *Engineering and Science*, pp. 362–74.

[19] Inter-Departmental Board, *Wireless Telegraphy* (Washington, D.C.: Government Printing Office, 1904), pp. 10, 11.

Thus, at this early stage a large regulatory role for government was anticipated. The justification for such regulation lay not in public service liberalism, but in the idea of cooperative planning to minimize interference. This is best illustrated through the example of vehicular traffic. The numerous traffic control indicators found in all but the most sparsely settled areas regulate motorists in order to prevent one person's vehicle use from interfering with another's. Obviously, without such signs as STOP, YIELD, or those indicating speed limits, the likelihood of automobile damage and personal injury would substantially increase. Some public authority or mechanism is needed to coordinate each person's vehicular use. Even at wireless's primitive stage in 1904, an authority or mechanism was needed to assign conditions of use such as frequency, time, and area of transmission. From this it followed that the authority also had to establish criteria that would rank applicants who otherwise would interfere with each other. Since national defense was considered the paramount use in 1904, licensing was, according to the report, the obvious mode of regulation followed by close supervision of the licensees.

This rationale persisted through the great changes that rapidly took place in the technology and primary uses of wireless transmission. Secretary of Commerce Herbert Hoover, who in the 1920s was more responsible than any other person for the public policy regime toward broadcasting, stated the prevailing view—one that would be embodied in the Communications Act of 1934: "We can no longer deal on the basis that there is room for everybody on the radio highways. There are more vehicles on the roads that can get by, and if they continue to jam in, all will be stopped."[20] As we will see, there were (and still are) other regulatory alternatives to the one taken, including toll broadcasting, the idea advanced by AT&T that likened radio to telephony.

The structure of the radio industry in the early 1920s can be largely attributed to government's role during World War I and a situation in which the patents of various corporations blocked each other in developing a complete wireless system. GE had become involved in vacuum tube research through its commercial interest in incandescent lamps and X-ray tubes. It had organized one of the earliest laboratories devoted to fundamental scientific and engineering research. Among its projects were the essential principles of illumination. Since illumination and electrical communication research lead down the same path, it was inevitable that the research interests of GE and AT&T would collide. In fact there were twenty patent interferences between the two companies between 1912

[20] Quoted in Federal Communications Commission, *Second Interim Report by the Office of Network Study: Television Network Program Procurement: Part 2* (Washington, D.C.: Government Printing Office, 1965), p. 79.

and 1926.[21] The patent situation was an incredible labyrinth that could be solved either through costly and protracted litigation or through some form of accommodation between the two companies and, indeed, others who held important patents.

To illustrate the extraordinary tangle, consider vacuum tube patents. The Federal Trade Commission's investigation of antitrust questions in the radio industry in 1923 concluded:

> It seems, therefore, that the best known form of vacuum tube could not be manufactured by the Marconi Company of America unless it had rights in the de Forest patents and in the patents on pure electric discharge tubes. The American Telephone and Telegraph Co. could not manufacture such tubes for radio purposes unless it acquired rights in the Fleming patent and cleared up the interferences of the Arnold application with the Langmuir application on pure electron discharge tubes, while the General Electric Company could not manufacture, sell or use such tubes for radio purposes unless it acquired rights in the Fleming, de Forest and Arnold Inventions.[22]

Nor was that the only labyrinth. Virtually every component of a radio system was subject to the same tangle.

These events occurred around the period of World War I, when patriotic feelings were running high, the navy had a deep concern about the future defense capabilities of radio, and professional planning was ascendant. Inevitably, such considerations led government officials both to seek a solution to the patent tangle and to assure that control of radio and its development were in American hands. Consequently, the Bureau of Steam Engineering wrote a letter to both GE and AT&T in 1920 stating that "the bureau has constantly held the view that all interests will be best served through some agreement among the several holders of permanent patents whereby the market can be freely supplied with tubes. . . . It is a public necessity that such an arrangement can be made without further delay."[23]

The U. S. government was in a strong position to dictate the future structure of the radio industry as a result of the war. As in the cases of communication and transportation, the Wilson administration had assumed control over radio in the war. Moreover, a bill was introduced in Congress in 1918 to retain government control of wireless. Notwithstanding the navy's support, the bill ran into considerable opposition, largely on the grounds that it would involve the navy in a commercial business. If the United States was not to play the major developmental role in ra-

[21] MacLaurin, *Invention to Innovation*, pp. 95–97.

[22] Federal Trade Commission, *Report on the Radio Industry* (Washington, D.C.: Government Printing Office, 1924), p. 27.

[23] Ibid., p. 29.

dio, government policymakers felt that American firms should. The strong patriotism that followed the war reinforced that feeling. GE, which had lost its substantial wartime market for expensive wireless equipment was negotiating the sale of the rights to its alternator (a critical radio component) to the American Marconi Company, a British subsidiary. Under pressure from the navy, GE changed its stance from seller to buyer.[24]

The director of Naval Communications proposed that GE should take the lead in forming a radio operating company. GE initially resisted, asserting that its business was in the manufacture and distribution of electric apparatus, not the operation of radio stations. The solution that the navy proposed was for GE to organize and sponsor a new operating company. This would simultaneously solve the navy's radio operation problem and provide GE with a market for radio equipment that would supplant the substantial wireless equipment sales lost because of the armistice.[25] But this solution, while attractive, ran head on into the patent blocking problem. Owen D. Young, GE's counsel and vice-president, observed: "It was utterly impossible for anybody to do anything in radio. . . at that time. The Westinghouse Company, the American Tel & Tel Company, the United Fruit Company and the General Electric Company all had patents but nobody had patents enough to make a system. And so there was a complete stalemate."[26]

All of these considerations required, first, that the new radio company should acquire the operating rights and pertinent patents of American Marconi. Second, it was important to assure that the new company would not be foreign dominated. Third, the new company should be structured so as to issue stock to the holders of the important patents in exchange for their contributing radio patents to a pool. But since a patent privilege grants a monopoly, the arrangement among the members of the patent pool would have to provide for a division of the product and service markets among the participants. For example, one company might make receivers, another transmitters, and so on.

With these considerations in mind the Radio Corporation of America was formed in 1919. Under the principal agreement RCA received title to all American Marconi assets (with certain exceptions). In exchange the Marconi company received 2 million shares of RCA's preferred stock. To assure that the navy was satisfied, the new corporation's charter required

[24] The material is recounted in Gleason L. Archer, *History of Radio to 1926* (New York: American Historical Society, 1938), pp. 156–67.

[25] In addition to Archer, *Radio to 1926*, pp. 164–67, see also Josephine Young Case and Everett Needham Case, *Owen D. Young and American Enterprise* (Boston: David R. Godine, 1982), pp. 173–80.

[26] Quoted in Archer, *Radio to 1926*, pp. 168, 169.

that only American citizens could become directors and officers. By the time the RCA-Marconi principal agreement was executed, RCA had entered into an agreement with GE, transferring all the GE radio patents to it and requiring GE to sell radio devices only to RCA and the U.S. government. RCA in turn was bound to acquire all of its radio equipment from GE.[27] Thus, at this stage of RCA's development, the navy was satisfied that radio development in the United States was in American hands, and GE now had a market for radio equipment to replace the lost wartime navy market. Only the patent problem remained.

AT&T Joins the Patent Pool

The key, of course, was to persuade AT&T to go along. If this could be done, the smaller patent holders would inevitably follow. Early in 1920 A. J. Hepburn, acting chief of the navy's Bureau of Steam Engineering, started the ball rolling by sending identical letters to AT&T and GE complaining that because of the patent situation, which has "prevented the marketing of tubes to the public such vessels are not able to communicate with greatest efficiency except with the shore. . . . It inevitably follows that the lives of crews and passengers are imperiled beyond reasonable necessity."[28] After putting the two companies on the spot, Hepburn directly urged them to remedy the situation forthwith. Although the navy sought little more than cross-licensing of vacuum tube patents, the agreements reached went beyond that. Indeed, they amounted to an allocation of exclusive and nonexclusive fields in virtually every phase of communications, including broadcasting, land telephone, and transoceanic telegraph.

Thus, before considering the July 1920 agreements, we should note that a conceptual revolution had taken place. The parties conceived of an overall field that we can call communication by wire and wireless. All of the skill that could be mustered by legal craftsmen was necessary to specify the "proper" boundaries between the companies. Since these agreements did much more than simply comply with the navy's request to unblock vacuum tube patents, a major inference of the comprehensive agreements was that without such an allocative agreement each firm would inevitably invade large parts of the others' fields.

As we will see, this indeed did happen in broadcasting, and the July 1, 1920, agreements, although carefully crafted, were soon subject to heated disputes between GE and AT&T. The agreements were ex-

[27] The agreements are reprinted in FTC, *Report on the Radio Industry*, pp. 118–30.

[28] The letter to GE is reprinted in Case and Case, *Owen D. Young*, pp. 209, 210.

tremely complex and consisted of a lengthy license agreement (the basic agreement between GE and AT&T); an extension agreement allowing GE to assign rights to RCA and AT&T to assign rights to Western Electric; and a termination agreement specifying the conditions under which the license agreement could expire before July 1, 1930, the termination date.[29] As carefully as the agreement was drawn, bitter disputes would soon erupt, largely because of ambiguities in the clauses concerning broadcasting.

But in 1920 none of this was foreseen. As the GE strategists anticipated, United Fruit, Westinghouse, and others joined the patent pool. United Fruit, which became involved in radio because it owned a fleet of ships, brought important patents for crystal sets and the loop antenna into the pool. Westinghouse brought many patents to the pool, including an important improvement for radio receiving sets (the feedback circuit). With these patents and others RCA was made the exclusive selling agent for radio sets that were to be manufactured by members of the pool. Of all the communications companies, only the domestic telegraph companies did not participate in the pool, although Postal Telegraph and RCA entered into a traffic agreement in 1922. Approached by GE, Western Union was reluctant to come to terms with RCA, then a much smaller competitor in transatlantic communications traffic.[30]

While the AT&T-GE basic agreement did not call for AT&T to become an RCA shareholder, GE's planners intended that the other participants in the pool should be enticed to enter into such a venture. Equity ownership by AT&T and other pool members would provide an infusion of capital and, therefore, an enlargement of RCA's resources. Second, the GE planners felt that such a commitment would enlarge the new companies' technological base by forging closer links with the other companies involved in radio research. Third, since considerable political reaction had developed over the claim that RCA was nothing but a GE front, the participation of other firms would partly deflect the charge. AT&T initially resisted an equity participation in RCA, but it eventually succumbed because GE's negotiators would not come to terms on any other basis.[31]

When a point is particularly salient to one party in negotiations and the other party is relatively indifferent, the indifferent party will usually make concessions. AT&T officials had not yet considered how important wireless would be and had not then formulated a clear view on how it

[29] These agreements are reprinted in FTC, *Report on the Radio Industry*, pp. 130–41.

[30] Hiram L. Jome, *Economics of the Radio Industry* (Chicago: A. W. Shaw, 1925), pp. 157–60 and Aitken, *Continuous Wave*, pp. 444–45.

[31] Aitken, *Continuous Wave*, pp. 446–49.

would fit into the company's overall business. For these reasons AT&T reluctantly purchased RCA stock. All appeared, for a short while, to be harmonious among the patent pool participants. In 1921 AT&T, GE, Westinghouse, and United Fruit held both common and preferred stock in RCA.[32] But in 1922 AT&T began selling off its RCA stock. The harmony would be short-lived.

Westinghouse and the Broadcasting Boom

When the navy had commissioned Westinghouse to undertake wireless research in World War I, Frank Conrad was the engineer in charge of the research. After the war Conrad began broadcasting as a hobby and discovered that a number of people in his area were picking up the broadcasts. In the spring of 1920 Conrad began broadcasting phonograph records on his station and soon began to entertain requests from his listeners. Later he devised a regular schedule for broadcasting records and, exhausting his own collection, borrowed records from a local record store in exchange for which its name was mentioned on the air. The record store discovered after a short while that its sales of the records played on the air were greater than those not played. Soon Conrad's sons, eager to join in, became announcers and introduced local Pittsburgh talent, who provided live entertainment. In turn, the increasing demand soon spurred entrepreneurs to make and sell "amateur wireless sets."[33]

The lessons of these events were not lost on Westinghouse policymakers, who drew the appropriate conclusions. H. P. Davis, a Westinghouse vice-president, stated a few years later:

> An advertisement of a local department store in a Pittsburgh newspaper, calling attention to a stock of radio receivers which could be used to receive the programs sent out by Dr. Conrad, caused the thought to come to me that the efforts that were being made to develop radio telephony as a confidential means of communication were wrong, and that instead its field was really one of wide publicity, in fact, the only means of instantaneous collective communication ever devised. . . . We became convinced that we had in our hands in this idea the instrument that would prove to be the greatest and most direct means of mass communication and mass education that had ever appeared.[34]

[32] Stock ownership data for 1921 and 1922 is contained in Gleason L. Archer, *Big Business and Radio* (New York: American Historical Co., 1939), pp. 7, 8; and Jome, *Economics of Radio*, pp. 56, 57.

[33] Archer, *Radio to 1926*, pp. 198–205; and Erik Barnouw, *A Tower in Babel* (New York: Oxford University Press, 1966), pp. 64–68.

[34] Quoted in Archer, *Radio to 1926*, pp. 200, 201.

Under Davis's guidance Westinghouse developed an ambitious plan consisting of erecting a powerful transmitter in the Pittsburgh area and developing a regular schedule for the station, which would broadcast every night. In this way listeners would develop the habit of routinely listening to the station in the same way that they routinely read newspapers. Obviously, listeners would require receivers that Westinghouse or a designee would produce. On October 27, 1920, the Department of Commerce licensed KDKA to Westinghouse. On Election Day—a day deliberately chosen by Westinghouse officials for dramatic impact—KDKA made history by broadcasting the 1920 presidential election returns. The acquisition of important patents during the same few months, together with the broadcasting achievement, made Westinghouse an important power in the wireless field. These factors led Owen Young to invite Westinghouse into the RCA agreements. And it also led the other wireless innovators to ask why they did not think of radio broadcasting first.[35]

The radio boom began almost immediately after KDKA began operating. In 1921 there was one licensed standard broadcast station (although there were others that were experimental). By 1922 there were 30, and by 1923 there were 556. In 1922 approximately 100,000 radio sets were produced. By 1923 the figure had jumped to 500,000, and by 1925 2 million sets were produced. Thus, a major industry had been created in less than five years. Perhaps most important, the promise appeared even greater. In 1922 60,000 households had radios. By 1923 the figure jumped almost sevenfold to 400,000. By 1925 2.75 million households had sets. But even this incredible increase barely touched the vast market of families that would almost certainly want one or more radio sets.[36]

AT&T and Toll Broadcasting

The temptation was too great to resist. AT&T wanted a part of at least some of the burgeoning broadcasting markets: transmitting equipment, transmission, or receiving equipment. Its strong patent position and participation in the RCA patent pool rendered the possibilities of entering one or more of these markets easier and, therefore, more tempting. But there were difficulties just as there were opportunities, the most important of which was AT&T's peculiar business status as a public service corporation. With its status as a protected telephone network manager it

[35] Barnouw, *Tower in Babel*, pp. 68–74 and Archer, *Radio to 1926*, pp. 200–204.

[36] The data is taken from Bureau of the Census, *Historical Statistics of the United States: Colonial Times to 1970* (Washington, D.C.: Government Printing House, 1975), p. 796.

could not simply enter any market it selected, even one in a related field of communications. Radio as a *supplementary* device in telephony to reach inaccessible places or to extend long distance raised no issues. But unless some rationale could be developed that showed the connection between radio broadcasting and AT&T's telephone obligations, which were imposed under the public service concept, and the specific undertakings AT&T had made over the years, the company could not defend its actions against the inevitable criticism that it was extending its monopoly.

We should realize that aside from the complaints about the "telephone trust," both Congress and the Federal Trade Commission (FTC) were deeply concerned about monopoly problems in radio. The FTC noted:

> In considering House Bill 13773, there was developed some evidence indicating that the radio art was being restricted by the acquisition of the basic patents on radio by closely affiliated interests. . . . The main cause of these complaints was no doubt the inability of manufacturers of receiving sets and dealers to obtain an adequate supply of vacuum tubes. With respect to the transmission of messages by radio, it was contended that exclusive rights had been granted certain parties which resulted in the elimination of competition in service to the localities covered by such contracts.[37]

The FTC went on to state that it received many complaints about the radio pool.

While AT&T's entry into broadcasting antedated the publication of the FTC report, criticism of the direction that the radio industry was taking began even before the construction of the patent pool. AT&T, aware of this, had much to lose if its moves into radio were used as "evidence" of its avarice in its main field — telephony. Its solution was to devise a rationale and mode of operation patterned on its telephone obligations with an equal "public service" obligation. Aside from this historical interest, AT&T's conception of toll broadcasting illustrates a path that the broadcasting industry might have taken that compares interestingly to the radio-television network-dominated system that gradually evolved in the 1920s and 1930s. Many of the complaints made about the banality and similarity of network television, and about the general public's inability to broadcast, *might* have been avoided if toll broadcasting had prevailed. Anyone, of course, can communicate by telephone; very few of us have ever been on television or radio.

Perhaps the best way to understand toll broadcasting is to compare wire and wireless transmission. In the former the telephone company supplied the transmission and receiving devices and distributed (includ-

[37] FTC, *Report on the Radio Industry*, p. 9.

ing switched) information between communicating parties. The telephone company cannot censor messages or determine who can use the system, subject, of course, to the payment of reasonable fees and other moderate regulations. In contrast to telephone, radio transmission equipment was very expensive and could not be supplied to subscribers on the same basis. Thus, unlike telephone users, the radio appeared to bar would-be users from transmitting what they wanted to communicate. They could receive messages because receiving equipment, even in the primitive stage of radio, was very cheap relative to transmitting equipment. Radio, from the perspective of one seeking to communicate with others, had only half the capability of wire telephony. Instead of intercommunication, broadcasting consisted of active transmission and dumb reception. Moreover, as the Westinghouse story illustrates, persons controlling transmission can effectively exclude anyone they want from providing messages or can censor what is transmitted. Toll broadcasting was intended to redress the differences between wire and wireless.

Although two-way radio and interactive cable transmission exist today, they are still unusual forms of broadcasting. Toll broadcasting was designed to remedy this problem. It is impossible to pinpoint the exact date that AT&T conceived the idea of toll broadcasting. Its 1921 Annual Report mentioned the *possibilities* of a one-way service consisting of a transmitter sending out news, music, and the like within a local area. But who would pay for this service was not set forth.[38] There were three principal methods to pay for a service in which waves spread out in many directions: you could license users of sets, have receiver manufacturers finance the programs, or have advertisers pay for time. The last method—commercial advertising—has prevailed in the United States, whereas the first method has been widely adopted in countries where public broadcasting prevails.[39] Toll broadcasting was most consistent with the third alternative.

The person primarily responsible for devising AT&T's broadcasting plans was its vice-president, Walter Gifford, who represented AT&T's interest in RCA and sat on RCA's board of directors. On January 26, 1922, Gifford approved the construction of broadcast equipment at the headquarters of the Long Lines department and the operation of an experimental station. The public announcement stated that AT&T "will provide no program of its own, but provide the channels through which anyone with whom it makes a contract can send out their own programs. Just as the company leases its long distance wire facilities for the use of

[38] AT&T, 1921 Annual Report, p. 20.

[39] William Peck Banning, *Commercial Broadcasting Pioneer* (Cambridge, Mass.: Harvard University Press, 1946), p. xx.

newspapers, banks and other concerns so it will lease its radio telephone facilities and will not provide the matter which is sent out from this station."[40] The announcement went on to describe the demand for such a service from newspapers, places of entertainment, and many kinds of businesses. It then noted the opportunity for market expansion; although the New York metropolitan area contained 11 million people, there were then only 35,000 receivers. Finally, if the system proved successful, AT&T envisioned a network of local stations connected by long distance wires so that the same program could be heard simultaneously in many places. In short, the idea of a network—but with major differences from current radio and television networks—was conceived by AT&T in its first experimental broadcasting venture.

Needless to say, the public announcement led some persons to charge that toll broadcasting was little more than a coverup for an attempt to extend the telephone monopoly into a new area. Equally important, it led to tensions between AT&T and the other members of the patent pool. Thus, RCA's president, J. G. Harbord, wrote to Gifford on September 7, 1923, charging that AT&T was going beyond the toll concept and was encroaching on RCA's nontoll broadcasting. RCA requested that AT&T refrain from doing so.[41] Gifford's reply further spelled out the conception of toll broadcasting. Under no circumstances would AT&T produce programs or otherwise engage in entertainment. But it reserved the right to fill up air time. Therefore, if a time slot was not sold, AT&T would provide the time to noncommercial interests. With a flourish, Gifford said AT&T's broadcast service would be available "for the use of *all* who desired to broadcast" (emphasis supplied).[42]

The canons of decency were established and, therefore, the word *all* is an exaggeration, but the system outlined certainly goes conceptually further toward opening up the airwaves to diversity than the system that eventually took root. If the toll conception had become the dominant form of broadcasting, stations would have become merely distributing agents for communication, leasing use of the costly transmitting devices. Because one receiver will not interfere with the use of another person (unlike wire telephony), there is no reason for the distributing common carrier to control, impose, or even suggest standards for the receiving instrument under the toll broadcasting system. Moreover, since even a relatively large block of broadcast time (for example, thirty minutes) can be subdivided (or wholesaled) into smaller units, there would have been a much greater possibility for diversity than under the system that has

[40] Quoted in ibid., p. 68.

[41] J. G. Harbord to W. S. Gifford, 7 September 1923, in AT&T Archives, Box 50.

[42] W. S. Gifford to J. G. Harbord, 12 September 1923, p. 2, in AT&T Archives, Box 50.

evolved. Like telephony, rates and other standards could have been regulated. But this was not to be. Control of broadcasting passed into the hands of radio networks.

The analogy with telephony was made even more manifest in a February 1922 AT&T press release. First, the release made clear that toll broadcasting was not only the company's plan for broadcasting, but a *model* for what the system of broadcasting generally should be. During this period, support for toll broadcasting, moreover, was coming from public officials who were reaching conclusions that drew close analogies between telephony and broadcasting. In March 1922 the Radio Telephone Technical Committee of the Department of Commerce held that "radio telephony" (by which they then meant broadcasting) was a "public utility" and that federal regulation was the only plausible solution to the interference that was increasingly cropping up as stations proliferated. Moreover, the committee recommended the division of the radio spectrum so that parts of it would be reserved for noncommercial and public uses. In that way there would be opportunity even for broadcasters who could not presumably afford the payment of a toll.[43]

AT&T obviously had much to gain from the widespread adoption of toll broadcasting. But guidelines would have to be spelled out on interconnection issues. Phone lines were necessary for several purposes in early radio: connecting the broadcasting studios to towers that were generally located outside of city limits, connecting mobile broadcasting facilities (such as those at football games) to the radio studio, and connecting stations in distant cities for simultaneous broadcast.[44]

Obviously, Bell System interconnection restrictions could cripple any broadcaster with more than the most modest ambitions, including another company that sought to use the toll system. In late 1921 an RCA memorandum expressed concern about AT&T's ability to arbitrarily deny telephone interconnection to broadcasters. "The Radio Corporation and the manufacturing companies will request the Telephone Company to waive their present restrictions with reference to connecting outside devices to its circuits so that private broadcasting stations of these companies may be connected to the regular wire lines of the Telephone Company."[45]

Another set of issues concerned AT&T's attitude toward the patents it controlled and the leverage it exercised through them. Anything short of a liberal policy of low royalty licensing to all reasonable applicants would

[43] "Urges Federal Rules over Radiotelephones," *New York Times*, March 11, 1922, p. 9.

[44] L.N. Stoskopf, "Telephoning Radio Programs to the Nation," *Bell Telephone Quarterly* 7 (January 1928): 5–16.

[45] *Memorandum to the Committee on Broadcasting Appointed by the Board of Directors, Radio Corporation of America*, December 29, 1921, p. 3, in AT&T Archives, Box 50.

be inconsistent with the public service company conception and, moreover, would provide ammunition to the company's enemies. On the other hand, it could not simply permit the infringement of its patents. Aside from the revenue loss, such a course of action would signal an abandonment of its patent rights and a tolerance toward other profit-seeking companies that would permit them to reap the fruits of AT&T's research and development without cost. Clearly, this was an equally unacceptable solution, especially as AT&T was embarking on the costly, risky, and—at least for a while—unprofitable venture in toll broadcasting.[46]

The equivocal approach AT&T took was to assert (but not litigate) its patent and interconnection rights. AT&T guarded against potential arbitrariness by adopting a liberal licensing and interconnection policy. Thus, in December 1921 Gifford wrote a letter to Westinghouse pointing out that the latter was interconnecting with Bell of Pennsylvania's wires. The letter continued: "I trust that you have not forgotten that the license agreement does not permit you to connect radio telephone equipment to the lines of any public service communication system. As you know we have this matter under consideration at the present time."[47] Yet GE was granted permission for a remote pickup that would be linked to its radio station. But AT&T pointedly asserted that this permission was not intended to establish a precedent. In early 1924 AT&T president H. B. Thayer, trying to justify AT&T's course of action, issued a statement to stockholders:

> We have recognized the fact that many broadcasters, in making wrongful use of our inventions, have been ignorant of their infringement. We have, therefore, established reasonable license fees, the payment of which, coupled with an agreement to refrain from further infringement, would liquidate any claims . . . and would give the broadcaster a legal right to the use of the patents. . . . The fees are so moderate as to represent a return far below the customary profits on unpatented electrical apparatus.[48]

Thayer claimed that four hundred radio stations (approximately 80 percent of those then in operation) were infringing on AT&T patents. The company, however, had not pursued its infringement claims vigorously and almost appeared to be pleading with broadcasters to take out a license. It was under attack as a "monopolist."[49] The tensions with its part-

[46] That AT&T considered toll broadcasting a high-risk experiment is clear from internal memoranda. See Memorandum, Edgar S. Bloom to W. S. Gifford, 9 November 1922, in AT&T Archives, Box 42.

[47] Quoted in Archer, *Big Business and Radio*, pp. 20, 21.

[48] H. B. Thayer, "The Radio Broadcasting Situation," *Bell Telephone Quarterly* 3 (April 1924): 115.

[49] See "Will Seek Control over Broadcasting," *New York Times*, March 7, 1924, pp. 1, 6.

ners in the RCA patent pool were on the rise, and the regulation that had been tried thus far was inadequate. In short, the broadcasting agreements that had developed so rapidly were highly unstable. They would soon fall apart. And toll broadcasting would collapse with it.

The Labyrinth of Disputes

AT&T's commercial station, WBAY, began broadcasting on August 3, 1922. Its initial broadcast was disappointing, its signal very weak. A few days later AT&T began to use Western Electric's transmitter, whose call letters were WEAF. The results were far superior, and pursuant to existing federal law that required a station's call letters to be identical to the call letters of its broadcasting towers, WBAY became WEAF. The preliminary announcement stated: "Anyone desiring to use these facilities for radio broadcasting should make arrangements with A. W. Drake, general commercial manager, long lines department . . . [who] can advise fully with reference to . . . periods of operation and the charges thereof."[50] Even though "anyone" could use the facilities of WEAF, the station did reserve the right to refuse such programs as the management thought to be undesirable. Nevertheless, the tenor of the policy statement indicates that the only restrictions would be based on standards of decency, not political or religious perspectives. From an internal 1923 policy statement, it is clear that the station's managers were more concerned with presenting a dignified tone than with dictating content.[51] It is obvious, however, that if the AT&T conception of toll broadcasting had continued, some formalized rules would have had to have been implemented in order to assure easy access to use the system.

But toll broadcasting never advanced that far. The earliest sign of trouble among the RCA partners was an April 1922 AT&T announcement that it had disposed of its RCA stock. The reason then advanced was that holding RCA stock contravened AT&T's policy to hold stock only in Bell System companies.[52] Perhaps that was the principal reason that AT&T disposed of RCA stock so soon after it was acquired. But AT&T must also have been concerned with a serious conflict of interest that would ensue from broadcasting in the New York area, for its principal rival there was WJZ, a joint venture of Westinghouse and RCA. Only one other station, WOR, operated by the Bamberger Department Store chain, was in the same power class as WJZ and WEAF. When WJZ applied to AT&T for

[50] Quoted in Archer, *Big Business and Radio*, p. 54.

[51] See "Radio Broadcasting in the Metropolitan Area," in AT&T Archives, Box 42. It is apparent that the memorandum was prepared in 1923.

[52] "Out of Radio Corporation," *New York Times*, April 22, 1922, p. 18.

use of leased wires in order to carry the 1922 World Series, the first time this would have been done, it was turned down. This event marked the opening of hostilities between what became known as, respectively, the "telephone group" and the "radio group."

RCA Chairman Young's complaint to AT&T's president met with a rebuff. President Thayer argued that the basic RCA license agreements forbade interconnection into the telephone network. In an October 13, 1922, letter, Thayer explicitly stated that it was the Bell System's intention to connect only its own broadcasting stations into the telephone network in those markets where it had stations. Where AT&T had "no broadcasting station, but one of the other parties to the license agreement has such a station, we have been ready to consider every case . . . and to give all possible consideration to meeting the needs of the situation, so that the public might obtain this service pending the period in which this whole broadcasting matter is being considered and brought to a definite policy."[53] On first impression the letter appears extremely arrogant, but the last portion of the letter must be considered in the overall context. AT&T was awaiting clarification of its broadcasting problems before it would adopt a definite policy. Until then it would defer any long-term commitments and would assert all of its rights.

As we have seen in previous chapters, public service companies are under a common law obligation to serve persons and businesses (except telephone companies) seeking connection into the system who are willing to abide by the telephone company's reasonable regulations. Although many states enacted compulsory telephone interconnection laws, it was not required if it worked a substantial detriment to the quality of service.[54] And that was the basic problem AT&T and others pointed to in radio. Stations interfered with one another, reducing service quality. We have already noted Secretary of Commerce Hoover's analogy of radio to automobile traffic. The quality of every broadcaster's transmission was being reduced by the interference caused by others.

Hoover endorsed the conclusions of the first conference of radio representatives in February 1922, which recommended government regulation, including frequency allocation assignment by the secretary of commerce to each station and a restriction on the same frequency assignment to stations within 750 miles of one another; but Congress did not act. Indeed, it did not act until 1927. Moreover, to illustrate the chaotic events at the time the above AT&T-RCA correspondence was taking place, one should consider Hoover's remarks at the second conference in

[53] Quoted in Archer, *Big Business and Radio*, p. 58.

[54] Ellsworth Nichols, *Public Utility Service and Discrimination* (Rochester, N.Y.: Public Utilities Reports, 1928), p. 652.

March 1923: that in one year the number of broadcasting stations had increased from 60 to 588. At the time only three frequencies were available for broadcasting.[55]

It is within this context that AT&T strongly endorsed legislative action to regulate radio. Its 1922 Annual Report asserted that interference destroys the value of radio, even as a source of entertainment. "It is the expectation that ultimately by national and international control interference will be obviated."[56] For the same reason at the second conference AT&T took the position that, given the small number of frequencies then available, the number of stations should be reduced. Under this view, the interests of listeners ought to be paramount. One licensee in each area should act as distributor for all persons wishing to broadcast. A federal authority would assign the distributor a sufficient number of wavelengths to assure that this obligation was consistent with high-quality service and noninterference. Thus, under the plan, AT&T advocated the then revolutionary idea of tuning receiving sets up and down the wave scales to listen to any program desired.[57] The company pointed out that under such a system one would not have to make a substantial investment in broadcasting equipment in order to be heard. The system, thus, would have encouraged diversity as well as the development of programs by a wide variety of groups. Further, the system contemplated that the information distributor would play no role in programming.

It is within the context of the potential for chaos, the lack of government regulation, and AT&T's toll plan that its behavior must be considered. Although AT&T obviously was very self-interested, that alone did not mean that its behavior was contrary to the public interest. Nor, in view of government's fumbling in the matter, could AT&T be accused of arrogance, especially since it continuously favored and actively sought government regulation of radio. But its position was bound to lead to tensions within as well as outside the patent pool. Moreover, the situation was complicated by the need to assert its patents against infringers. If it failed to do so, as an AT&T bulletin to its affiliates stated, it "may prove embarrassing from a patent standpoint."[58]

Attempts to mediate the disputes between the telephone group and the radio group in 1922 failed on the interconnection issue as well as on issues involving interpretation of the license contracts application to broadcasting. In late 1923 tensions between the parties were further

[55] Laurence F. Schmeckebier, *The Federal Radio Commission: Its History, Activities and Organization* (Washington, D.C.: Brookings Institution, 1932), pp. 4–6.

[56] AT&T, 1922 Annual Report, p. 19.

[57] "Wants Less Broadcasting," *New York Times*, March 23, 1923, p. 17.

[58] Quoted in Federal Communications Commission, *Investigation of the Telephone Industry in the United States* (Washington, D.C.: Government Printing Office, 1939), p. 390.

heightened, and the matters were referred to an arbitrator, Roland Boyden, a highly respected Boston attorney. In early 1924 a draft of his decision was prepared that favored the radio group on virtually every disputed point. One of the most important questions was whether the phrase *wire telephony* included the furnishing of wire facilities to broadcasters. If so, it would have been exclusively within the province of AT&T. The radio group claimed, however, that it was intrinsically a part of broadcasting, a contention with which the arbitrator agreed.[59]

Although the setback to AT&T was enormous, the radio group also suffered several disappointments.[60] But AT&T still had a strategy that it would use before the arbitrator's decision became final. In a December 1924 report RCA's general sales manager summarized AT&T's new argument and the quandary it presented to the radio group.

> Finally, the Telephone Group attacks the referee's decision from a new angle pointing out in the second brief . . . that the contract, as interpreted, becomes an agreement by the parties for non-use in certain fields of inventions of one or both groups . . . "in other words, it forces both the Radio Group and the Telephone Group into an agreement for suppression or non-use of inventions in certain fields." . . . Such an agreement is illegal and comes within the scope of the Sherman Antitrust Act.[61]

In short, according to AT&T's view, the license agreement was illegal and, therefore, unenforceable. The arbitrator's decision, then, could not lawfully be obeyed. AT&T employed John W. Davis, the 1924 Democratic presidential candidate and an outstanding legal scholar, to assess the situation. In essence he supported the AT&T position that the license agreement was unenforceable. One should note that in January 1924 the FTC had brought an action against eight members of the radio pool, charging a conspiracy in violation of the FTC Act, section 5. In essence the FTC, whose charge tended to support Davis's argument, claimed that the conspirators had engaged in seven major unfair methods of competition.[62] Given these factors as well as legislative interest in radio, the best strategy for both the radio group and the telephone group was to privately settle their disputes and not to pursue legal remedies with the attendant publicity. Thus far the dispute had been kept relatively quiet; both the referee's draft report and the Davis opinion, for example, had

[59] The arguments and the arbitrator's conclusions are extensively quoted in FCC Special Investigation Docket 1, Exhibit 289, *Bell System Policies and Practices in Radio Broadcasting* (1936), pp. 22–28.

[60] See the 15 November 1924 letter from Albert G. Davis to GE President Gerard Swope summarizing the draft, in Archer, *Big Business and Radio*, pp. 170–72.

[61] Quoted in ibid., p. 173.

[62] See the summary in Jome, *Economics of Radio*, pp. 222, 223.

been circulated *only* to the disputing parties. If the tensions were further exacerbated through formal legal proceedings, the entire affair could become a political football match in which both sides would be losers.

While the ensuing negotiations were taking place, both sides were active on other fronts. From the AT&T perspective the most important one was the pursuit of the radio network idea, in which radio stations would be linked by telephone lines so that the joined stations could simultaneously broadcast the same program. Toll broadcasting was thus further enlarged, since local, regional, and national programs would be possible. As early as January 1923 AT&T conducted an experimental network broadcast. The first regular network of modest proportions began during the summer of 1923. By the fall of 1923 the AT&T network consisted of three East Coast stations, and the concept had been firmly established.[63] Gradually AT&T adopted a plan for a widespread network toll system consisting of owned and licensed stations.[64] When one considers the advantages that a network had, even in the 1920s, over purely local stations and AT&T's ascendant position as telephone system network manager, it is apparent that this provided AT&T with a great advantage in broadcasting compared to the radio group or other competitors.

Of course, these competitors considered a number of ways to circumvent connecting into the telephone network. One alternative was to use the telegraph lines of Western Union or Postal Telegraph. While this could be done, trial results were very unsatisfactory. Telegraph lines, as we noted earlier, were inadequate for telephonic speech, and especially bad for the transmission of the range of speech and music that radio required. Moreover, transmission was often interrupted by the dots and dashes of regular telegraph transmission.[65] The high risks and costs of upgrading made improvement of the telegraph lines an unlikely undertaking.

The other alternative, forcefully advocated by RCA General Manager David Sarnoff as early as 1922, was to avoid the telephone lines almost entirely by resorting to superpower transmission that would cover large areas. GE and RCA had already begun the operation of experimental superpower stations, and the dispute between the telephone group and the radio group spurred RCA to make further improvements in superpower transmission. Although promising, superpower presented enormous obstacles compared to a telephone linked system. Long wave superpower was encumbered with severe static noise, power requirements were very high, and the antennas needed were large and expensive

[63] Banning, *Broadcasting Pioneer*, chaps. 11, 12.

[64] Barnouw, *Tower in Babel*, p. 145.

[65] MacLaurin, *Invention to Innovation*, pp. 114, 115.

structures. Short wave superpower, while solving some of these problems, had serious difficulties of its own. Interference from such sources as automobile ignition systems was severe. Not all frequencies were usable at all times; rather, a frequency would be usable for a few hours and then would have to be replaced with another. Further, such changes occurred frequently. Fading and distortion were major problems, and service could disappear entirely because of magnetic storms caused by sunspot activity.[66]

While the radio and telephone groups were engaged in their dispute, other firms had moved into the various broadcasting markets and drew the attention of both groups. AT&T was caught in the quandary of whether its public service obligations or its patent rights should prevail in cases of conflict. Since it asserted patent claims in connection with broadcasting transmitters, it had to either enforce or waive these claims. Many stations were infringing on the claims because transmitters could easily be assembled from readily purchasable components. Fearful of adverse publicity if it enforced the claims, AT&T at first ignored the alleged infringements. AT&T policymakers in 1923 adopted the path of forgiving any station its past infringement if it agreed to pay a license fee ranging from five hundred to two thousand dollars, depending on the power of the station. Refusal to take out a license would subject the alleged infringer to a lawsuit. But more important, telephone interconnection would be denied infringers.[67] Such denial may have been inconsistent with public service obligations. Still stations did not comply.

A 1924 AT&T internal memorandum strongly recommended against a *general* attempt to force compliance. "Any general attempt to force nonlicensed broadcasting stations to either obtain a license under our patents or to stop broadcasting is probably impracticable and undesirable from a public relations standpoint."[68] Instead, AT&T took the path of selective enforcement, choosing WHN of New York as its first defendant. WHN's transmitter included parts that were covered by patents controlled by AT&T and other pool members. Under advice from patent counsel, it was necessary to assert patent rights. On March 5, 1924, AT&T requested the federal district court to enjoin WHN from broadcasting. AT&T asserted that if the case against WHN was successful, it would pursue others. Decrying the AT&T monopoly and the concept of toll

66 Fagen, *Engineering and Science*, p. 408; Archer, *Radio to 1926*, pp. 264, 371; Banning, *Broadcasting Pioneer*, pp. 179–81; and Archer, *Big Business and Radio*, pp. 51, 52, 96–98.

67 Barnouw, *Tower in Babel*, pp. 117, 118; Banning, *Broadcasting Pioneer*, pp. 131–40; and FCC, *Bell System Policies*, p. 95.

68 Quoted in FCC, *Bell System Policies* p. 98.

broadcasting, WHN vowed to fight to the bitter end.[69] The David-versus-Goliath aspect of the battle stoked up charges of monopoly against AT&T.[70] The telephone group defended itself by claiming that it only sought its due under the patent laws and would license all applicants on reasonable terms. It did not desire a radio monopoly.[71]

While the WHN suit was pending, AT&T also had to defend itself against charges of monopoly in a House committee hearing on a bill, part of which would deny a license to a station that was unlawfully monopolizing or seeking to monopolize broadcasting in any market. Under all of these circumstances it is not surprising that the WHN suit ended quickly. In early April 1924 both sides announced a settlement of the suit. WHN agreed to take out a license from AT&T for two thousand dollars. Interestingly, although WHN officials originally attacked the toll concept, the agreement authorized the station to engage in toll broadcasting.[72] Even though this affair was settled, AT&T faced a more fundamental issue. Large as the company was, it still had scarce resources. Should those resources be placed in the booming field of wire telephony? Should AT&T commit substantial resources to radio, which thus far had caused it much trouble?

The End of Toll Broadcasting

Even though AT&T had enjoyed success in broadcasting, there were doubters in the company all along. A February 1924 confidential memorandum from AT&T Assistant Vice-President A. G. Harkness to Vice-President E. S. Bloom spelled out many of the difficulties AT&T would face in broadcasting and proposed several *tentative* solutions. While arguing that no firm is better equipped to carry out broadcasting than AT&T, Harkness nevertheless said: "It may be undesirable for many reasons, and from the standpoint of both organization and public relations, for the Bell System to develop and carry on radio broadcasting as an integral part of telephone service."[73] One possibility was a separate corporation that would be owned by AT&T and that, in turn, would own and operate stations. Other suggestions were made, including the defensive

[69] "WEAF Begins Suit to Silence a Rival," *New York Times*, March 6, 1924, p. 15. On AT&T's need to enforce patent claims, see *Brief of Bell System Companies on Commission Walker's Proposed Report on the Telephone Investigation* (1938), pp. 198–202.

[70] "Will Seek Control over Broadcasting," *New York Times*, March 7, 1924, pp. 1, 6.

[71] "AT&T Heads Deny Monopoly of Radio," *New York Times*, March 13, 1924, p. 21.

[72] Ibid., and "WHN Gets License Ending Radio Suit," *New York Times*, April 11, 1924, p. 19.

[73] A. G. Harkness to E. S. Bloom, 2 February 1924, p. 4, in AT&T Archives, Box 42.

one of Western Electric refusing to sell high-power transmitters to others until a definite policy was established. A separate subsidiary would go some way toward solving the inherent contradiction between public service responsibilities and the inherent right of other companies to select their customers, and hence refuse to deal with some firms. But in a company the size of AT&T the claim of a subsidiary operating independently would probably not have been believed. The slightest evidence of wrongdoing by the subsidiary would have resulted in fingers pointed at AT&T.

It is next to impossible to determine which event finally led to a settlement between the radio group and the telephone group and AT&T's withdrawal from broadcasting. Nevertheless, several investigators have pointed to the H. V. Kaltenborn episode as critical. The incident certainly took the soul out of the toll broadcasting concept. The *Brooklyn Eagle* newspaper had been invited to present a series of talks on WEAF about world affairs. The paper chose its associate editor, the outspoken Kaltenborn, to present his views. Early in 1924 Kaltenborn sharply criticized Secretary of State Charles E. Hughes's outright rejection of Soviet overtures to establish diplomatic relations with the United States. Hughes, who was tuned to the broadcast with several prominent guests in attendance, was extremely angry, called a Washington AT&T representative, and "laid down the law to him."[74] The word relayed to AT&T's New York headquarters was that Kaltenborn should not be permitted to criticize public officials.

The political pressure mounted when President Coolidge expressed his displeasure at the controversial broadcast.[75] According to Edgar H. Felix, a WEAF publicity director, "The Kaltenborn episode and other disputes gave the telephone executives 'nightmares.' They had originally embraced the 'toll' conception with the beguiling thought that they could lease facilities without responsibility; this seemed a sound telephone approach. Now they were enmeshed in agonizing policy problems."[76] Later in the career of broadcasting, commentators would routinely criticize government officials and policies, but in the medium's infancy the boundaries of permissible commentary were not clear. A public service company highly regulated at both the federal and state levels had to feel particularly vulnerable to the potential wrath of government. Although telephone service was expanding rapidly during the 1920s, criticism of the "telephone trust" continued unabated, and many modern liberals of the day recommended dismemberment of AT&T, public ownership, and more.

[74] Quoted in Barnouw, *Tower in Babel*, p. 140.

[75] Banning, *Broadcasting Pioneer*, p. 156.

[76] The quote is from Barnouw, *Tower in Babel*, p. 141, and is a paraphrase of Felix's reminiscences in the Columbia University Oral History Collection, New York.

A few examples will illustrate the continual barrage of political criticism faced by AT&T in the telephone field. In 1924 New York Representative John J. O'Connor demanded that Congress investigate the "complete monopoly" AT&T. With the perspicacity one often finds in legislative "experts," O'Connor claimed that excessive rate increases were due to the installation of the dial system, which he described as "a nuisance [which] may ultimately have to be abandoned."[77] In September 1926 New York City, Boston, Los Angeles, and other cities complained to the ICC that AT&T was a monopoly in violation of the antitrust laws. Rate increases had triggered the municipalities' complaints.[78] The barrage of charges in the telephone part of the business was trouble enough for AT&T. Alone it may not have been enough to persuade AT&T to leave the broadcasting business, but taken together with the Kaltenborn controversy, the battle with the radio group, and other radio problems, the company decided to leave broadcasting.

After protracted negotiations both sides came to agreement, signing the twelve documents that made peace between the radio and telephone groups in July 1926. AT&T agreed to withdraw from broadcasting and sell its existing stations to RCA. In its public statement announcing the sale of WEAF to RCA, AT&T said: "The further the experiment was carried the more evident it became that while the technical principle was similar to that of a telephone system, the objective of a broadcasting station was quite different from that of a telephone system. Consequently . . . the broadcasting station which we built up might be more suitably operated by other interests."[79] As a result of the agreement RCA controlled the two best stations in the nation's largest metropolitan area. In the Washington D.C., area, AT&T agreed to discontinue WCAP and sell its air time to WRC, an RCA affiliate. With these steps RCA was well on its way to the establishment of a network. The National Broadcasting Corporation (NBC) was incorporated in September 1926 and sent AT&T a check for $1 million to take title to WEAF, a key station in the new network.[80]

In statements announcing the formation of NBC, RCA asserted that it approved the concept of network broadcasting, which AT&T had pioneered. But though it employed the phrase "toll broadcasting" for a short while, gradually dropping it, RCA's conception—formulated largely by

[77] "Again Asks House for Phone Inquiry," *New York Times*, May 28, 1924, p. 36.

[78] "Cities Join to Fight Higher Phone Rates," *New York Times*, September 4, 1926, p. 1. See also "Gifford Again Denies Illegal Monopoly," *New York Times*, September 8, 1926, p. 29.

[79] Quoted in "WEAF to Merge with Radio Station WJZ. AT&T Will Quit the Broadcasting Field," *New York Times*, July 22, 1926, p. 1.

[80] Barnouw, *Tower in Babel*, p. 186.

David Sarnoff—differed in critical respects from AT&T's. The telephone company envisioned a broadcasting station as a big telephone booth that anyone might use after waiting in line and paying a reasonable fee. If lines were too long, more "telephone booths" in the form of additional stations would be added. In this way a maximum of cultural, social, and political expression would be aired. As we saw, government frightened AT&T slightly away from the original conception. But it still remained the ideal. From the outset RCA set a very different tone for broadcasting—a tone that exists to the present. NBC would control what went on the air. It would determine and often produce entertainment and "news." Only those persons and entertainments that met NBC's approval would be permitted on its network.[81] AT&T's toll idea, which would have encouraged diversity, was dead.

Of course, AT&T gained much from the agreements or it would not have signed them. The original license agreement was modified so that furnishing wire circuits for broadcasting would explicitly be considered telephony and, therefore, under the control of AT&T, which received an exclusive license under GE's patents for furnishing wire service. If AT&T failed to provide them, GE and RCA would have the right to do so. AT&T, however, agreed to furnish the broadcasting group with wire facilities. Both the telephone group and the radio group were allowed to manufacture transmitting equipment under the pooled patents. Superpower transmission, once advocated by Sarnoff, was forgotten; wire transmission would be the mode employed in broadcasting. Members of the radio group, which had to some extent been using telegraph lines, would shift entirely to AT&T. And though the agreements permitted AT&T to reenter broadcasting after July 1, 1936, the company left the field forever, aiding RCA and NBC, through its interconnection policies and patent pooling, to achieve dominance in the field.[82] It was clear from internal memoranda that AT&T intended to withdraw entirely from all aspects of broadcasting, even the manufacture of radios, which the agreements permitted.[83]

But while the agreements left the radio group and AT&T satisfied (or, at least, not dissatisfied), many other participants in the radio industry who were not members of either group were very disturbed—for example, companies formed during the 1920s, notably set manufacturers such as Zenith, Philco, and Emerson; independent radio stations and CBS, a new network begun in 1927. CBS, for example, charged that AT&T was

[81] For details, see ibid, chap. 4.

[82] Summarized from FCC, *Bell System Policies*, sec. 3.

[83] Edgar S. Bloom to all presidents, 6 August 1926, in AT&T Archives, Box 42.

reluctant to furnish it with wire facilities.[84] These companies' complaints would be translated into government action and would eventually contribute to the founding of the Federal Communications Commission. Radio regulation had wandered far from the path of public service liberalism. Under modern liberalism it ironically combined business firms responsible to no one but their shareholders with a powerful government licensing agency that could supervise the content of programs. The public's ability to use the new medium was beside the point.

[84] On the independent set manufacturers, see MacLaurin, *Invention to Innovation*, chap. 7. On CBS charge, see Barnouw, *Tower in Babel*, pp. 193–95.

9

The End of the Old Deal

Prelude to the New Deal

THE NEW interventionist philosophy, ascendant during World War I, retreated from the end of that war until the Great Depression. But as the case of radio regulation illustrated, the new philosophy, although no longer at center stage, continued to play a significant role in American public policymaking. It would have to await the crisis of the Great Depression before it became the dominant theory of public policymaking. Nevertheless, even during the three Republican presidencies in the 1920s, radio regulation was hardly an aberration. In numerous areas government intervened in ways inconsistent with public service liberalism, even as the older system flourished under the guidance of public utility commissions. Indeed, a single agency—the ICC—would embrace old and new theories of public policymaking.

Consider only the following measures adopted during the 1920s. The most important regulatory statute enacted during the era was the Transportation Act of 1920. Although it manifested compromises between shipper and railroad interests, the single most important aspect of the statute was the ICC's mandate to prepare a plan for a new national network based on the consolidation of weak roads and strong ones within the framework of competition. Central to the statute was the idea that the ICC would employ scientific management techniques to produce a highly efficient system. For the same reason the agency was empowered to fix rates that a scientific and efficient management would; income based on rates in excess of those deemed fair under this standard would revert to government under the so-called recapture clause. Even service details and securities matters had to be approved by the government agency.[1] In short, under the 1920 act, instead of the cooperation and division of responsibility under public service liberalism, government planners had the upper hand.

[1] See Robert L. Cushman, *The Independent Regulatory Commissions* (New York: Oxford University Press, 1941), pp. 116–29; I. L. Sharfman, *The Interstate Commerce Commission* (New York: Commonwealth Fund, 1931), pt. 1, pp. 177–244; and Stephen Skowroneck, *Building a New American State* (Cambridge: Cambridge University Press, 1982), pp. 248, 281, 282, 286. For a somewhat different view, see K. Austin Kerr, *American Railroad Politics* (Pittsburgh: University of Pittsburgh Press, 1982).

Banking policy, too, underwent a subtle transformation during the 1920s that would have important ramifications during the New Deal and after. Although macroeconomic policy has existed in a rudimentary form for centuries, a major breakthrough occurred in the 1920s when Benjamin Strong, the governor of the Federal Reserve Bank of New York, came to dominate the Federal Reserve System. His ascent signaled a major shift in emphasis in Federal Reserve policy, from regulating banks for depositor protection to macroeconomic policymaking. In the 1920s, while not ignoring traditional regulatory functions, Strong employed open market operations in government securities as a means to plan high and stable levels of business activity and to control inflationary tendencies.[2] Just as an important precursor of New Deal macroeconomic policy existed in the 1920s, so also the New Deal's modern liberal version of agricultural policy was firmly rooted in 1920s policy. Whereas President Calvin Coolidge vetoed legislation that had no other purpose than to protect farm incomes, President Herbert Hoover offered the Agricultural Marketing Act, which committed the federal government to raise farm prices by purchasing surplus crops. Of course, this was a minor effort compared to the elaborate set of agricultural programs that began during the New Deal and have continued into the 1990s.

As these examples show, radio policy was not unique. Modern liberalism continued from the wartime period and came to dominate American public policymaking in the New Deal. Its central hypothesis was that government could plan, sometimes in conjunction with the behavior stemming from the operation of markets, as in the case of radio, sometimes supplanting market mechanisms, as in the case of agriculture. But whether we look at the microsocietal or macrosocietal levels, or whether we consider such pre–New Deal policies as open market operations or such New Deal policies as the National Industrial Recovery Act or the development of fiscal policy, planning through governmental action has been a critical part of modern liberalism. No longer is the model of public service liberalism employed: set public goals, ask whether the unaided market can properly achieve them, and if it cannot, devise a public-private collaborative arrangement to attain them, with the private sector in charge of attaining goals subject to government supervision. Under modern liberalism, we look to government institutions to plan the attainment of goals.

Planning leads to the second key characteristic of modern liberalism: a new mandarinate of professional administrators is needed to engage in the various planning processes undertaken by government. Professional

[2] This summary is based on Lester V. Chandler, *Benjamin Strong, Central Banker* (Washington, D.C.: Brookings Institution, 1958), passim.

economists, accountants, public administrators, scientific and technical experts, and most of all lawyers began to staff the positions of power in the increasingly large federal and state bureaucracies. While the language of scientific management atrophied, faith in it did not; rather, it increased. Thus, even in the period from 1921 to 1931 the number of federal civilian employees increased from approximately 561,000 to 610,000. But by 1940 it was in excess of 1 million.[3] And that, of course, was only the beginning.

Planning implies control and compulsion. In order to be effective, administrators must have the ability to control the recalcitrant through sanctions. Government always has been able to compel obedience through sanctions. But there is an important difference between compulsion and control under public service liberalism and what it involves under modern liberalism. The spirit of the former lies in cooperation or collaboration to achieve such goals as universal telephone service, for example. Lines of responsibility between private and public sectors are clear; the technical experts who formulate the plans *to achieve* goals are in the private sectors. Government agents monitor, looking over the shoulders of those in the private sector, prodding where necessary. Compulsion or antagonism is a last resort, signaling the breakdown of collaboration. In contrast, planner-bureaucrats are at center stage under modern liberalism, essentially giving orders to achieve *their* goals. Not surprisingly, almost from the outset of the New Deal, conflict between administrators and those under their control supplanted cooperation. And regardless of whether successor presidential administrations have been Republican or Democratic, the antagonism has continued unabated; in business circles (and others as well) *bureaucrat* has become an epithet, not a technical description of an administrator.

Goals, whether under public service liberalism or the modern type, are numerous. But asking the basic questions in the manner advanced by John Stuart Mill acts to restrain the zeal that societal groups manifest to transfer resources or privileges to them through government programs. Public goals will always remain numerous and flexible, but modern liberalism has no pattern of restraint comparable to that delineated under public service liberalism. Accordingly, more and more groups looked primarily to the state to achieve their desired ends, and planning administrators, eager to enlarge their jurisdictions, were more than happy to oblige. The enormous explosion in government-run transfer mechanisms and entitlements that began to increase dramatically during the Lyndon Johnson administration is a direct outgrowth of modern liberalism's ear-

[3] U.S. Bureau of the Census, *Historical Statistics of the United States: Colonial Times to 1970* (Washington, D.C.: Government Printing Office, 1975), p. 1102.

lier career. But it was during the New Deal that modern liberalism became the predominant philosophy of public policy in America.

The Coming of the New Deal

The assumption of Franklin D. Roosevelt to the presidency in 1933 after a landslide victory in 1932 was unquestionably one of the paramount events in American political-economic history. There was considerable policy disagreement among various groups during the New Deal period. Few defenders of public service liberalism as a *general principle* remained, although many of its public institutions, such as public utilities commissions, stayed intact and a few new programs were based on it. In large part, dissatisfaction with the failings of the old economic order, rather than with an explicit and detailed philosophy of what should replace it, led to the adoption of modern liberalism. John Stuart Mill had not been overruled by superior wisdom; he had merely been shunted aside in favor of a new style of public policymaking.

The New Deal's policies cannot be reduced to a single coherent philosophy. The New Dealers themselves were of several schools. As historian E. W. Hawley shows, by early 1938 there were four discernible trends: antitrusters who sought to restructure the American economy, budget balancers, spenders, and planners.[4] All the prevailing trends were inconsistent with public service liberalism. The Great Depression created conditions that were beyond the control of the public. Without fault on their part, workers became unemployed, businesses went bankrupt, and farmers lost their homesteads. Thrift, hard work, and the other virtues that were supposed to lead to material rewards failed to do so. The policy changes that took place are concisely summarized in the words of Rexford Guy Tugwell, one of the premier New Deal brain trusters:

> Men and women do not ask much from this world in which they find themselves—not more, at least, than ought to be guaranteed them by our resources and achievements. They ask security—security of access to the goods of simple living, security of employment, security in ill-health and old age. . . .
>
> If we ask why it is, with an equipment adequate to this purpose, this provision of a minimum of security still remains conspicuously lacking, the answer must be discovered in the kind of economic and political organization which dominates our industrial machine.[5]

[4] E. W. Hawley, *The New Deal and the Problem of Monopoly* (Princeton: Princeton University Press, 1966), p. 398.

[5] Rexford G. Tugwell, *The Industrial Discipline and the Governmental Arts* (New York: Columbia University Press, 1933), p. 199.

The concept of "security" clearly has an appeal, but it is one latent with ambiguity. Escaping the vicissitudes of the market for business people, of weather for farmers, of unemployment for workers, was only the easily understood beginning of attempts to end all uncertainty. Security, in short, is extraordinarily elastic in application. It led to the attempt at government-dominated comprehensive economic planning in the National Industrial Recovery Act, the shift of planning power to government from the banking industry in the 1933 and 1935 banking statutes, and a host of other statutes that enlarged American state power. In this way government (first federal, then state and local) adopted the role of problem solver. If there was a "problem"—inevitably a source of insecurity—the habit took root of first looking to government to solve it. In contrast, abandoned was the old system of asking whether the market could satisfy public goals and only after a negative finding bringing government in.

Government during the New Deal sponsored such disparate legislation as minimum wage setting and the National Labor Relations Act, the Social Security Act, agricultural price support programs, and small business protective statutes such as the Robinson-Patman price discrimination law and fair trade laws. Demands for the solution of problems came not only from interest groups outside government, but from the burgeoning administration within it. As we will see, the complex of factors that shaped the New Deal generally led to the enactment of the 1934 Communications Act, but with one additional element: it embraced—at least in part—the public service principle. The Act included elements of the new philosophy as well as a continuation of the past in telephone regulation.

The Origins of the Federal Communications Commission

At the most obvious level the FCC was an institution that combined the radio regulatory functions of the Federal Radio Commission and the telephone regulatory functions of the ICC. The important "Roper Report," prepared in 1934 under the direction of President Franklin D. Roosevelt by an interdepartmental committee under the leadership of Secretary of Commerce Daniel C. Roper, stated:

> The present diversity in the regulation of means of communication should be ended in the interest of economy, convenience and efficiency. . . . There should be one place in the Federal Government to which one could go with any complaint or seeking redress of any grievance over which the Federal Government would properly have jurisdiction against these monopolies [telephone and telegraph]. . . . There is growing discontent with certain features of radio

> broadcasting. . . . The companies using these agencies—the telephone, the telegraph and broadcasting—should be regulated by one body.[6]

Similarly, President Roosevelt, in transmitting a communications bill to Congress, stated: "I have long felt that for the sake of clarity and effectiveness the relationship of the Federal Government to . . . 'utilities' should be divided into three fields: transportation, power and communications. . . . In the field of communications, however, there is today no single agency charged with broad authority."[7] Accordingly, the president recommended the creation of such a unified body, to be known as the Federal Communications Commission.

The House of Representatives report went further, speaking of the ill effects of "disjointed regulation." It stated that the communications bill was largely based on existing legislation and except for the change of administrative authority "does not very greatly change or add to existing law."[8] Nevertheless, there was more involved than administrative consolidation. The report envisioned a future report by the new FCC on such topics as whether AT&T ought to be compelled to engage in competitive bidding for equipment instead of relying on Western Electric. Further, the House report complained that the statutes administered by the ICC contained important gaps that hindered effective regulation. The law establishing the FCC was intended to be relatively uncontroversial. But on the other hand, it was expected to be a prelude to more drastic regulation that would follow an FCC investigation of telephone practices. The FCC investigation did take place, but the subsequent legislation never did.

AT&T's opposition to the enactment of what became the Communications Act of 1934 stemmed, in part, from the fear that worse would follow. Certainly there was much in the political atmosphere of the early New Deal to support the company's anxiety. Within one hundred days after Franklin D. Roosevelt's inauguration the president had guided a revolution in government-business relations through Congress—and he promised more. Among the statutes that rapidly sailed through Congress were those that: divorced commercial banking from investment banking and imposed new layers of regulation on commercial banks, set up agencies to regulate agriculture, established federal control over new securities issues, created the position of federal coordinator of transportation, and most controversially, created the National Recovery Administration

[6] U.S. Senate, *A Study of Communications by an Interdepartmental Committee* (Washington, D.C.: Government Printing Office, 1934), pp. 25, 26.

[7] The president's message is contained in U.S. House of Representatives, *Report to Accompany S.3285* (June 1, 1934), reprinted in Bernard Schwartz, ed., *The Economic Regulation of Business and Industry* (New York: Chelsea House, 1973), 4:2437.

[8] Ibid., 4:2438.

(NRA), which was to establish federal supervision over virtually every activity in almost every industry.[9] From AT&T's perspective the federal government had in many industries marched well beyond the boundaries between private management and regulation that public service liberalism contemplated. It did not take too much foresight to envision that, given the public mood toward business, the establishment of the FCC would likely be a prelude to overly stringent regulation—or even nationalization.

Even more ominous was the experience of the electric power and gas industries—the sister public utilities—during this period. During the 1920s holding companies were established in these industries, the alleged purposes of which were to evade state regulation by incorporating out of the state or states in which the operating companies supplied their services. The claim was that this rendered state-mandated utility accounting practices virtually ineffective and siphoned funds from operating companies.[10] The public utility scandals that erupted after the 1929 stock market crash, a lengthy FTC investigation of gas and electric utility holding companies, and the focus of public anger on utility holding company magnate Samuel Insull for his business indiscretions made the charge of being a "holding company" about as insulting and dangerous as being called a Communist during Senator Joseph McCarthy's brief reign of notoriety.[11] Although the Public Utility Holding Company Act, which, among other things, compelled the restructuring and simplification of gas and electric public utility holding companies, was not enacted until 1935, the near certainty of such a new statute was clear well before then. To many people looking for a scapegoat upon whom the Great Depression could be blamed, public utility holding companies were ideal candidates.

It is within this context that AT&T's strategy and attitude toward the establishment of the FCC must be considered. The congressional study that paralleled the executive branch Roper study was clearly hostile to AT&T. The congressional report, published in 1934, commonly known as the Splawn Report, asserted: "The holding company has been found as a result of this investigation to be as prolific of abuses in the field of communications as in other utilities already studied." After citing the *single* instance of the Associated Telephone Utilities Company as the example of widespread abuses, the report went on: "Moreover, American Tele-

[9] See Arthur M. Schlesinger, Jr., *The Coming of the New Deal* (Boston: Houghton Mifflin, 1959), pp. 20–21.

[10] Details of the charges against public utility holding companies are contained in many sources. One of the briefest and best is the New York PSC report in *New York State Electric and Gas Corporation*, PUR 1932E, 1, 3–4.

[11] The best summary of the public utility holding company scandals is Forrest McDonald, *Insull* (Chicago: University of Chicago Press, 1962), pp. 214–339.

phone and Telegraph Company which is both a holding and an operating company is more powerful and skilled than any State government with which it has to deal. A bill regulating communications in interstate commerce will fall far short of being effective unless it first restricts the use of the holding company to what is absolutely essential and necessary."[12]

The "evidence"—if one can use that word—to describe the Splawn Report's sweeping assertions was nonexistent. With the single exception cited above, no evidence came to light that suggested rampant abuse in the telephone industry. Certainly nothing in either the Splawn or Roper investigations, or in the legislative inquiries, supported any charges of abuse against the telephone industry's largest company. Nevertheless, as any student of politics realizes, fictions are often as important as facts. Scapegoats serve the function of simply explaining complex issues.

Thus, when AT&T president Walter S. Gifford appeared before Congress in 1934, he argued that the FCC would be an unnecessary creation. His major argument before congressional committees was that neither AT&T nor any other telephone interest (with the one exception already noted) were holding companies in the abusive sense. Most important, neither the ICC nor any other governmental body had received complaints about interstate rates, which, in any event, had been declining in real terms since 1926. Neither of the major governmental reports nor Congress had specifically criticized AT&T's rates or service. Gifford argued that only vague paranoid suspicions that AT&T was concealing something motivated the proponents of strong federal telephone control. Further, neither the ICC nor any state PUC was charged with dereliction of duty. Finally, telephone regulation was largely a state matter and not a federal one, since 98.5 percent of telephone traffic was local or intrastate long distance. A new agency, AT&T charged, would disturb the delicate structure of company-PUC relations that had grown up because it might preempt state regulation.[13]

Only the independent telephone operating companies, speaking through the United States Independent Telephone Association, of the major interests testifying before Congress, fully supported AT&T's position. Other interests complained about various facets of the bill to create the FCC. But even when no enthusiasm could be worked up, the prevailing attitude is captured in the remarks of Western Union's president: "We can adjust our practices to conform with its requirements without

[12] U.S. House of Representatives, *Preliminary Report on Communications Companies* (Washington, D.C.: Government Printing Office, 1934), pp. xxx, xxxi.

[13] See U.S. House of Representatives, Committee on Interstate and Foreign Commerce, *Federal Communications Commission, Hearings* (Washington, D.C.: Government Printing Office, 1934), pp. 165–202.

much difficulty and without many changes."[14] The principal business witness who enthusiastically endorsed the unification of communications regulation and the prompt creation of the FCC was David Sarnoff of RCA. As the most powerful radio network, it had much to gain from the establishment of a permanent commission that would license all new applicants for radio stations. From RCA's perspective anything that increased entry barriers into radio was to the good. Moreover, radio, unlike telephony, would not be subject to rate regulation. Finally, whereas AT&T's nontelephonic activities would be subject to FCC scrutiny, RCA's (such as its ownership of a phonograph record company) would not.

The idea of a federal agency that combined the disparate jurisdictions of the various communications regulatory authorities was not a New Deal invention. In preparing his report Secretary Roper remarked that his committee was heavily influenced by a bill introduced in 1929, upon which extensive hearings were held.[15] Yet before the change in government philosophy to one strongly anti–big business and before the near hysteria over holding companies, about which AT&T had reason to fear, the company had opposed the creation of the FCC. This, of course, indicates that AT&T had still other reasons to oppose the creation of such an agency. AT&T was anxious to avoid again becoming embroiled in radio controversies, which it felt was inevitable if communications regulation was integrated. For this reason AT&T had opposed such an agency during the preceding Republican administration as it did during the New Deal.

The 1929–1930 hearings were conducted at a considerably more leisurely pace than those in 1934, lasting from May to early February. The hearings were, thus, far more extensive and, one should add, more dispassionate; they were far more an inquiry than an inquisition. Like the 1934 act, the 1929 bill's principal announced purpose was the unification of federal communications regulation. Latent in this recommendation was criticism of radio regulation, which was then entrusted to the Federal Radio Commission (FRC). The establishment of the FRC had been proposed as a way of solving the interference problem through station licensing. At the same time, vesting the power in an *independent* regulatory commission assuaged those legislators who were fearful of placing full control over radio in the hands of Secretary of Commerce Herbert Hoover. A delicate compromise was reached in which jurisdiction over radio was divided in a complex manner between the secretary and the

[14] Ibid., p. 203.

[15] "Radio, Telegraph Mergers: Federal Communications Commission," *Telephony* (November 18, 1933): 25; and "Federal Communications Rule Proposed," *Telephony* (December 23, 1933): 12.

newly created FRC.[16] AT&T had withdrawn from broadcasting in part to avoid becoming embroiled in a hot political dispute. Since part of the compromise was to charter the FRC for a temporary period, Congress would inevitably debate communications again. And if prior debates were a guide, later ones would be heated, with various participants charging that opponents were nothing more than surreptitious agents of the "telephone trust"—the ultimate insult. AT&T wished to avoid becoming involved in such controversies.

The developing structure of broadcasting promised to make future disputes over communications regulation even more antagonistic than past ones. Controversies between independents and NBC continued unabated after the FRC's creation. But networking was so successful that RCA established two—the Red and the Blue. Meanwhile, a group of promoters who were rebuffed by NBC when they sought to produce shows for it created their own network—Columbia Broadcasting System (CBS). The new network, which needed telephone lines to link stations that would broadcast programs simultaneously and to link studios (often located outside city business districts) with remote broadcasting locations, charged that an NBC-AT&T conspiracy was denying them interconnection. The FRC lacked the power to order interconnection. Given the high transaction costs of lobbying each state PUC or legislature to require mandatory interconnection under these circumstances, the independents and CBS had a clear incentive to promote the establishment of a new federal agency with comprehensive jurisdiction over wire and wireless transmission and the boundaries between them.

Was there substance to the CBS claim? We have seen that when AT&T operated broadcasting stations it sometimes refused interconnection. But the evidence indicates that, within the limits of the system's capacity and prior commitments, the policy changed after AT&T left broadcasting. After internal discussion AT&T agreed to furnish such interconnection as long as it either supplied or approved the equipment and wiring connected into the telephone network.[17] Nevertheless, AT&T, unable to fully keep up with demand, sometimes delayed provision of lines, most importantly to CBS in 1927.[18] Notably, in its testimony CBS did not complain about the problem. Owen D. Young, then RCA's chairman, did advocate that the proposed new commission should have the power to

[16] Philip T. Rosen, *The Modern Stentors* (Westport, Conn.: Greenwood Press, 1980), pp. 11, 102–6; and "Agree on Measure for Radio Control," *New York Times*, January 23, 1927, p. 20.

[17] FCC Special Investigation Docket 1, Exhibit 289, *Bell System Policies and Practices in Radio Broadcasting* (1936), pp. 137–63.

[18] Erik Barnouw, *A Tower in Babel* (New York: Oxford University Press, 1966), pp. 195, 201.

compel interconnection. Although he did not charge AT&T with any impediment, he wanted the security of a formal rule.[19]

Many independent broadcasters who feared their demise in the face of the growth of the networks favored an FCC for different reasons. According to one study, "Aggrieved station owners challenged every aspect of the 1927 law as well as specific FRC rulings."[20] The FRC was effectively lending the imprimatur of government approval to the new system of network broadcasting in which the networks controlled program production (even if they did not necessarily produce them) and transmission—a system quite unlike AT&T's toll system. With RCA's control of patents in every phase of wireless transmission, the FRC was viewed with suspicion or hostility by some radio stations. For example, in January 1929 the president of the independent New Jersey State Broadcasters Association charged that the FRC was incompetent and had "been used to effect a monopoly in the radio business. Almost all the publicity emanating from the commission has worked hardships on small stations."[21]

The critics of the FRC, however, did not call for the end of regulation, but rather what they conceived to be superior regulation. For example, radio independents approved of portions of the 1929–1930 bill that enlarged an aggrieved party's right to appeal an adverse commission decision to the courts. They also favored expanding the right of intervention in agency proceedings.[22] Thus, both the independents and the networks supported the idea of an FCC, but for very different reasons. AT&T and the independent telephone interests were caught in a defensive posture against the apparently reasonable surface argument that all interstate communications regulation should be vested in one body.

Advocates of a unified agency also felt compelled to aver that ICC telephone regulation was ineffective. Even though the highly respected ICC commissioner Joseph B. Eastman and others argued that the ICC was a good telephone regulator, the reports of both congressional investigations *without any evidence to support their position* argued otherwise, using the vague distrust of an enterprise as large as AT&T as a major weapon. Under intense questioning Eastman stated that only 1.36 percent of total Bell System messages, constituting less than 10 percent of the system's revenues, were interstate. Directly asked whether the ICC was so busy with its railroad work that it ignored telephone work—consistently a favored theme of FCC advocates—Eastman issued an emphatic denial: "I think it is due to the fact that the information coming in to the commis-

[19] U.S. Senate, Committee on Interstate Commerce, *Commission on Communications, Hearings*, part 9 (Washington, D.C.: Government Printing Office, 1929), pp. 1094, 1095.

[20] Rosen, *Modern Stentors*, p. 13.

[21] "Says Radio Board Fosters Monopoly," *New York Times*, January 11, 1929, p. 16.

[22] "New Radio Bill Nears Senate," *New York Times*, March 30, 1930, sec. 4, p. 15.

sion in the form of complaints from users of these services, or in the informal protests has not seemed to justify any greater activity than the commission has exercised."[23] In turn, this was attributable to declining long distance rates coupled with continually improved technology. But his argument fell on deaf congressional ears.

Another factor that cemented an alliance in favor of a new commission was the settlement of the disputes between the independent station operators and independent receiver manufacturers, on the one hand, and the members of the RCA patent pool, on the other. In January 1924 the FTC issued a complaint against the members of the RCA patent pool. But because of adverse Supreme Court decisions, the agency was compelled to conclude that it could not grant relief in the matter and, accordingly, in 1928 dropped the case. In 1929 a concurrent resolution of Congress requested the Department of Justice to take action under the antitrust laws using the the evidence gathered by the FTC, which included almost seventeen thousand pages of testimony.[24] After investigating the matter, the Antitrust Division brought suit in 1931, seeking to divorce RCA from GE and the other companies that had participated in the pool.[25]

For its part, AT&T was fearful of the resentment that an antitrust judgment against it could unleash—especially in a severe depression in which many people sought convenient scapegoats. Accordingly, it urgently sought to settle the matter.[26] Adopting a position at odds with GE, RCA, and the Department of Justice, AT&T opted for a quick, reasonable settlement, hoping to stave off a backlash in the form of harsher antitrust treatment.[27] After lengthy negotiations, the Department of Justice officially approved a consent decree, the most important feature of which was that RCA was divorced from GE, which received RCA debentures and various property in return. The newly independent RCA could engage in broadcasting, receiver manufacturing, and maritime communications. All cross-licensing agreements were rewritten to render them nonexclusive so that competing set makers would receive licenses. RCA's network subsidiary, NBC, assumed the operations of various Westinghouse and GE stations.[28]

With the final settlement of the patent-licensing dispute the way was

[23] U.S. Senate, *Commission on Communications, Hearings*, p. 1585.

[24] "Attempt to Reopen Radio Trust Case," *New York Times*, January 12, 1929, p. 14.

[25] "Seeks Separation of Radio Companies," *New York Times*, April 30, 1931, p. 28.

[26] See, for example, AT&T to the Attorney General, 23 October 1931, and J. E. Otterson to C. P. Cooper, 17 June 1931, in AT&T Archives, Box 44.

[27] J. E. Otterson to C. P. Cooper, 10 June 1931, in AT&T Archives, Box 41.

[28] Hugh G. J. Aitken, *The Continuous Wave* (Princeton: Princeton University Press, 1985), pp. 504–9; and Rosen, *Modern Stentors*, pp. 152, 153.

clear for the radio interests to fight for the new regulatory agency, to which the political atmosphere of the early 1930s was most conducive. Even though there was objection to NBC having two networks, the *system* of networks furnishing programs, selling time to advertisers, and distributing the programs to network affiliates was essentially approved. At the same time a place was reserved in the new structure of radio for independent broadcasters and set makers. As a result of the antitrust settlement, their support was assured.

Public Service Liberalism in Retreat

The new statute creating the FCC, as one discerning commentator notes, contained an important, albeit ambiguous, statement in its first section.[29] A purpose of the new statute was "to make available, so far as possible, to *all* the people of the United States a rapid, efficient, nationwide and worldwide wire and radio communication service with adequate facilities at reasonable charges."[30] This language was not discussed in the hearings or reports accompanying the legislation. The Senate Interstate Commerce Committee Report simply stated that the language was intended to assure an adequate communication system for the United States. The counterpart House committee report is equally terse. The legislative debate is silent on the meaning of the language. In view of this silence, one must necessarily be cautious and not provide a strained meaning to the phrase. But in its paraphrase of Theodore Vail's proclamation of universal service as a central goal of the Bell System, it is hard not to read in the 1934 act's language a tribute to both AT&T's accomplishments and public service liberalism.

Nevertheless, times had changed. The Great Depression was perceived by many as a failure of the capitalist system and its policy formulations, including public service liberalism's rebuttable presumption in favor of free markets. Coupled with the Great Depression were major scandals and failures in such heavily regulated industries as electric power and banking. Not surprisingly, the public policies of American capitalism generally and those governing the public service industries were in retreat. Nevertheless, it would be a mistake to think that the old order was swept away in the way the 1917 Russian Revolution swept away all vestiges of the czarist regime. Rather, the old order was gradually eroded. In the area of communications the 1934 Communications Act

[29] G. Hamilton Loeb, *The Communications Act Policy toward Competition: A Failure to Communicate* (Cambridge, Mass.: Harvard University Center for Information Policy Research, 1977), pp. 61–64.

[30] Communications Act of 1934, title I, sec. 1.

both continued older public service principles and supplanted them, most dramatically on the radio side of its jurisdiction.

The End of Public Service Liberalism

After the New Deal, public service liberalism continued to operate in the traditional public utility industries and several others, such as the savings and loan industry, whose primary function was to serve a set of values associated with private home ownership. The independent regulatory commission was the principal governmental institution that assured the performance of public service goals. Most other economic sectors in the United States were in private hands, but legislators and the state bureaucracy imposed few new statutes or rules on them until the presidency of Lyndon Johnson. In large part this stemmed from the widespread satisfaction with economic progress made from the end of World War II through 1963. Not only did GNP (in constant dollars) grow 4.7 percent in the 1948–1956 period and 3.0 percent in the 1957–1963 period, but per capita personal income and other indicia of economic well-being rose accordingly.[31] Although there were dips and minor recessions, a much feared new Great Depression with massive unemployment never materialized. Andrew Shonfield, writing in 1965, summarized the main characteristics of the extraordinary postwar boom in the United States and the other advanced industrial countries of the West. Economic growth was much steadier than in the past. The growth of production was extremely rapid and the benefits of the new prosperity were widely diffused.[32]

But even in the midst of this self-congratulatory period, there were rumblings of discontent. Regardless of economic progress or the lack of it, black people and numerous white sympathizers were deeply distraught about social inequality. Then during the Johnson administration, a major change took place. The harmonious consensus was over. A variety of demands for social change became widespread, and policymakers appreciated that the wellsprings of dissatisfaction ran deep and wide. Environmental concerns, consumer dissatisfaction, and other rumblings of discontent soon reached the policy agenda and spawned movements or, at least, political entrepreneurs claiming to speak on behalf of dissatisfied interests. Many of these concerns, such as air pollution, can be translated into economic terms, but treating them entirely in this framework ob-

[31] See Alvin H. Hansen, *The Postwar American Economy: Performance and Problems* (New York: Norton, 1964), for the best description of the American economy through 1963.

[32] Andrew Shonfield, *Modern Capitalism* (New York: Oxford University Press, 1965), p. 61.

scures a critical point. Values other than efficiency were at stake. The political pressures led to a major sea change in American public policy. Cumulatively, they led to a massive increase in government programs and the large number of so-called social programs.

These programs, diverse as they are, have at least one important element in common: they were undertaken in the spirit of modern liberalism. Plans were made to alter certain conditions, regulations were imposed, and resources redistributed; no inquiry was made as to whether market-type arrangements could achieve better results or the same results at lower cost. Further, the relationship between the legislators who sponsored programs and the bureaucracies in charge of administering them, on the one hand, and the affected business interests, on the other, was usually one of conflict and control, not collaboration. From 1965 to 1985 total nondefense federal spending had almost tripled. But many of the dissatisfactions remained. Poverty did not disappear, educational results were unsatisfactory, welfare clientelism remained rampant, many persons did not receive reasonable health care, civil rights still remained a festering issue, and environmental quality remained low in many densely populated parts of the country. In short, after enormous public expenditures, the results achieved under modern liberalism were unsatisfactory.

As we saw at the beginning of this book the inevitable backlash occurred. It began in the mid-1970s and continues into the 1990s. The new conservatives advocate free competition, market mechanisms, deregulation, and privatization as the answer to virtually all problems. They point not only to the failure of modern liberalism in America to solve important problems, but also to the extraordinary harms to liberty, economic progress, and social goals that Communist-bloc massive state intervention wreaked on that unfortunate part of the world. The gradual encroachment of the state into more and more areas, they warn, could surreptitiously transform the United States and other Western nations into a replica of the Soviet Union. But critics of the new conservatives decry their lack of concern about so many social problems and the American future generally, calling for a step up in planning through industrial policy, massive environmental planning, and the like. Only government action, they argue, can solve such long-range problems.

A revival of public service liberalism can, however, meet the objections of both schools. It begins with a presumption in favor of private enterprise and free markets and recognition of the dangers of excessive government intervention. But there are many values in addition to economic efficiency, and if the unaided market cannot satisfy them, state intervention can frequently help to do so. Besides, as the telephone industry history until 1934 illustrates, heavily regulated enterprises can operate efficiently even when they are monopolies. Public service liberalism, moreover, makes the corporate charter a meaningful document again and even without government compulsion would direct firms to be attentive to goals in addition to profit maximization. Further, privately owned but publicly chartered companies might be employed to attain such goals as educating children or providing medical care. Such traditional tools of public service liberalism as the chartering of monopolies, restrictive entry, and cross-subsidization of rates can be employed to assure that virtually everyone is able to obtain a service. Thus, pressing issues could be solved with state involvement limited to cooperative supervision. Obviously, such suggestions are sketchy and intended only to provoke thinking along new lines. But as I have sought to show in this book, public service liberalism is an alternative that has worked.

Index

Adams, Charles F., 124
Adams, Henry C., 32
Africa, 52
Agricultural Marketing Act, 272
Alexanderson, Ernst F., 245
Allen, Walter S., 135
American Bell Telephone Company, 87–89, 95–96; annual report, 99, 100, 113; and commission regulation, 159; and independents, 135; and Keystone, 125–26; and license contracts, 111, 113, 114, 115; and long distance, 96–103; and patent monopolies, 127–28, 129; and Western Electric, 115–21
American Marconi Company, 246, 250–51
American Medical Association, 5
American Revolution, 34
American Speaking Telephone Company, 65, 93, 94–95, 116. *See also* Bell patents; Western Union
American Tel & Tel Company, 250
Anderson, George W., 163
Antitrust laws, 121, 184, 186–91, 282; and Bell acquisitions, 179–80; and competition, 165, 188, 189; and leasing issues, 104; and natural monopoly, 184–85; and the radio industry, 249; and small-scale enterprise, return to, 167; and the Willis-Graham Act, 201
Antwerp, port of, 52
Asia, 52
Associated Press, 40
Associated Telephone Utilities Company, 277–78
Astor, John J., 155
AT&T (American Telephone and Telegraph Company), 79, 84–85, 101, 146–48, 283; acquisition maneuvers by, 154–58, 175–78, 186–91; and the American Bell Telephone Company, 87–89, 99, 101–3; breakup of, 51, 88, 140, 174, 180, 203, 233; and end-to-end service, 221, 225; and the FCC, 276, 277–82; foundation of, 94, 99–103; and horizontal integration, 88, 109–15; and independents, rise of, 130–40; and licensing, 108, 109–15, 131–32; Long Lines Department of, 88, 102, 256; and the National Bell Telephone Company, creation of, 94, 96; and the natural monopoly concept, 143–45; and the new political economy, 165–204; and the patent pool, 251–53; political strategy of, 140–43; during the Progressive Era, 126, 128–65; and PUCs, 221, 225, 226, 233, 235–36; and the regulated network manager system, 165, 172, 174–76, 194–95, 199, 254–55; and the Splawn Report, 277; and toll broadcasting, 240, 254–70; and *United States v. AT&T*, 104, 154, 221; vertical structure of, 88–89, 115–21; and wireless transmission, 241–70. *See also* Local service; Long distance service
Averch-Johnson (AJ) effect, 67

Bamberger Department store chain, 260
Banking, 14, 17, 18, 19, 34, 179; and the invention of the telegraph, 25, 40, 44; and the New Deal, 272, 275, 276; and the Panic of 1907, 169–72
Bartlett, Charles L., 185
Barton, Enos, 116
Baruch, Bernard, 7
Batteries, 80–83, 107, 115, 223, 231
Baughman, James P., 79, 105–6
Bell, Alexander Graham, 38, 39, 54–67, 71; attitude toward invention and profit, 68; conception of the telephone, as an instrument of mass communication, 74–75; demonstration-lectures of, 90–91; 1876 patent, 57, 63–65, 90; and the first transcontinental telephone line, opening ceremony for, 150, 242; and the transmission of speech through light waves, 240
Bell, Daniel, 154
Bell Laboratories, 76, 88, 115–16, 121, 246
Bell Patent Association, 59, 90, 109, 116

Bell patents, 55–57, 62–67, 93–95; basic, expiration of, 110, 111, 112; and licensing, 110, 111; and the New England Telephone Company, agreement with, 91; original 1876 patent, 57, 63–65, 90; and research, 68–71. *See also* Patent(s)

Bell System, 44, 55, 62, 188, 281; and acquisitions, 175, 177, 178, 182, 183, 191; and the American Bell Telephone Company, 87; and automatic switching, 76–78; and battery systems, 80–83; and complementarity and integration, 78–80; enlargement of, 70–71, 87–89, 130; genesis of, 89–96; horizontal structure of, 93; and the ICC, 201; and independents, 130–40, 174, 179–80; and the Kingsbury Commitment, 192–95, 199; and leasing of telephones, 103–9; and the Merchants Association finding, 162; and the Panic of 1907, 169; and patent monopoly, 124–29; stock in, holding of, policies regarding, 260; as universal, Vail's summarization of, 151, 199, 283; and Western Union, acquisition of, 182, 183; and wireless transmission, 260, 261, 266. *See also* Licensees, Bell System

Bell Telephone Company (corporation), 91, 92, 93, 94

Bell Telephone Company (Massachusetts Trust), 68–71, 91, 92, 128

Bell Telephone Company of Missouri, 111

Bell Telephone Company of Pennsylvania, 88

Bell Telephone Company of Philadelphia, 120

Berliner microphone, 128

Bernstein, Neil N., 232, 233

Black Death, 29

Blake, Francis J., 68, 93, 128

Bloom, E. S., 266

Board of Public Utilities, 200

Board of Railroad Commissioners, 159

Boettinger, Henry, 81

Bonbright, James C., 234

Bourseul, Charles, 57

Brandeis, Louis, 6, 165, 167–68

Brass v. Stoeser, 27

Britain, 127. *See also* England

Broadcasting. *See* Radio industry; Television industry; Toll broadcasting

Brock, Gerald, 195

Brooklyn Eagle, 267

BTU standards, 230

Bureaucracy, 273

Bureau of the Census, U.S., 176

Bureau of Public Roads, 238

Bureau of Steam Engineering, 249, 251

Burleson, Albert S., 197, 198, 199

Cable car companies, 24, 28

California Railroad Commission, 108, 126, 189, 218, 220

Campbell, George A., 72, 100

Canada, 127

Canals, 19–20, 38, 49, 86

Capitalism, 3, 12, 29, 51–54, 170; and "creative destruction," 51; and the Great Depression, linkage of, 283

"Capture theory," 48

Carty, John J., 150, 246

CBS (Columbia Broadcasting Service), 280–81

Centennial Exhibition, 61

Central New York Telephone and Telegraph Company, 114

Central Union Telephone Company, 143, 188

Chandler, Alfred D., 24–25, 165, 166

Charles River Bridge case, 87

Charles Williams shop, 118

Chicago Telephone Company, 99

Christianity, 85

Cincinnati Bell, 88, 113

Civil War, 9, 21, 54, 58, 94

Cleveland Telephone, 154

Coase, R. H., 206

Colpitts, E. H., 245

Commerce Department, 7, 211, 248; and the Roper Report, 275–76; and wireless transmission, 258, 261, 279

Commercial and Financial Chronicle, 139, 140, 142, 152

"Common calling," 29

"Common control" system, 78

Common law, 12–13, 15–17, 28–29, 31, 83, 217

Communications Act (1934), 213, 248, 275, 276, 283

Communism, 285

Competition, 33, 67, 115, 175, 199, 240;

and antitrust laws, 165, 188, 189; and AT&T's consolidation strategy, 172; belief in, 3–4; and capital markets, 102; decline of, and the new political economy, 165; and efficiency, 102; foreign, 83; free, 3, 4, 14, 45–47, 167, 285; Gary on, 6; and the "great good," 31; and interconnection, 193–94; and long distance service, 193, 195; and nationalization, 167; and railroads, influence of, 32; vs. regulation, 200; Schumpeter on, 46, 51; and technological progress, 9. *See also* Monopoly
Complementarity, 78–80
Computer II decision, 90
Congress, 93, 200, 268; and complaints about high rates, 236–37; and the FCC, origins of, 276, 278, 280, 283; Gifford's appearance before, 278; Justice Department report to (1914), 193; and the Mann-Elkins Act, 184, 185, 186; and monopoly problems in radio, 255; and patent laws, 54; and the Roper Report, 275–76, 277, 278, 279; and the Senate Interstate Commerce Committee Report, 283; and the Splawn Report, 277, 278; and the telegraph, 40; and the "telephone trust" charge, 236–37, 280; and wireless transmission, 249–50, 261. *See also specific acts*
Conrad, Frank, 253
Constitution, U.S., 16, 28, 42, 44, 54; commerce clause of, 185; and federalism, 89; Fourth Amendment to, 28; and the public interest concept, 28–29; and PUCs, 207, 208–9
Continental Telephone, 231
Contoocook Valley Telephone Company, 139
Coolidge, Calvin, 267, 272
Cotton gins, 35
Coy, George W., 75
"Creative destruction," concept of, 46–47, 51
Credit, 34, 45, 166
Croly, Herbert, 6, 165
Cross subsidies, 50, 225–26, 228–29, 234, 286
Cullom Report (1886), 32
Cunningham, William J., 168
Cushman, Sylvanus D., 56, 63
Customer premises equipment, 82, 118, 223–24

Daggett, Stuart, 202
Dallas Telephone Company, 178
Davis, H. P., 253–54
Davis, John W., 263
Dean Company, 129
Declaration of Trust, 91
Defense Department, 211
Democratic party, 185, 199, 263, 273
Department of Justice, 184
Deregulation, 3–4, 285
De Wolf, Wallace L., 156–57
District Telephone Company of New Haven, 75
Dolbear, Amos, 56, 65
Douglas, Stephen, 5
Drake, A. W., 260
Drawbaugh, Daniel, 64, 127
Du Pont, 166

Eastman, Joseph B., 281–82
Edison, Thomas A., 43, 68, 69
Electricity, 24, 38, 205, 206, 237; and public service obligations, 45; and the telegraph, development of, 39; and telephone research and development, 70
Ely, Richard T., 145, 161
Emerson, 269
Emery, James J., Jr., 91
End-to-end responsibility, 108, 220–25, 232
England, 4, 27, 29–30, 84–85; and the closing of the port of Antwerp, 52; common law in, 12–13, 15–17, 28–29, 31, 83, 217; industrial supremacy of, 9; laissez-faire policy in, 10–12; legal system in, 16, 17, 29; nationalization in, 42, 141; nineteenth-century political economy in, 9–12, 15–16; patent laws in, 53–54; regulation of rates in, 41–42; the telegraph in, 26, 41–42. *See also* Britain
Eric Canal, 19–20
Erie Telegraph and Telephone, 154–55
Europe, 4, 25, 52. *See also* England

"Fairness" issue, and rate structures, 234
Farmer lines, 177

Farmers, 32
Farrar, Edward, 56
FCC (Federal Communications Commission), 76, 88, 125–26; *Computer II* decision, 90; data, on AT&T voting stock, 114; founding of, 270, 275–83; during the Progressive Era, 125–26, 154; and service standards, 232; and the Willis-Graham Act, 201
Federal Highway Act, 238
Federalism, 89
Federalist Papers, 207
Federal Reserve Bank of New York, 272
Federal Reserve System, 272
Federal Telephone Company, 127
Felix, Edgar H., 267
Ferries, 35, 38
Fessenden, Reginald, 245
Field, Stephen J., 123
Fish, Frederick, 149, 153; and acquisition maneuvers, 156, 158; and commission regulation, 160; and financial problems, 169, 170–71; resignation of, 169, 171; and wireless transmission, 241, 245
Forbes, William H., 89, 94, 95, 113–14, 118
Ford, Gerald R., 8
Ford Model T, 166
Forest, Lee de, 241, 242, 243, 246
Fortune, 89
Fowler, J. A., 190
FRC (Federal Radio Commission), 279–80, 281
Friedman, Milton, 3
FTC (Federal Trade Commission), 249, 255, 263, 277, 282

Gallatin, Albert, 19
Gary, Elbert, 6
Gary, Theodore, 179
Gas industry, 24, 33, 37, 45, 206, 237
GE (General Electric Company), 156, 245, 246, 249–53, 264, 282
General Accounting Office, 211
Germany, 4
Gifford, George, 66
Gifford, Walter, 256, 257, 259, 278
Gilliland, E. T., 100
Gilliland Electric Manufacturing Company, 117, 118, 119
GNP (gross national product), 8, 23–24, 169, 284
Gordon MacKay Shoe Machinery Company, 104, 106
Gould, George, 183
Gould, Jay, 65–66, 119
Grampp, William D., 11
Granger complaint, 49
Granger movement, 32
Gray, Elisha, 55, 56, 58–62, 65, 68; and the "graded multiple," 78; and the multiplex problem, 58; and Western Electric, 116, 117
Great Britain, 127. *See also* England
Great Depression, 130, 178, 239, 271, 283; blame for, 277; and fear of future depressions, 284. *See also* New Deal
Great Fire of London, 14
Great Northern Railroad, 27
Greece, 85
Griswold, A. H., 244

Haber, Samuel, 168
Haines, Henry S., 20
Hall, E. J., 80, 114–15, 173, 204
Hall Memorandum, 204
Hamilton, Alexander, 17
Handlin, Mary, 17
Handlin, Oscar, 17
Harbord, J. G., 257
Harkness, A. G., 266
Hartz, Louis, 19
Harvard University, 69
Hawley, Ellis W., 7, 198, 274
Hayek, Friedrich A., 3, 10
Hayes, Hammond V., 69, 81–82, 241, 245
Heath, Milton S., 18
Henry, Joseph, 59
Hepburn, A. J., 251
Hertz, Heinrich, 241
Himmelberg, Robert F., 8
Hockett v. State, 44–45
Hofstadter, Richard, 122
Holcomb, Alfred G., 63
Holmes, E. T., 90
Holmes Burglar Alarm Company, 74–75
Home Telephone Company, 178, 200
Hoover, Herbert, 248, 261–62, 279; and the Agricultural Marketing Act, 272; and the Commerce Department, 7
Horizontal integration, 88, 93, 109–15

Hovey, William, 103
Hubbard, Gardiner, 59, 65, 90, 91; and American Bell Telephone Company, formation of, 95; and leasing of telephones, 103, 104, 106; and license contracts, 109
Hughes, Charles E., 267
Huntington, Samuel, 48

ICC (Interstate Commerce Commission), 8, 20, 126–27, 146, 268, 278; complaints about, in the *Commercial and Financial Chronicle*, 142; establishment of, 158, 159, 184–86; and the FCC, origins of, 281; and the Justice Department, 189; and the Kingsbury Commitment, 193; and mergers, 201–3; and Northwestern, 191; and the postmaster general's call for new legislation, 199; and rates, 236–37, 278; and the Roper Report, 276, 271
Illinois Commerce Commission, 228
Illinois Railroad and Warehouse Commission, 33, 48, 158
Inflation, 3
Information revolution, 25
Insull, Samuel, 277
Insurance, 34–35
Interconnection, 176–77, 179, 193–95, 280–81; mandatory, 230; to noncompeting independents, 193; between rival systems, as a technological failure, 200
Interest groups, 50
Interstate Commerce Act (1887), 32, 49, 184
Interstate Telephone & Telegraph, 188
Interstate Telephone Company, 97–98
Iowa Telephone Company, 112
Iowa Union, 135
Irrigation, 33–34

Jackson, Andrew, 87
Jacksonianism, 87
Jacob, Frank, 181
Janicke v. Washington Mutual Telephone Co., 200
Japan, 4, 46, 83
Jefferson, Thomas, 19
Johnson, Paul B., 201
Johnson, Lyndon B., 273, 284
J. P. Morgan & Company, 152, 154, 170, 179, 188
Justice Department, 179, 180, 282; and AT&T's Western Union stock acquisition, 186–91; and end-to-end service, 221; as a "guardian" of the public interest, 211; and the Kingsbury Commitment, 191–93; and lease-only policy, 104; and the Mann-Elkins Act, 186; and the Sherman Act, 184, 188, 192; and *United States v. AT&T* (1974), 154

Kahn, Alfred, 137, 233–34
Kaltenborn, H. V., 267
Kaysen, Carl, 106
Kellogg, Milo G., 156–58
Kellogg Switchboard and Supply, 129, 154, 156–58
Keynes, John Maynard, 3, 12
Keystone Telephone Company, 124, 125–26, 129, 177
Kidder, Peabody and Company, 155
Kingsbury, Nathan, 187, 191–92
Kingsbury Commitment, 178, 180, 191–95, 199; activity leading to, 184; and Lewis, 197; and the Northwestern case, 190; and the Willis-Graham Act, 201; Wilson and, 198
Kitson Clark, G.S.R., 12
Kolko, Gabriel, 122

Labor: disputes, 11, 225; division of, 25; legislation, 15; unions, 211
Laissez-faire, 4, 9–12, 13, 14–15, 31, 52–53
La Porte trial (1895), 77
Lease-only policy, 85, 103–8, 223–24
Leverett, George, 153
Lewis, David J., 197
Liberalism: economic, 9–12; interest group, 5; structural, 84–121
Liberalism, modern, 4–5, 7–8, 165, 167–68, 180, 191; failure of, and the new conservatives, 285; and the Kingsbury Commitment, 195; and the New Deal, 272–74; and the radio industry, 137, 270; and World War I, 198
Liberal party, 11
Libertarianism, 4, 15
Licensees, Bell System, 89, 108, 118, 177–78; and antitrust lawsuits, 184; and horizontal integration, 109–15; and interconnection, 177; and license contracts, 109–15; and long distance service, 101, 108; and standards, 231; stock in, 95, 96; and

Licensees (*cont.*)
the toll business, 97; and Vail's consolidation strategy, 172, 173, 175; and vertical integration, 117–19. *See also* Licensing; *specific companies*
Licensing, 17, 18; and toll broadcasting, 258, 259. *See also* Licensees, Bell System
"Life cycle" theory, 206
Lincoln, Abraham, 5
Link, Arthur, 239
Little Rock Telephone Company, 178
Livesay, Harold, 120
Loading coil, 82
Local service, 82, 88; and the Bell System, 70, 89; and end-to-end service, 221; and license contracts, 110, 111; and telephone technology, 72
Lockwood, Thomas, 69, 77, 97, 112, 129, 241
Lodge, Oliver, 241
Long distance service, 70, 88, 94, 101–2, 185, 186, 191; and AT&T, foundation of, 96–103; and the Bell System, 89; and competition, 193, 195; control of, 97; and end-to-end service, 221; and the ICC, 203; and licensees, 114–15; and local monopolies, 193
Lowe, T.S.C., 37
Lowi, Theodore J., 5
LT&T (Lincoln Telephone and Telegraph), 162–63
Lumber companies, 224

Macaulay, T. B., 11
McCarthy, Joseph, 277
McCraw, Thomas K., 124
McCurdy, Charles W., 21
McDonough Telephone and Telegraph Company, 63
Mackay, Clarence, 183, 197
Mackay Companies, 183–84, 197
MacMeal, Harry B., 132, 157
McReynolds, James C., 190, 191, 192, 193
Madison, James, 20, 207
Magneto systems, 231
Mann-Elkins Act (1910), 159, 184, 185, 186
Marconi, Guglielmo, 241
Marshall, John, 18
Massachusetts Bay Company, 85
Massachusetts Gas and Electric Commission, 158
Massachusetts Institute of Technology (MIT), 69, 72, 90
Maxwell, James Clerk, 240
Merchants Association of New York, 160, 161–63
Mergers, 166, 199, 201
Mexico, 100
Michigan Railroad Commission, 187
Michigan Telephone, 154
Microeconomics, 12
Mill, John Stuart, 4, 12, 14–15, 273; and the natural monopoly concept, 144; on patent monopolies, 54
Minnesota Telephone Company, 112
Missouri Public Service Commission, 127, 189
Molecular Telephone Company, 63
Molina, E. C., 78
Monopoly, 21–22, 67, 83, 104, 183, 259; and antitrust suits, 188, 189; charges of, and the term "telephone trust," 183, 236–37, 267, 280; copyright, 13; and corporate charters, 85, 86–87, 96, 286; local, 200, 219; and the public service classification, 27, 28, 29, 30, 34; and PUCs, 218–20; and the radio industry, 265–66; regulated, 189; Smith on, 13–14; and the telegraph industry, 66
Monopoly, natural, 201; categorization of industries into, and regulation, 115; and the telephone industry, 27, 184–85, 189, 198, 218. *See also* Competition
Montgomery Ward, 139
Moon, John A., 199
Morality, 11–12, 14, 54
Morrill Act, 83
Morse, Samuel F. B., 25, 39, 40–41
Morse Code, 43, 241
Motion pictures, 80, 238
Muckrakers, 212
Muecci, Antonio, 56
Municipal League, 200
Municipal ownership, 165, 195–96, 197. *See also* Public ownership

Nader, Ralph, 211, 236
NARUC (National Association of Railway & Utilities Commissioners), 203–4, 211, 232

National Bell Telephone Company, 65, 94–95, 128
National Good Roads Association, 166
National Independent Telephone Association, 185
National Industrial Recovery Act, 272, 275
Nationalization, 41–42, 165, 167, 180, 191, 195–96. *See also* Public ownership
National Labor Relations Act, 275
Navy, U.S., 247–48, 249–50
NBC (National Broadcasting Company), 268, 269, 280, 282–83
NCF (National Civic Federation), 142, 160
Nebbia v. New York, 28
Nebraska Commission, 189
Nelson, John R., 17
New Deal, 5, 8, 271–84; coming of, 274–75; and modern liberalism, 272–74. *See also* Great Depression
New England Telephone & Telegraph, 225
New England Telephone Company, 91–92, 94, 164
New Jersey Bell, 228
New Jersey Board of Public Utility Commissioners, 107
New Jersey Court of Errors and Appeals, 34
New Jersey General Incorporation Law (1875), 95
New Jersey State Broadcasters Association, 281
New Mexico Bell, 184
New World, 85–86
New York Central Securities case, 208
New York Gas Light Company, 37
New York Telephone Company, 146, 148, 160, 183
New York Times, The, 128, 171, 180; on independent telephone interests, 193; on nationalization, 197; on rates, 236–37
Noble Prize, 76
Northwestern Long Distance Company, 190–91
Northwestern Telephone, 154
NRA (National Recovery Administration), 276–77

O'Connor, John J., 268
Office of Management and Budget, 211
Olcott v. Supervisors of Fond Du Lac Co., 35
Orton, William, 65

Panic of 1907, 169, 245
Parliament, 12, 53
Party line service, 115, 226, 227–28
Patent Office, U.S., 90, 128
Patent(s), 48, 51–52, 57, 63–65, 83–84; automatic exchange, 76–77; Edison, 68, 69; and lease-only policy, 103; monopolies, 13, 51–54; Morse, 40–41; pools, 247–53, 262, 265, 269; and the radio industry, 247–53, 262, 265, 269, 281–83; and rate regulation, 45; and research, 67–71; vacuum tube, 249, 251. *See also* Bell patents
PBX (private branch exchange), 125, 224
Pennsylvania Public Utilities Commission, 177
Pennsylvania Railroad, 149
Petrina, M., 56
"Phantom circuit," 180–83
Philadelphia Centennial Exposition, 64, 90
Philadelphia Electric, 125
Philco, 269
Pioneer Telephone Company, 97
Populism, 4
Porter, Patrick, 120
Posner, Richard, 211, 234
Postal Telegraph Company, 41, 125, 183, 184, 189; and Fessenden's patents, 245; and radio, 252, 264
Post Office Department, U.S., 8, 40, 59, 93, 185, 197–99
Price-fixing statutes, 34
Primm, James N., 19
Principles of Political Economy (Mill), 15
Private lines, 73–74, 104
Private telephone companies, 35
Privatization, 3–4, 285
Privy Council, 53
Profit, 51, 85, 172, 286; incentive, 89; and licensees, 114, 117; maximization, 104–5; and monopoly privileges, 219; and toll broadcasting, 259; and vertical integration, 121
Progressive Era, 4, 5, 22, 122–64, 167
PSC (public service commission), 219, 225, 227
PT&T (Pacific Telephone and Telegraph Company), 47, 49, 190, 191, 200, 218, 231

Public ownership, 195–99, 267. *See also* Municipal ownership; Nationalization
Public Utility Act (1907), 164
Public Utility Holding Company Act, 277
PUCs (public utility commissions), 177, 208–37, 239, 278; and the Constitution, 207, 208–9; and end-to-end responsibility, 220–25, 232; and industry cooperation, 209–12; and information and procedures, 215–17; lobbying of, 280; and monoploy privilege, 218; and the quality and quantity of service, 225–32; and rates, 232–37; regulation by, overview of, 212–15
Pupin, Michael I., 72, 149
Pupin coil, 177, 242
PURs (Public Utilities Reports), 217

Radio industry, 5, 8, 67, 239–70, 279–83; and modern liberalism, 237; and patent pools, 247–53; and public policy before 1920, 242–44; and Westinghouse, 253–54. *See also* Radio stations
Radio stations, 260–61, 265–66, 267, 268
Railroads, 15–16, 42–43, 166, 167–68; and the Cullom Report (1886), 32–33; and the ICC, 48, 167, 168, 184–85; investment in, 86; as a public service, in the nineteenth century, 20; Supreme Court decisions regarding, 35; and telephone wires, 182; and the Transportation Act, 7–8, 200
Railway Mail Service, 93, 113
Rate(s): "reasonable," concept of, 50, 213, 233; structures, 232-37
RCA (Radio Corporation of America), 247, 250–57, 260–64, 268–69; and the FCC, 279; and patent pools, 282
Reis, Philipp, 56, 57, 60
Reis telephone, 56, 57, 60
Report on Manufactures (Hamilton), 17
Republican party, 185, 199, 271, 273, 279
Ricardo, David, 12
Roads, public, 8, 18–19, 49, 166, 238
Robinson-Patman Act, 275
Rochester Telephone, 178
Roman Catholic church, 85
Roosevelt, Franklin D., 274, 275, 276
Roosevelt, Theodore, 6
Roper, Daniel C., 275–76. *See also* Roper Report
Roper Report, 275–76, 277, 278, 279
Rosenberg, Nathan, 67
Rotogravure, four-color, 238
Royal Society, 240
Russia, 85, 283. *See also* Soviet Union
Russia Company, 85

Sanders, Thomas, 59, 65, 90–92, 95
Sarnoff, David, 264, 269, 279
Scherer, F. M., 53
Schmookler, Jacob, 57
Schumpeter, Joseph, 10, 13, 46, 51
Scientific American, 56
Scientific management, 6–7, 168–69, 239, 273
Scribner, Charles E., 118
Sears, Roebuck, 125
Second Atlantic Cable, 25
Senate Interstate Commerce Committee, 283
Sharfman, I. L., 213
Sherman, John, 42
Sherman Act, 42, 180, 187, 192
Shoe machinery, 104–5, 106
Shonfield, Andrew, 284
Slaughter-house cases (1873), 21
Slavery, 5
Smith, Adam, 9, 12–14, 19, 25
Smith, George David, 69
Smithsonian Institution, 59
Socialism, 3, 4, 7, 165
Social Security Act, 275
South Dakota Board of Railroad Commissioners, 222
Southern Bell, 227
Southern New England Telephone Company (SNET), 88, 98–99
South Korea, 4, 83
Southwestern Telegraph and Telephone, 154
Soviet Union, 195, 284. *See also* Russia
Spencer, Herbert, 10–11, 15
Splawn Report, 277, 278
Stager, Anson, 116, 118–19
Standardization, 70, 173
Standard Oil Company, 179
Standards, 229–32, 258
State Department, 211
State v. Nebraska Telephone Co., 44

Statute of Labourers (1349), 29
Sterling Company, 129
Stigler, George, 3
Stone, John, 246
Street lighting, 36, 37
Stromberg-Carlson, 129, 139, 154, 156
Strong, Benjamin, 272
Strowger, Almon B., 76–77, 78, 115
Suburban Bell, 113
Supplemental Telephone Report, 161–63
Supreme Court, U.S., 16, 26, 42, 87, 123; and Bell patents, 56, 62, 63–65; and the "clothed with a public interest" idea, 28–29; on competition, 164; and end-to-end service, 221, 225; and patent monopoly, 127, 128, 282; rulings on gas and electricity companies, 37; rulings on rate discrimination, 36–37; rulings on tobacco warehouses, 33; and the sale of Kellogg stock, 157. *See also specific cases*
SWB (Southwestern Bell Telephone Company), 88, 177–78
Switchboards, 92, 112, 200
Switching, 74–79, 81, 82; automatic, 76–79, 115; and license contracts, 109; and long distance service, 101

Taft, William Howard, 26, 142, 185
Taney, Roger, 87
Tariffs, 13, 40, 83
Taxation, 11, 84, 86, 234
Taylor, Arthur J., 11
Taylor, Frederick, 6, 168
Taylor, George R., 10
Taylor, Lloyd W., 56
Telegraph, 8, 9, 24, 38, 185; centrality of, in business life, 33; frequencies used for, 98, 182; "harmonic," 60–61, 62, 90; invention of, 25–26; lines, 101; nationalization of, 165, 197; stations, and railroads, 182; and switching systems, 74; and the telephone, development of, comparison of, 38–43; use of relative to that of the telephone, 105; and Western Union, 59, 66; wires, and interference in telephone systems, 72. *See also* Morse Code; Western Union
Telephone: adoption of, rate of, 38–39; advantages of, 43–45; booths, 240, 269; cables, 82–83; centrality of, in business life, 33; directories, 75; invention of, 23, 38, 54–62, 180; as a public service, 8, 9, 24–50; technology, essential, 71–74; and the telegraph, development of, comparison of, 38–43; use of, relative to that of the telegraph, 105. *See also specific companies; specific services*
Telephone, Telegraph and Cable Company (TTCC), 139, 152
Telephony, 138, 156, 160
Television industry, 239, 240, 255, 257
Ten Hours Bill debate, 11
Thayer, H. B., 259, 261
Third World, 3
"Thumper," 72
Tobacco industry, 33–34
Toll broadcasting, 203, 240, 254–60, 266–70, 281
Toll service. *See* Long distance service
Tory party, 11
Totalitarianism, 195
Transportation Act of 1920, 7, 200–201, 208, 213, 271
Trunk lines, 75–76, 78, 81, 200, 226
Tugwell, Rexford Guy, 36, 274
Twenty-four hour service, 230
Typesetting, photoelectric, 238
Typewriter, 68

Unemployment, 3
Union Pacific Railroad, 93
United Fruit Company, 250, 252, 253
United States Independent Telephone Association, 278
United States Steel Corporation, 6, 152
United States v. AT&T, 104, 154, 221
Universal service, 176, 236; and PUCs, 225 226; Vail and, 151, 199, 283
Urbanization, 23–24
USITA (U.S. Independent Telephone Association), 203, 204
USITC (U.S. Independent Telephone Company), 156
Utopia, 15

Vail, Theodore, 89, 92–93, 97, 118; and acquisitions, 175–77, 182–83; and Carty, appointment of, 150; and centrally integrated systems, 113–14; and commission regulation, 160; and Fish, 149; Hall and, 114; and independents, 132, 135, 174, 179; and license contracts, 110, 111, 112;

Vail, Theodore (*cont.*)
and National Bell, 94; and the Panic of 1907, 169–70; and patents, 110; as president of AT&T, 130, 140, 146, 150–54, 160, 169–75, 245; and public ownership, 198; on regulation, 213–14; retirement of, 141, 204; and universal service, 151, 199, 283; and vertical integration, 120; and Western Union's initial attitude toward the telephone, 244; and the Williams shop, 117; and wireless transmission, 246
Valentine, Robert G., 6
Value-of-service pricing, 235–36
Vanderbilt, William H., 119
Vertical integration, 88, 115–21, 166
Viner, Jacob, 13
Virginia Company, 85

Wall Street Journal, 170
War Industries Board, 7, 198
Washington Department of Public Service, 47, 231
Water distribution, 23, 24, 25, 206, 237; and irrigation, 33–34; Supreme Court and, 36–37
Watson, Thomas A., 60, 61, 117; and the American Bell Telephone Company, 95; and the Bell Patent Association, 90; demonstration-lectures of, 90–91; and the first transcontinental telephone line, opening ceremony for, 150, 242; and the Interstate Telephone Company, 97–98; and the invention of the "Thumper," 72
Watt, James, 53
Wealth of Nations, The (Smith), 9, 14
Western Electric, 58, 136, 276; and AT&T's vertical structure, 88–89, 115–21; and automatic switching, 78; and Bell Labs, 89, 115–16; and competition, 136, 139; and the patent pool, 252; subsidiaries of, 88–89; and wireless transmission, 260, 267
Western Electric Telephone Company, 136
Western Union, 41, 43, 48, 55, 65–66, 68; AT&T's acquisition of, 154, 166, 175, 180–84, 186–91; and AT&T's position on the FCC, 278–79; attitude of towards the telephone, initial, 244; and the Bell Telephone Company, financial strain on, 92; and Fessenden's patents, 245; and Hubbard, 59, 90; and the New Haven switching system, 75; and the patent pool, 252; patent suit against, 94–95; and Postal Telegraph, 125; and radio networks, 264; "settlement," 110; transmission of news messages by, 96; Vail and, 93, 94; and Western Electric, 68, 116–17, 118, 119; withdrawal of, from the telephone business, 119
Westinghouse, 156, 246, 250, 282; and the broadcasting boom, 253–54; and interconnection policy, 259; and the patent pool, 252, 253
Wheatstone, Charles, 57
White, Samuel, 58–59, 61
Whitehead, Alfred North, 57
Wickersham, George W., 180, 186–89
Wiebe, Robert, 5
Williams, Charles, Jr., 74, 117, 118
Willis-Graham Act, 193, 201
Wilson, James Q., 211
Wilson, Woodrow, 6, 168, 190, 192, 199; and government control of wireless transmission, 249–50; and the Kingsbury Commitment, 198; and nationalization, 197
Wisconsin Railroad Commission, 108, 164, 196
Wisconsin Telephone Company, 154, 164
Woods, Frank H., 179
World Series, 261
World War I, 4, 5, 7, 198, 205; development of four-color rotogravure after, 238; economic progress after, 284; and interventionist philosophy, 271; and the radio industry, 244
World War II, 4, 103, 205, 224, 226; growth after, 130; long distance service before and after, 232
Wright, John H., 188–89
Wright Brothers, 166

Young, Owen D., 250, 254, 261, 280

Zenith, 269